# As the Condor Soars

# As the Condor Soars

## Conserving and Restoring Oregon's Birds

Edited by
Susan M. Haig,
Daniel D. Roby, and
Tashi A. Haig

Illustrated by
Ram Papish

Foreword by
John A. Wiens

CONTRIBUTING AUTHORS

John D. Alexander
Bob Altman
Matthew G. Betts
M. Ralph Browning
Charlie Bruce
Alan L. Contreras
Jesse D'Elia
Katie M. Dugger
Elise Elliott-Smith
Greg Green
Joan C. Hagar
Christian Hagen
Charles Henny
Frank B. Isaacs
Gary Ivey
Patricia L. Kennedy
Donald Kroodsma

David J. Lauten
Joe Liebezeit
Roy Lowe
Donald Lyons
Jeffrey S. Marks
Christopher Mathews
Brenda McComb
Randy Moore
Michael T. Murphy
S. Kim Nelson
M. Cathy Nowak
Lewis Oring
Peter Sanzenbacher
Mark Stern
Oriane Taft
Pepper Trail
David Wiens

Oregon State University Press   *Corvallis*

Publication of this book would not have been possible without the support of the Division of Migratory Birds, US Fish and Wildlife Service (Nanette Seto); Department of Fisheries, Wildlife and Conservation Sciences, Oregon State University (Selina Heppell); and the Oregon Cooperative Fish and Wildlife Research Unit (Katie Dugger).

Library of Congress Cataloging-in-Publication Data

Names: Haig, Susan M, editor.

Title: As the condor soars : conserving and restoring Oregon's birds / edited by Susan M. Haig,
    Daniel D. Roby, and Tashi A. Haig ; illustrated by Ram Papish ; foreword by John A. Wiens.

Description: Corvallis, OR : Oregon State University Press, 2022. | Includes bibliographical references
    and index.

Identifiers: LCCN 2022038409 | ISBN 9780870712166 (trade paperback) | ISBN 9780870712173 (ebook)

Subjects: LCSH: Birds—Conservation—Oregon—History. | Wildlife conservation—
    Oregon—History.

Classification: LCC QL684.O6 A88 2022 | DDC 598.09795—dc23/eng/20220816

LC record available at https://lccn.loc.gov/2022038409

∞ This paper meets the requirements of ANSI/NISO Z39.48-1992
    (Permanence of Paper).

Oregon State University Press

121 The Valley Library

Corvallis OR 97331–4501

541–737-3166 • fax 541–737-3170

www.osupress.oregonstate.edu

*Frontispiece:* California Condor. This endangered species was extirpated from Oregon early in the 20th century, but in May 2022 was reintroduced just south of the Oregon border by the Yurok Tribe. It is highlighted in the title of the book as a species for which we have great hope for its future recovery in Oregon. While the individual pictured here, Bird 463, tragically died of lead poisoning in 2020, hunters using non-lead ammunition will prevent this situation from occurring in Oregon which will save the species. Photo by Tim Huntington, webnectar.com

# DEDICATION

For the many ornithologists who have spent their lives contributing to the conservation of Oregon's avifauna. Let us hope we have all inspired another generation to carry on our efforts.

For Margaret Abigail Haig (1935–2020), our mother and Nanna, whose love of nature enveloped us and guided us to care for nature throughout our lives.

   SUSAN HAIG AND TASHI HAIG

For Juliet Dulany Roby (1916–2004) and Ross Roby (1914–2002), my parents, who supported and encouraged my childhood obsession with birds and other wildlife and tolerated me when I went off the deep-end with that obsession.

   DAN ROBY

## LAND ACKNOWLEDGMENT

The contributors to this book come from across the state called Oregon. Together we share a love and appreciation of the land it encompasses and concern for its past and future well-being. We offer our profound gratitude for those who have cared for it throughout the generations and for the opportunity to spend our lives traversing its diverse habitats and fragile ecosystems. We respectfully acknowledge and honor all Indigenous communities throughout time and into the future and are grateful for their vital presence. We extend our respect to the nine federally recognized Indigenous Nations of Oregon: the Klamath; Burns Paiute; Coquille; Confederated Tribes of Grand Ronde; Cow Creek Band of Umpqua Indians; Confederated Tribes of Umatilla; Confederated Tribes of Siletz; Confederated Tribes of Coos, Lower Umpqua, and Siuslaw; and the Confederated Tribes of Warm Springs. Additionally, we acknowledge all other tribes who have traditional connections to these lands, and any displaced Indigenous peoples who reside in Oregon. We recognize that Native communities were forcibly removed from the land where we now reside, teach, and carry out research. Thus it is our deepest hope that our lifetime commitments to conservation will demonstrate our deep respect for this land, which we recognize should by rights be under the stewardship of these Native communities.

# CONTENTS

# FIGURES

# TABLES

# AGENCY ABBREVIATIONS USED IN THIS BOOK

BLM         Bureau of Land Management

NMFS        National Marine FIsheries Service

NOAA        National Atmospheric and Oceanic Administration

NWR         National Wildlife Refuge

ODFW        Oregon Department of Fish and Wildlife

OSU         Oregon State University

USACE       US Army Corps of Engineers

USDA        US Department of Agriculture

USDI        US Department of the Interior

USFS        US Forest Service

USFWS       US Fish and Wildlife Service

USGS        US Geological Survey

# FOREWORD

Oregon is a geographically and ecologically diverse state. Its habitats range from the fertile farmlands of the Willamette Valley to the barren alkali flats of the Alvord Desert; from the moss-laden forests of the Coast Range to the sagebrush sea of the Great Basin; from the depths of Hells Canyon to the icy slopes and alpine meadows of the Cascade peaks. This variety is matched by a rich and varied avifauna. More than 540 bird species have been recorded in the state, a number surpassed by only a few other states.

This diversity of habitats and birds has attracted the attention of bird students—ornithologists—for well over a century. In Contreras et al.'s (2022) recent *History of Oregon Ornithology*, he describes the beginnings of ornithological investigations in Oregon, a period that saw the flowering of natural history: exploration, collection, description, observation, and management. Following the Second World War, ornithology began to change and grow as a scientific discipline. Natural history was still the foundation (as it always should be), but now there was an increased emphasis on quantification, statistical analysis, computer modeling, and genetics. Questions were now framed as hypotheses, and studies of birds were increasingly used to address theoretical constructs in other disciplines beyond ornithology, such as ecology, behavior, and evolution. As studies of birds grew in the scientific realm, public interest of birds also expanded: bird clubs proliferated, and "citizen science" blurred the distinction between formal science and bird-watching. Birds also became the focus of intense public controversies, pitting the lives of birds against the livelihoods of people. The impacts of Spotted Owls on forest practices in western Oregon or of waterfowl on water allocations in the Klamath Basin are but two examples.

All of this is part of the history of ornithology as a discipline. But history is made by people and therefore is just as much about the individuals who have fostered this disciplinary growth. People have built the natural history foundation of ornithology; brought new approaches, technologies, and questions to our studies of birds; enriched our knowledge about how birds live and how they affect our lives; and engaged the public to enhance appreciation of birds and promote their conservation. So, this book is about people.

But history is more than a telling of who did what and when. To understand how ornithology has grown and prospered in Oregon, one must also appreciate how people went about studying birds, and why they did so. The following pages relate many stories about many people. They illustrate the threads that have led Oregon to become, in Susan Haig's words, "a mecca for ornithology." Even so, they are but a sliver of the ornithological richness of the state. Many more people have helped to build our knowledge of the ecology, behavior, and conservation of Oregon birds. Here's a personal example. During the time I was on the faculty at OSU, from the mid-1960s to the late 1970s, my graduate students took full advantage of the diversity of birds and habitats that Oregon offered. They studied Common Murres and Brandt's Cormorants at Yaquina Head (Mike Scott), bird communities in oak woodlands of the Willamette Valley (Stan Anderson), water physiology of Sagebrush Sparrows in the Great Basin (Ralph Moldenhauer), breeding behavior of American Avocets at Summer Lake (Flash Gibson), song learning in Bewick's Wrens near Corvallis (Don Kroodsma), community structure of shrub-steppe birds north of Fort Rock (John Rotenberry), and prey localization by Northern Harriers at Malhuer (Bill Rice). The stories told here are just the tip of an iceberg of ornithology in Oregon that runs deep.

History, of course, is never finished; time's arrow always points to the future. The birds and habitats that people have been studying scientifically (and enjoying watching) for decades are dynamic, and the changes are becoming more profound every year. New species are making their way into the state, often with cascading consequences; the impacts of Barred Owls on Northern Spotted Owls are just one example. Other birds are changing their seasonal occurrence. When I first moved to Oregon, Anna's Hummingbirds rarely stuck around for the winter. Now they usually do, perhaps as a consequence of the year-round availability of food in feeders (I now need to be careful to keep a feeder heated during freezing weather). And the abundance of many species is declining.

There is a growing awareness that biodiversity is vital not just to the quality of our own lives, but to life itself. Conservation is essential, and ornithology is an important part of that conservation. Ornithology, in turn, will continue to be an unfolding history of the people who study birds, manage them, seek to conserve them, or simply devote their free time and energy to observing and enjoying them.

JOHN WIENS

## John Wiens

John Wiens joined the Department of Zoology at Oregon State University in 1966 and was chairman of the department from 1976 to 1978. He left OSU in 1978 to teach at the University of New Mexico and eventually Colorado State University. In 2002, Wiens's focus shifted from academia and ecological research to conservation. He joined The Nature Conservancy as chief scientist and then served as chief scientist for Point Blue (PRBO) Conservation Science from 2008 to 2012, when he returned to Oregon to retire.

Wiens's research in Oregon emphasized the ecology of birds in sagebrush shrub-steppe. More broadly, his research has focused on community ecology and behavior of birds and insects in semiarid environments and on spatial relationships, particularly landscape ecology. He also spent 25 years documenting the effects of the Exxon Valdez oil spill on seabirds.

Wiens served as president of the International Association for Landscape Ecology, and for eight years he edited *The Auk*. Wiens was recognized for his contributions to ornithological research as the recipient of the Elliott Coues Award from the American Ornithologists' Union and the Loye and Alden Miller Research Award for lifetime achievement from the Cooper Ornithological Society. He is a fellow of the American Ornithological Society.

*Photo by Ann Wiens*

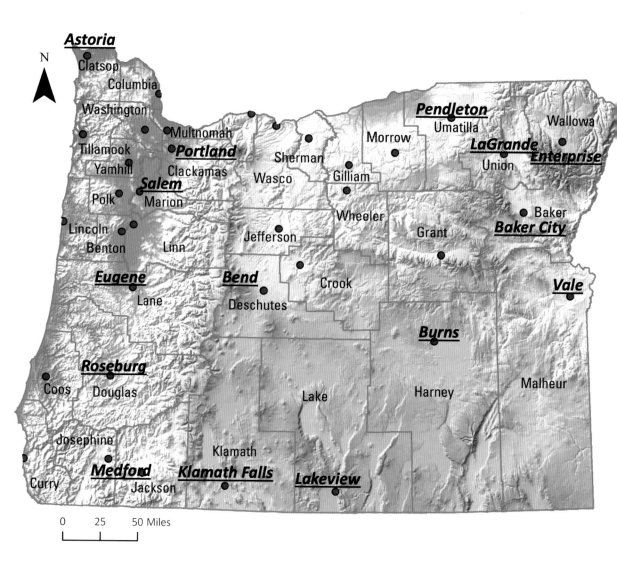

FIGURE 1. A topographic map of Oregon, with counties and key cities indicated.
Patti Haggerty, US Geological Survey, Corvallis, Oregon

A pair of Western Grebes "rushing" during their courtship display, one of the most complex and spectacular displays by a North American bird. This species nests in the wetlands and marshes of southern and eastern Oregon and winters along the coast of Oregon and on the Columbia River. Photo by Mick Thompson

# PREFACE

The deep and colorful history of early Oregon ornithologists provides enough of a contribution to the emerging field of ornithology to put the state in a place of prominence among pioneering avian endeavors in North America. From Native tribes' worship of California Condors and Bald Eagles to the early expeditions of Lewis and Clark and John Kirk Townsend up through William Finley, C. D. Littlefield, and David Marshall—the grand diversity of Oregon's varied ecosystems attracted a stunning array of brilliant minds to witness the numbers, diversity, ecology, and behavior of the state's birds.

Their initial work paved the way for a different breed of ornithologists, moving from descriptive to hypothesis-testing approaches and from appreciation of birds to deep concern over their future. Oregon has blossomed with ornithologists over the past 50 years, and their work has become more important with time. In this volume, we examine the development and emerging importance of ornithologists and their roles in defining and working to mediate some of the state's and region's most critical environmental issues. The various essays and chapters are written by those who worked directly on the issues they describe. Thus the perspectives are uniquely "straight from the horse's mouth" at a time when many are retired or late in their careers and have had time to reflect on the good and bad of what transpired.

Oregon's outstanding natural resources put the state's environmental crises on a national scale, and the history of substantive and credible research in support of management provides us with the opportunity to lead and provide best practices for others to follow. For example, for decades, newspapers across North America have highlighted the plight of the Northern Spotted Owl and related timber management issues. Restricted water rights owing to changing climates in the Klamath Basin involve multiple states and tribes. Furthermore, the ongoing discussion about renewable energy technologies clearly highlights the need to understand their effects on wildlife, especially birds. And now, in 2022, California Condors have been successfully released back to where they once soared over Oregon skies less than 100 years ago.

While this book describes the history of key environmental challenges Oregon birds have faced, it also reminds us of the deep personal ties ornithologists form with each other as we put our mark on history. For example, there are familial ties. Howard Marshall Wight came to OSU in 1916 (when it was Oregon Agricultural College), got his undergraduate and MS degrees in zoology, and went on to become a pioneer in the field of wildlife biology. Likewise, his son, Howard Morgan Wight, got his degree at OSU and returned in the 1960s to join the wildlife faculty and later lead the Oregon Cooperative Fish and Wildlife Research Unit.

Former OSU Professor of Zoology John Wiens and his son David Wiens have a similar story. Dave is an OSU alum who is currently a wildlife ecologist with the US Geological Survey (USGS) and a faculty member in OSU's Department of Fisheries, Wildlife, and Conservation Sciences. Or Chris Mathews, retired head of OSU's Biochemistry Department and former president of the Audubon Society of Corvallis, and his grandfather Robert Cushman Murphy. Murphy was an ornithologist with the American Museum of Natural History, former president of the American Ornithologists' Union, and an early Antarctic explorer for whom Murphy's Petrel was named. Or even Joan Hagar's uncle Joseph Hagar, who raised her father and who was the state ornithologist in Massachusetts. Her father, Donald Hagar, followed in his path to be an ornithologist and served on the Peregrine Falcon Recovery Team of the US Fish and Wildlife Service (USFWS) when DDT was discovered to cause eggshell-thinning in raptors and other birds throughout North America. Joan is currently a forest avian ecologist for USGS and is on the faculty in the College of Forestry at OSU.

When the early bird painter and ornithologist George Miksch Sutton spent time in Oregon, how could he have imagined that in the coming decades five generations of Oregon ornithologists would spring from his efforts? His undergraduate student John Wiens would become a faculty member at OSU, followed by his son David. Another Sutton student was my major professor, Lewis Oring, from the University of Nevada-Reno. We spent more than 20 years working on shorebird and wetland issues in southeastern Oregon. I have now spent 25 years with my students and postdocs working across the state and beyond addressing various endangered species and wetland issues. One of those students, Hope Draheim, joined the Genetics Department at the USFWS National Wildlife Forensic Lab in Ashland, Oregon, and another, Jesse D'Elia, leads endangered species recovery efforts for the USFWS in Portland. Among other things, Jesse has worked to bring about the return of the California Condor to Oregon. They now are training their own generation of avian conservation biologists.

This book is a thumbnail sketch of some of the most complex environmental issues that our state has faced, with birds at the center of the debate. And it is a celebration of what has become a mecca for ornithology, a tribute to some of the many ornithologists who have dedicated their lives to conservation of our natural resources. May we learn from this history as we face our most serious threat to Oregon's birds and the environment to date: rapid anthropogenic climate change.

SUSAN HAIG

Western Screech-Owl in a tree cavity used as a daytime roost. Western Screech-Owls are vulnerable to predation from Barred Owls, a new arrival to the Pacific Northwest from eastern North America. Photo by Alan Dyck

# ACKNOWLEDGMENTS

This book was an enormous undertaking on the part of all the authors, photographers, artist Ram Papish, editors, and reviewers. Without each person coming through with their essays, photos, and more, the book would not have delivered cogent assessments about the critical conservation efforts undertaken in Oregon over the past 50-plus years. Thus, we gratefully acknowledge all who contributed.

Text editor Susan Haig thanks the following for helpful discussions related to ornithology and conservation in Oregon: Alan Contreras, Bruce Dugger, Eric Forsman, Chuck Henny, Molly Monroe, Maura Naughton, Lew Oring, Carrie Phillips, Fred Ramsey, Carol Schuler, and Mark Stern. We are especially grateful to Eric Forsman for his dogged editing efforts and to the following for reviewing all or parts of the book: Blair Csuti, Alan Franklin, Barb Haig, Garth Herring, Bruce Marcot, Barbara Ralston, and J. Michael Scott.

As photo editor, Dan Roby had unprecedented support from the photographers who selflessly donated use of their images for free. The list of photographers and their photos can be found on page 319. We further appreciate the use of archival photos from The Evergreen State College, the US Fish and Wildlife Service, and Malheur Field Station. We are particularly grateful for technical help with photographic issues from Tim Lawes. Biography editor Tashi Haig had tremendous support in tracking down histories and photos from the authors contributing to the book (list is found on page 317) and many others.

Oregon's Audubon chapters were great about digging up and sharing their histories: Marcia Cutler and Jim Fairchild (Audubon Society of Corvallis); Bob Sallinger (Portland Audubon); Mary Shivell (East Cascades Audubon Society); Maeve Sowles (Lane County Audubon); Erin Ulrich (Rogue River Audubon); and Dawn Villaescusa (Audubon Society of Lincoln City).

We sincerely thank Patti Haggerty (USGS, Corvallis, Oregon) for creating the topographic maps of Oregon, Molly Jensen Haig for creating the Oregon wetlands map, and David Ziolkowski (USGS, Laurel, Maryland), who created the map of the current Oregon Breeding Bird Survey routes. Kelly Haig Huber produced the Literature Cited section and helped with other editing. Ez Shaughnessy Jandrasi produced the Appendix, and Devon Thomas created the Index.

We are enormously grateful to Katie Dugger and the USGS Oregon Cooperative Fish and Wildlife Research Unit; Selina Heppell and the OSU Department of Fisheries, Wildlife, and Conservation Sciences; and Nanette Seto of the Migratory Birds and Habitat Program of the US Fish and Wildlife Service for providing the funds to cover printing costs for this book.

We have had great support from OSU Press and are particularly grateful to Kim Hogeland, Marty Brown, and Micki Reaman for their help throughout the process of developing and publishing this book. We thank Ashleigh McKown for copyediting and Erin Kirk for layout and design of the book.

For contributors with USGS affiliations (Katie Dugger, Elise Elliott-Smith, Joan Hagar, Chuck Henny, and David Wiens), we provide this required statement: Any use of trade, product, or firm names is for descriptive purposes only and does not imply endorsement by the US government.

SUSAN HAIG, DAN ROBY, AND TASHI HAIG

# THE BIRDS OF OREGON
## PEPPER TRAIL

The bald eagle and the beautiful buzzard of the Columbia still continue with us.
—MERIWETHER LEWIS, 3 January 1806

Lewis and Clark began to write this book
coming down the Columbia, first to report the news
of the marvels familiar to the Native peoples
the "great numbers of wild fowls of various kinds"
the Lewis's Woodpeckers in the big pines
the Clark's Nutcrackers in the mountain heights
the Bald Eagle and the beautiful buzzard
—the California Condor—
soaring over the tumble and roar of Celilo Falls

Celilo is silent now, drowned behind a dam
And condors have not carved the air of Oregon
For a century or more, but our birds are a marvel still
Their voices echoing in every wild place
The meadowlarks' sweet songs our valley music
Oystercatchers yelping louder than the crashing surf
Sage Grouse booming in the high desert dawn
Rosy-finch calls cutting the Steens' cold wind
Spotted Owls moaning to each other in the dark

The Bald Eagle embodies nature's bright possibility
Nearly gone, now returned to fierce abundance
Others—endangered murrelet and owl—
Remain our task to preserve for those to come
While the "beautiful buzzard" is a soaring symbol
Of wild hope, the great condor soon to be returned
to Oregon's skies, the dream of many years—
for dreams, like birds, have wings, and together
we fly.

OPPOSITE: California Condor; also known to the Lewis & Clark Expedition as
"the beautiful buzzard." Photo by Tim Huntington, webnectar.com

N

Lewis & Clark
NWR

Wa

0    25    50 Kilometers

0         25        50 Miles

Three Arches
NWR

Cape Meares
NWR

Wapato Lk
NWR

Tualatin Lk
NWR

Nestucca Bay
NWR

Baskett
Slough
NWR

Willamette R.

Siletz Bay
NWR

Ankeny
NWR

Santiam R.

Santiam R.

Deschutes R.

John

Coast Range

Cascade Range

Pacific Ocean

WL Finley
NWR

McKenzie R.

Fern Ridge
Reservoir

Willamette R.

Crooked R.

Harney

Umpqua R.

Bandon Marsh
NWR

Crater Lk

Klamath Marsh
NWR

Summ

U. Klamath
NWR

Sycan R.

Sprague R.

Rogue R.

Oregon Islands
NWR

Klamath
Mtns.

Klamath Lk.

L. Klamath
NWR

Goose Lk.

California

ngton

Umatilla
NWR

McNary
NWR

Cold Springs
NWR

Columbia R.

McKay Ck.
NWR

Zumwalt
Prairie

Blue Mts

Wallowa Mts

Snake R.

Columbia Plateau

Idaho

Strawberry Mts

Deer Flat
NWR

Warm Springs
Res.

Malheur R.

Lk, Owyhee

Basin

Malheur
NWR

Malheur Lk.

Harney Lk

Owyhee R.

Steens Mts

Lk Abert

Warner Wetlands

Hart Mtn.
NWR

FIGURE 2. A topo-
graphic map of Oregon,
with significant
natural features and
sites of importance to
birds indicated. Patti
Haggerty, US Geological
Survey, Corvallis,
Oregon

Sheldon
NWR

Nevada

# Early History

SUSAN HAIG

The publication of Aldo Leopold's *Game Management* in 1933 and *A Sand County Almanac* in 1949 helped transform the field of ornithology from identifying birds and building museum specimen collections to thinking about the habitats that birds occupied, their essential needs for survival throughout the annual cycle, and threats they faced. Leopold went a step further and suggested that land could be manipulated or managed to benefit a species or group of species. At that time, most attention was focused on waterfowl and upland game birds (e.g., quail, grouse, turkeys). Even though the Migratory Bird Treaty Act was passed in 1918 and prohibited hunting of nongame birds, it was not until the 1980s that attention became focused on management efforts to benefit nongame birds (Table 1). It was yet another decade before the emerging field of conservation biology came

William Finley in the field, with mule and tripod.

into play and ornithologists began thinking about maintaining biodiversity and genetic issues related to small populations. In Oregon, the 1993 Northwest Forest Plan was another turning point for birds in the state. And we are now in an age where changing climates dictate our focus. Thus, in the past 50 years, we have advanced our thinking about birds more than the previous millennia. Sadly, we have also lost more habitat and bird diversity in the past 50 years than during the previous millennia.

Oregonians have always been at the forefront of issues related to ornithology. Our incredibly diverse state provides for enormous biodiversity, yet each ecosystem has its fragilities, of which Oregon birds are usually the best indicators (see the Appendix for a list of the regularly occurring bird species in Oregon, including their scientific names, as well as those discussed in this book). In this book we examine the roles that birds and Oregon ornithologists have played in the ever-changing paradigms of environmental conservation. The efforts of state, federal, and tribal biologists—along with university researchers, bird conservation groups, and private citizens—are inextricably interwoven into the recent history of protecting and restoring Oregon's avifauna. Here we attempt to untangle the complex web of issues, research, management, conservation, and Oregonians' love of birds as they influence the fortunes of bird populations throughout the state.

In this book we also highlight some of the noteworthy Oregon ornithologists who have spent their careers dedicated to a better understanding of Oregon birds and their conservation. There are many others who trained at Oregon universities and went on to accomplish so much in other locations (e.g., Jerry Bertrand, Gordon Gullion, Richard Reynolds, Michael Scott, Noah Strycker), but we simply do not have the room to highlight them all. It is a tragedy of riches.

David Marshall.

To put the book into perspective, we must briefly revisit the contributions of the most influential ornithologists in Oregon's history: William L. Finley (1876–1953) and David B. Marshall (1926–2011). William L. Finley (his uncle William A. Finley was the first president of Oregon State University) was born of a pioneering family in California in 1876 and moved to Portland when he was 10 years old. His love of nature, travel, and photography sent him on incredible journeys to Alaska, Mexico, and beyond. Eventually, he settled in Oregon, and his photos persuaded President Theodore Roosevelt to protect Tule Lake and Lower Klamath Lake. He also was credited with convincing Roosevelt to protect the Three Arch Rocks area on the Oregon Coast, where one of the rocks is named for Finley. Finley served on the board of what became the National Audubon Society, was the second president of what became Portland Audubon, and was twice appointed to the Oregon Wildlife Commission. To honor his accomplishments, the William L. Finley National Wildlife Refuge near Corvallis was named for him.

David Marshall was an Oregonian through and through—born in Portland, graduated from OSU in wildlife, and spent his life protecting endangered species and establishing federal wildlife refuges, including Ankeny, Baskett Slough, and Finley National Wildlife Refuges (Grover and Grover 2000, Marshall 2008). He received his first job (as a technician at Malheur National Wildlife Refuge) from Stanley Jewett (of Gabrielson and Jewett [1940] fame). Later, during a brief stint in Washington, DC, he became chief of endangered species for the US Fish and Wildlife Service (USFWS) and helped develop the Endangered Species Acts of 1966, 1969, and 1973. He returned to Oregon to assume a similar role in the USFWS regional office in Portland. While proposing land to be protected for the Finley refuge, Marshall invited Ira Gabrielson (then president of the Wildlife Management Institute in Washington, DC) to investigate controversial development raised by the Government Accounting Office about land acquisition. The issues magically disappeared after his visit. Even in retirement, Marshall was keen to help with bird conservation measures and served as a consultant to the Oregon Department of Fish and Wildlife (ODFW). In 2003, he coauthored *Birds of Oregon* with Alan Contreras and Matt Hunter. Dave's books and papers are housed in the David B. Marshall Library, at the Columbia River Intertribal Fish Commission's Streamlet Library in Portland.

Stanley Jewett. Courtesy of the American Ornithologists' Union

Bureau of Biological Survey chief Ira Gabrielson (1889–1977) casts a line off a vessel of the Upper Mississippi River National Wildlife and Fish Refuge during a 1937 congress. Courtesy of the US Fish and Wildlife Service

OPPOSITE: Northern Spotted Owl. The twin threats of loss of old-growth late successional forest habitat and displacement by the invasive Barred Owl are driving the continued population decline of this threatened species in Oregon. Photo by Jared Hobbs

**TABLE 1. State and federal activities of importance to Oregon birds**

| Date | Event |
|------|-------|
| 1805 | The journals kept by Meriwether Lewis and William Clark refer to abundant fish, elk, and other wildlife, including California Condors, in the area that is now Oregon. |
| 1848 | Oregon becomes an official territory of the United States. |
| 1850 | California Condors are no longer regularly seen along the Columbia River, although they are seen elsewhere in interior Oregon. |
| 1859 | Oregon is admitted to the Union. |
| 1865 | The first fish cannery is built on the Columbia River; it packs 4,000 cases of salmon during its first year of operation. |
| 1871 | The Oregon Legislature passes the first game laws; April–July is closed for the taking and selling of swans and certain ducks. |
| 1882 | Ring-necked Pheasants from China are successfully introduced in the Willamette Valley of Oregon, the first successful introduction of this gamebird species in the United States. |
| 1885 | The Section of Economic Ornithology is established within the USDA Division of Entomology. |
| 1891 | 30,000 Ring-necked Pheasants are harvested in Linn County, nine years after the first successful introduction of pheasants in the United States. |
| 1893 | Hollister McGuire is appointed the first State Fish and Game Protector, making him the first employee of the agency that eventually becomes the Oregon Department of Fish and Wildlife. A total of $2,000 was appropriated for salaries and office expenses. |
| 1896 | The federal Division of Economic Ornithology and Mammalogy is renamed the Division of Biological Survey. |
| 1899 | The Fish Commission and the Office of the Game and Forestry Warden are established as separate federal agencies. The Fish Commission regulates commercial fishing. The Game and Forestry Warden oversees game, game fish, and forestry laws. |
| 1900 | The Lacey Act of 1900 is passed by Congress. It became the first federal law protecting wildlife, establishing civil and criminal penalties for the illegal trade of wild animals and plants across state lines. |
|      | Gray Partridges from central Europe are successfully introduced to western Oregon. |
| 1901 | L. B. W. Quimby is hired as the first Game and Forestry Warden. He prints 5,000 copies of the fish and game laws at his own expense, stating that "People can't abide by these laws if they don't know what they are." |
|      | Oregon bag limits are set at 50 ducks per day. |
| 1902 | A Rainier lumber mill receives the state's first pollution fine ($50), for dumping sawdust into the Columbia River and damaging commercial fishing. |
| 1904 | Last reliable observation of wild California Condors in Oregon, near Drain. |
| 1905 | An Oregon resident hunting license costing $1 is established, along with the first State Game Fund for the management of fish and wildlife. Selling of game is prohibited. |
|      | The federal Division of Biological Survey is renamed the Bureau of Biological Survey. |
| 1913 | Geese are included in a 30 waterfowl per week bag limit in Oregon. |

1918    Congress approves the Migratory Bird Treaty Act, whereby Canada, Mexico, and the United States work together to protect migratory birds. The act grants authority to the US Department of the Interior to establish regulations for hunting, transport, possession, sale, and killing of migratory birds. It also grants authority to the states to implement laws for hunting seasons and protection, as long as the state laws are consistent with, or more restrictive than, the federal law.

1921    The State Board of Fish and Game Commissioners is split into two agencies, the Fish Commission of Oregon and the Oregon State Game Commission.

1929    The Migratory Bird Conservation Act creates a commission to consider areas to be purchased and sets prices. The act made way for development of the National Wildlife Refuge System in the United States.

1931    Responsibility for fish and game law enforcement in Oregon is transferred to the Oregon State Police.

1934    Congress passes the federal Migratory Bird Hunting and Conservation Stamp Act (Duck Stamp Act).

1937    The first students from the Oregon State University Fish and Wildlife program are graduated.

        The US Federal Aid in Wildlife Restoration Act (Pittman-Robertson Act) passes and provides funding to the states for wildlife restoration programs.

1938    Construction of Bonneville Dam on the Lower Columbia River is completed.

1939    The federal Bureau of Biological Survey and the Bureau of Fisheries are transferred to the Department of the Interior.

1940    The federal Bald Eagle Protection Act is passed.

        The federal Bureau of Biological Survey and the Bureau of Fisheries are merged to become the US Fish and Wildlife Service.

1941    The Oregon Legislature delegates authority to the Oregon Game Commission to set seasons and bag limits.

1944    Summer Lake Wildlife Area becomes the first state wildlife area, with the aid of funds from the Pittman-Robertson Act.

1951    Chukar partridges are successfully introduced to Oregon from India.

1961    Merriam's Wild Turkey is successfully introduced to Oregon.

        The federal Wetlands Loan Act is passed by Congress.

1962    The federal Golden Eagle Protection Act is appended to the Bald Eagle Protection Act of 1940.

1966    The National Wildlife Refuge System Administration Act is passed by Congress.

1970    The National Environmental Policy Act (NEPA) is passed by Congress. It is a major environmental law that promotes the enhancement of the environment through analysis of the impacts of federal projects and establishes the president's Council on Environmental Quality (CEQ).

        The federal Clean Air Act is passed.

1971    The Oregon Legislature delegates responsibility for 235 nongame wildlife species to the Game Commission.

| 1972 | The federal Marine Mammal Protection Act is passed. |
|------|------|
| | The federal Clean Water Act is passed. |
| | The sale or use of DDT is prohibited in the United States by the Environmental Protection Agency. |
| | The Migratory Bird Treaty Act is amended to include protection for an additional 32 families of birds, including eagles, hawks, owls, cormorants, and corvids (crows and jays). |
| 1973 | The US Endangered Species Act is passed by Congress. |
| | The Oregon Game Commission is renamed the Oregon Wildlife Commission. |
| 1975 | The Oregon Fish Commission and the Oregon Wildlife Commission are merged to form the Oregon Department of Fish and Wildlife (ODFW). |
| | CITES (Convention on International Trade in Endangered Species of Wild Flora and Fauna) enters into force. |
| 1976 | Oregon wildlife violations are elevated to Class C felonies. |
| 1979 | Oregon begins the Wildlife Checkoff option on state income tax forms, earning $337,000 for management of nongame wildlife. |
| 1980 | The federal Northwest Power Planning Act gives fish and wildlife resources equal footing with power generation, irrigation, and flood control in the Columbia River Basin. |
| 1983 | The first Oregon Migratory Waterfowl Stamp is issued by ODFW. |
| 1985 | Peregrine Falcons are reintroduced to the Columbia River Gorge for the first time in 30 years. |
| | A portion of Oregon's share of income from the first Oregon Migratory Waterfowl Stamp goes to Ducks Unlimited to build 400 nesting islands on the Copper River Delta in Alaska for Dusky Canada Geese that winter in the Willamette Valley. |
| 1986 | The International Convention on Wetlands (RAMSAR) enters into force. |
| 1987 | The Oregon Legislature passes the Oregon Endangered Species Act. |
| 1988 | The Intergovernmental Panel on Climate Change (IPCC), the United Nations body for assessing the science related to climate change, is formed by the World Meteorological Organization and the UN Environment Programme. |
| 1989 | The North American Wetlands Conservation Act (NAWCA) enters into force. |
| | An Oregon Upland Game Bird stamp is approved. |
| 1990 | Five new Peregrine Falcon nests are found, bringing the state total to 16. ODFW hacks 26 young Peregrine Falcons at five additional sites and places two foster chicks in a southwest Oregon nest. |
| | The Northern Spotted Owl is listed as a Threatened species under the federal Endangered Species Act. |
| 1992 | The Marbled Murrelet is listed as a Threatened species in California, Oregon, and Washington under the federal Endangered Species Act. |
| 1993 | The federal Forest Ecosystem Management Assessment Team (FEMAT) issues its report on Forest Ecosystem Management. |
| | The Western Snowy Plover is listed as a Threatened species under the federal Endangered Species Act. |
| | Oregon Department of Fish and Wildlife celebrates its 100th anniversary. |

| 1994 | The federal government adopts the Northwest Forest Plan. |
|---|---|
| 1999 | A 639-foot wood chip vessel named the *New Carissa* runs aground near Coos Bay with 400,000 gallons of bunker and diesel fuel aboard. Up to 70,000 gallons of fuel are spilled into the water and on the beach. ODFW and other responders document 200 dead birds, including 2 Snowy Plovers and 24 Marbled Murrelets, which are protected under Endangered Species Acts at the state and federal level. About 40 percent of the population of 100 Western Snowy Plovers in Oregon is found to have traces of oil. |
| | The Peregrine Falcon is delisted (removed) from the federal list of threatened and endangered species under the Endangered Species Act. |
| | The Oregon Legislature elevates the previously named Governor's Watershed Enhancement Board to a new independent department named the Oregon Watershed Enhancement Board (OWEB). |
| | OWEB receives lottery proceeds and General Fund revenue to fund watershed health projects and coordinate salmon restoration efforts among all state agencies and local watershed councils. |
| 2000 | The Neotropical Migratory Bird Conservation Act establishes a matching grants program to fund projects that promote the conservation of migratory birds in the Caribbean, Latin America, and the United States. |
| 2007 | The Peregrine Falcon is delisted (removed) from the state's list of threatened and endangered species under the Oregon Endangered Species Act. |
| | The Bald Eagle is delisted from the federal list of threatened and endangered species under the Endangered Species Act. |
| 2008 | The Lacey Act is amended to regulate the importation of any wild species protected by international or domestic law and prohibits the spread of invasive or non-native species of flora and fauna. |
| 2009 | The California Brown Pelican is delisted from the federal list of threatened and endangered species under the Endangered Species Act. |
| 2017 | The Trump administration issues a new interpretation of the Migratory Bird Treaty Act that eliminates industry liability for incidental take of birds protected under the act. |
| 2021 | In January, the US Fish and Wildlife Service publishes a final rule that eliminates industry liability for incidental take of protected migratory birds. Nine months later the USFWS revokes that rule and reinstates prohibitions on incidental take. |
| 2022 | An initial cohort of California Condors are released in the Bald Hills of Redwood National Park within Yurok Ancestral Territory (~8 miles south of the Oregon border) with the hope that they will soon return to their ancestral home in Oregon. |

# Professional Ornithologists

SUSAN HAIG

From the time of Finley to Marshall, there was generally little emphasis on academics for ornithologists or wildlife biologists. People with college degrees were often looked down upon, as it was assumed that they did not have the practical experience to handle the job. That changed in the 1950s and 1960s when a college degree, if not an advanced degree, became more the norm. Perhaps it was the influence of *A Sand County Almanac* and/or the availability of funds to attend college via the GI Bill for returning World War II veterans, but it was during this time that various Oregon institutions of higher learning began to offer training in wildlife ecology and ornithology.

## Oregon State University

Oregon State University has been the state's flagship university for bird study from the late 1940s onward, training outstanding ornithologists like the ones outlined in this book. Perhaps the greatest growth in ornithological expertise has been in OSU's Department of Fisheries, Wildlife, and Conservation Sciences, although ornithologists have also found their way into the College of Forestry (Matt Betts, Joan Hagar, Brenda McComb, and Jim Rivers), the Range Department (Jonathan Dinkins), the Department of Integrative Biology (formerly the Zoology Department; Jamie Cornelius, Meta Landys, Douglas Warrick, and John Wiens), and the Department of Statistics (Fred Ramsey). Taken together, today, OSU has more doctoral-level ornithologists studying wild birds than almost any other place in the country, with the exception of Cornell University.

The growing influx of avian-related faculty members at OSU came as the result of a strategic decision by OSU in the 1970s to involve federal scientists as courtesy faculty members and to allow them to participate in training graduate students and teaching classes. This brought several federal research units to campus and resulted in millions of dollars of research funding awarded to OSU, training of many more graduate students, more classes being offered, job opportunities for undergraduate and graduate students, and greater contributions to Oregon conservation efforts than would ever have been possible otherwise. This all came at little cost to OSU or, more accurately, with great benefit to OSU, not only through the training opportunities for students but also through the millions of dollars of overhead awarded to the university from contracts brought in by federal scientists who serve as unpaid faculty members. This model has been highly successful for OSU, which has been recognized as offering one of the top graduate programs in conservation biology in the country and having one of the most productive wildlife faculty in the country.

The Department of Fisheries, Wildlife, and Conservation Sciences was established in 1935 as the Department of Fish, Game, and Fur Animal Management (OSU Department of Fisheries and Wildlife Comprehensive Review 1935–85). That year, the first ornithologist hired was Arthur Einarson, who wrote *The Gulls (Laridae) of the World; Their Plumages, Moults, Variations, Relationships and Distribution* (1925). Until retirement in 1957, Einarson and his students studied waterfowl and Ring-necked Pheasants. His work on waterfowl was carried on by Robert Jarvis when he joined the OSU faculty in 1971. Upland gamebird research was conducted by John Crawford and his students from 1974 to 2010. Upon the retirement of Jarvis and Crawford, OSU replaced them with Bruce Dugger (filling Bob Jarvis's position as a waterfowl and wetlands ecologist) and avian

## Howard Marshall Wight and Howard Morgan Wight

Prior to Howard Marshall Wight coming to Oregon in 1916, he had been a soldier in World War I, where he was awarded several medals for valor. Shortly thereafter, he arrived in Corvallis to pursue his M.S. degree while teaching in zoology. He attained his M.S. in 1918 and left Oregon State University in 1928 to join the faculty at the University of Michigan. There he rose to be an associate professor of forest zoology and a pioneer in the emerging field of wildlife management. His book *Field and Laboratory Technic in Wildlife Management* was one of the first wildlife technique manuals written. He died in 1942 at age 53.

His son, Howard Morgan Wight, was born in Corvallis in 1923. His family moved to Michigan when he was five years old. He began college at Hillsdale College in

Michigan and then joined the army during World War II. After the war, he finished his B.S. degree at OSU and went on to get his M.S. at Penn State. Several years later, in the 1960s, he joined the wildlife faculty at OSU, where he studied migratory birds. Upon reopening of OSU's Oregon Cooperative Fish and Wildlife Research Unit in 1971, Wight became the unit leader until his death in 1975. During his tenure at OSU, he directed graduate studies for notable avian ecologists Eric Forsman, Brad Griffith, Chuck Henny, Richard Reynolds, and Mike Vaughn, among others.

forest ecologist W. Douglas Robinson in 2002. Susan Haig was the first female ornithologist to join the wildlife faculty in 1994. Patricia Kennedy became the first OSU-paid (non-federal) female ornithologist when she joined the wildlife faculty in 2002. The first female Ph.D. student graduated in 1993: Sue Shaeffer defended her doctoral dissertation on the population ecology of Dusky Canada Geese (*Branta canadensis occidentalis*) under Bob Jarvis.

Einarson was a federal biologist and began what is known today as the US Geological Survey's (USGS) Oregon Cooperative Fish and Wildlife Research Unit. An effort by the federal

## Arthur Einarson

Arthur Einarsen (1897–1965) was a field biologist for the state of Washington's Department of Fisheries and rose to become chief biologist for the state's Department of Game in 1932. In 1934, he moved to Corvallis, where he served as the first unit leader of the then USFWS's Oregon Cooperative Wildlife Research Unit from 1935 to 1957. Upon joining the cooperative research unit, Einarsen directed graduate studies on waterfowl and antelope, and authored a variety of papers on species specific to the Pacific Northwest: "Ring-necked Pheasant" (1945), "Pronghorn Antelope" (1948), "Deer of North America" (1956), and "Black Brant: Sea Goose of the Pacific Northwest" (1965). After Einarsen's death in 1965, the Northwest Section of the Wildlife Society established the Arthur S. Einarsen Award in 1966 to recognize outstanding professionalism in services dedicated to wildlife of the Pacific Northwest. Awardees have included ornithologists Jack Ward Thomas (1981), Charles Meslow (1991), John Crawford (2001), Joe Lint (2010), and Dan Edge (2016).

government to provide graduate training for wildlife scientists, the Cooperative Research Unit program began nationally in 1935, and the Oregon Unit was the second one to be established. The Oregon Unit was closed in 1959 but restarted in 1971 under the leadership of Howard Wight, already an OSU faculty member since the 1960s. Wight died of cancer in 1976 and was succeeded by Charles Meslow in 1975. It was under Wight's and then Meslow's supervision that Eric Forsman began his pioneering and monumental studies of the Northern Spotted Owl (*Strix occidentalis caurina*). Eric's background gave him the ability to see many sides of the complex owl issues in the Pacific Northwest. Chuck Meslow was joined in the unit by Robert Anthony as assistant unit leader in 1977. Upon Meslow's retirement in 1995, Anthony became unit leader in 1995, and seabird biologist and physiological ecologist Daniel Roby joined the unit and faculty. He was joined by quantitative ecologist Katie Dugger upon Anthony's retirement in 2010. Bob Anthony and Katie Dugger's involvement in the enormous undertaking of evaluating Northern Spotted Owl demography and status every five years continues to be the largest such effort ever conducted for a bird species.

A second federal research center of significance to Oregon ornithology was the (now) USGS Forest and Rangeland Ecosystem Science Center (FRESC). FRESC was conceived by raptor biologist Michael Collopy in 1992 and was loosely modeled after the Cooperative Research Unit program. FRESC scientists are housed on the OSU campus and hold faculty appointments in various OSU departments, although principally in the Department of Fisheries, Wildlife, and Conservation Sciences. Susan Haig was the first ornithologist hired by FRESC (in 1994). Soon, raptor ecologist and contaminant biologist Charles Henny, an Oregon native, joined FRESC. He was followed by Joan Hagar (forest birds), Collin

Eagles-Smith (avian contaminant biologist), and David Wiens (raptor ecologist).

A third federal research center on campus is the US Forest Service Pacific Northwest Forest Experiment Station. It was there that Eric Forsman spent much of his career working on Northern Spotted Owls.

OSU's avian ecology program has expanded, so that currently the OSU AVES group (avian faculty and students) and the OSU Bird Nerds (undergraduate bird club) have annually produced the Willamette Valley Bird Symposium (a community outreach effort) since 2014. The undergraduate OSU Bird Nerds were the first official student bird club recognized by the American Ornithological Society.

OSU has produced influential ornithologists from early on, and while ornithology and wildlife ecology have traditionally been male-dominated fields, the early women attaining their advanced degrees in wildlife ecology at OSU went on to make significant contributions to Oregon and beyond. Gay Simpson (M.S., OSU 1976) was the first woman waterfowl biologist for South Dakota Game and Fish. Following her untimely death at age 40, the Central Flyway Council created the Gay Simpson Award for Waterfowl Students at South Dakota State University. Rebecca Goggans (M.S., OSU 1987) was the first woman wildlife biologist hired by ODFW and worked on nongame issues. Maura Naughton (M.S., OSU 1997) was a USFWS seabird biologist, USFWS shorebird coordinator, wildlife biologist at Finley Refuge, and became acting director of migratory birds for USFWS in Portland. Kim Nelson (M.S., OSU 1989) went on to lead research and conservation efforts for Marbled Murrelets for more than 30 years. And Carol Schuler (M.S., OSU 1987) went on to lead many programs within the USFWS and the USGS at the local, regional, and national levels, including testifying before the US Congress.

## Fred Ramsey

Fred Ramsey grew up in Portland birding with his brother from a young age. He attended the University of Oregon and Iowa State University before returning to a faculty position in the Department of Statistics at Oregon State University. Over a 40-year period, there were few ornithology graduate students at OSU who did not have a helpful Fred Ramsey on their committee. He may have taken the back door to ornithology by becoming a statistician, but he is responsible for the success of many graduate students (and their major professors). When not advising students, Fred spent a great deal of time consulting on environmental issues around the world. He was a key advisor and consultant to the US Fish and Wildlife Service and the EPA in Oregon and beyond. He is largely credited with the development of statistical methods for analyzing data from avian point-count surveys.

Fred's birding passion led him to be one of Oregon's top birders, publishing his *Birding Oregon* in 1979. Fred was a founding member and president of the Audubon Society of Corvallis. He has been the field trip chair, which includes stopovers at the Oregon Shakespeare Theatre, for the ASC for decades.

*Courtesy of the Audubon Society of Corvallis*

## E. Charles "Chuck" Meslow

Chuck Meslow has been a fixture in Oregon avian research since he moved to Oregon in 1971 and subsequently became the assistant unit leader of the Oregon Cooperative Wildlife Research Unit at Oregon State University in 1976. He mentored at least 31 graduate students who worked on a broad range of birds from Spotted Owls to House Wrens. His long list of graduate students includes many people who made significant contributions to avian conservation in Oregon and elsewhere, including but not limited to Evelyn Bull, Eric Forsman, Kit Hershey, Ruth Wilson Jacobs, David Johnson, Bill Mannan, Bruce Marcot, Tom McCabe, Kim Mellen, Gary Miller, Michael Morrison, Richard Reynolds, Keith Swindle, and Brian Woodbridge. Chuck initiated two of the long-term Spotted Owl demographic studies in Oregon, and as graduate advisor to Eric Forsman in the 1970s, he helped to develop the first radiotelemetry study of Spotted Owls in Oregon. Chuck served on the Interagency Scientific Committee and the Forest Ecosystem Management Assessment Teams, which eventually became the Northwest Forest Plan. During his long career, Chuck was an outspoken advocate for research on forest birds and mammals. He was also an active member of The Wildlife Society, for which he served as president in 1985 and 1986. Chuck retired from OSU in 1994 and then spent an additional five years as the western field representative for the Wildlife Management Institute. He was awarded the David B. Marshall lifetime achievement award by the Oregon Chapter of the Wildlife Society in 1991 and the prestigious Aldo Leopold Award by The Wildlife Society in 2005.

*Photo by Eric Meslow*

## Other Academics

As a land-grant school, OSU is bound by law to carry out research in the interest of the state. Without that mandate, other colleges and universities in the state have the freedom to develop their own curriculum. Most avian faculty at these institutions have focused on training undergraduates in avian ecology so they can go on to more focused graduate programs. For example, Michael T. Murphy joined the Portland State University faculty in 2000. Stewart Janes recently retired as an ornithologist on the Southern Oregon University biology faculty (see the Birdsong and Acoustic Studies section on p. 33). David Craig teaches biology at Willamette University, and Katie O'Reilly is an eco-physiologist who studies shorebirds and seabirds at the University of Portland (see the Bird Behavior section on p. 36).

# Ornithological Laboratories

## National Fish and Wildlife Forensic Laboratory

PEPPER TRAIL

The National Fish and Wildlife Forensic Laboratory in Ashland, Oregon, is a unique facility that has earned a place in the history of Oregon ornithology. The lab is part of the USFWS's Office of Law Enforcement, and its mission is to provide scientific analysis in support of federal wildlife crime investigations. It is the only federal wildlife forensic facility in the United States and provides training and analytical support for international partners through joint enforcement of the Convention on International Trade in Endangered Species (CITES) of Wild Fauna and Flora. The lab's rather surprising location in the small southern Oregon town of Ashland was achieved through the offices of Oregon Senators Mark Hatfield and Bob Packwood, who supported the lab's creation.

The National Fish and Wildlife Forensic Laboratory opened in 1989, and one of the first staff members hired was ornithologist Beth Ann Sabo. Sabo came to the lab from the Smithsonian Institution, where she trained with Roxie Laybourne, the legendary pioneer in the field of feather identification. This meticulous training in the details of feather structure and appearance was ideal preparation for work with bird evidence in wildlife crime investigations. Such evidence is often partial, modified, or degraded and frequently consists of nothing more than loose feathers.

Sabo established the lab's bird specimen collection and developed many of the identification protocols that are still in use. The lab's bird collection is tailored to the needs of wildlife law enforcement and now includes more than 1,600 species, representing virtually all birds regularly encountered in casework. In addition to standard skin and skeleton specimens, the collection includes artifacts with feathers, sets of loose flight feathers from hundreds of species, and hundreds of glass slides with feather barbs mounted for microscopic examination.

Sabo left the lab in 1998, and Pepper Trail was hired as the new ornithologist. Although Trail came to the position with significant ornithological research experience, he lacked formal training in feather identification. His first major case involved thousands of mixed feathers from a variety of North American raptor species, providing an excellent—if daunting—introduction to the most important group in forensic ornithology casework. In 2017, the lab's ornithology team gained a second member with the hiring of Ariel Gaffney.

From 1989 to 2020, ornithologists at the National Fish and Wildlife Forensic Laboratory analyzed bird evidence in more than 2,500 investigations, completing well over 100,000 identifications. More than 900 bird species have been identified in lab casework, everything from cassowaries to penguins to birds of paradise to 423 species protected by the Migratory Bird Treaty Act (MBTA) of 1918.

Through their support of law enforcement investigations, lab ornithologists have made significant contributions to avian conservation. An early case in the 1990s targeted a major falcon-smuggling ring, and recent investigations have focused on the burgeoning trade in live songbirds from the Caribbean, Indonesia, and Southeast Asia. The lab provided identification of oiled bird carcasses from the *Exxon Valdez* and *Deepwater Horizon* oil spills, which resulted in millions of dollars in fines. A series of investigations of avian mortality at oil pits resulted in the elimination of hundreds of these hazardous sites. The trade in bird feathers—a prime reason for passage of the MBTA more than a century ago—continues today, and lab ornithologists have identified thousands of feathers in such investigations, with the most significant species represented being Bald and Golden Eagles.

A major contribution of the National Fish and Wildlife Forensic Laboratory to ornithology has been the Feather Atlas of North American Birds, which was initiated by Trail in 2007 as a service to USFWS law enforcement personnel. This website, hosted by the lab, provides image scans of flight feathers (primaries, secondaries, and rectrices), along with feather measurements and associated specimen data, and distinguishes age- and sex-related plumage types for many species. The Feather Atlas has grown to include more than 1,900 scans representing 425 species, and it continues to expand. The site is now used by researchers,

birders, and artists as well as by law enforcement, and receives more than 1.5 million visits per year.

The focus of wildlife law enforcement continually evolves, as new aspects of illegal trade develop or are uncovered. Identification work at the lab has raised awareness about the ongoing trade in the skulls of hornbill species and the need for further protections of some threatened African species. A recent priority has been investigation of the illegal trade in hummingbird love charms (*chuparrosas*) coming from Mexico. Lab ornithologists have documented the presence of thirteen MBTA-protected hummingbird species in these objects, which incorporate whole dried carcasses of hummingbirds. The most frequently identified species is the Ruby-throated Hummingbird (*Archilochus colubris*). Because that species does not breed in Mexico but only winters there, each Ruby-throated Hummingbird killed for these love charms is a bird from the US breeding population.

In addition to the species identification work carried out in the lab's morphology section, important contributions to ornithology have been made by the lab's geneticists and pathologists. The genetics section provides species identification for evidence from items not suitable for morphological identification; for example, blood on raptor pole traps and breast meat from hunted waterfowl. The section has also compiled a continent-wide population genetics database for Bald and Golden Eagles. This enables analysts to determine the number of individual birds represented in a mass of evidence and whether separate items (e.g., a headless eagle carcass in a freezer and an eagle head on a crafted item) originated from the same bird. Such determinations have important implications for law enforcement investigations. In one case, genetic analyses determined that 112 eagles were represented

(39 Bald Eagles and 73 Golden Eagles), which was more than three times the number of individual birds that could be verified morphologically. This greatly increased the penalties in the case.

The lab's pathology section is tasked with determining the cause of death from animal carcasses submitted as evidence. Over the past 10 years, more than 4,700 bird necropsies have been performed by lab pathologists, including 680 Bald Eagles and 705 Golden Eagles. Overall, the most frequent causes of death among birds submitted for pathology examination were gunshot, pesticide poisoning, lead poisoning, and wind turbine collisions. The high-profile guilty plea by Duke Energy for bird deaths at wind energy facilities in Wyoming resulted in part from eagle necropsies performed at the lab. Necropsies at the lab also documented significant avian mortality at "power tower" solar facilities, as birds were burned flying through the solar flux. Another focus of lab pathologists has been working with the California Condor recovery program to document causes of death from carcasses of that highly endangered species. In 136 condor necropsies, 62 deaths (46%) were found to be related to lead poisoning.

From the small town of Ashland, Oregon, ornithologists at the National Fish and Wildlife Forensic Laboratory continue to contribute to bird conservation in every corner of the world.

## Pepper Trail

Pepper Trail was the senior ornithologist at the US Fish and Wildlife Service National Fish and Wildlife Forensic Laboratory in Ashland, Oregon, from 1998 to 2021. A leading expert in feather identification, Trail identifies bird remains submitted as evidence in federal wildlife crime investigations. One of his proudest contributions is the Feather Atlas of North American Birds. Since its founding, the Feather Atlas has catalogued over 425 species with more than 1,900 scanned images of flight feathers.

Trail is passionate about conservation, and when not fighting bird crime, he is known for his work addressing conservation issues with many local environmental groups in southern Oregon. He was especially involved in the efforts that led to the creation and expansion of the Cascade-Siskiyou National Monument. Trail has led many ornithological trips to every corner of the globe and enjoys writing about his avian sources of inspiration, having contributed to more than 20 peer-reviewed publications and over 100 species accounts of pigeons and doves for the *Handbook of Birds of the World* series. His poetry collection *Cascade-Siskiyou: Poems* earned him a spot as a finalist for the Oregon Book Award in Poetry. Trail is also a fellow of the American Ornithological Society.

*Photo by Susan Sawyer*

# Klamath Bird Observatory

JOHN ALEXANDER

Klamath Bird Observatory (KBO) grew out of the Partners in Flight International Bird Conservation Initiative and the Klamath Bird Monitoring Network that was started in 1992 by C. J. Ralph and Kim Hollinger of the US Forest Service Pacific Southwest Research Station Redwood Sciences Laboratory (Alexander et al. 2004a, Alexander 2011). Ralph and Hollinger invited John Alexander to help them, and over the next several years with support from many partners, they expanded the network, realizing a comprehensive regional bird monitoring and conservation program throughout the Klamath-Siskiyou Ecoregion of northern California and southern Oregon.

In 1996, in collaboration with Point Blue Conservation Science (formerly Point Reyes Bird Observatory), KBO's Upper Klamath Field Station near Fort Klamath, Oregon, opened with support from the USFWS, Bureau of Land Management (BLM), US Forest Service, and many other partners. From an expanding regional bird monitoring program, KBO fledged as an institution supporting observation-based science, using scientific results from bird research and monitoring efforts to inform conservation and land management. KBO incorporated in 2000 with their headquarters in Ashland, Oregon, where they continue to build the diverse partnerships that have been essential to their success. In that year, they established their mission to advance bird conservation through science, education, and partnerships. Jaime Stephens joined KBO in 2002, quickly growing into her current capacity as KBO's science director.

Implementing the Partners in Flight science-based approach that involves using birds as ecological indicators (Alexander 2011, Rosenberg et al. 2016), KBO employs a conservation planning approach that is fueled by (1) long-term monitoring that provides information about changes in bird populations, (2) in-depth theoretical research that advances an understanding of bird distributions and movements, and (3) applied ecology that addresses natural resource management challenges. KBO's conservation science is applied at local, regional, and international scales. KBO also manages an interactive data center and decision support system— the Avian Knowledge Northwest (www. AvianKnowledgeNorthwest.net) node of the Avian Knowledge Network—that provides scientific resources across the Pacific Northwest.

KBO brings a rich body of ornithological science to partners. For example, KBO contributed to OSU's efforts to develop a state-of-the-art species-centered habitat distribution modeling approach (Shirley et al. 2013, Betts et al. 2014) that Halstead et al. (2019) used to describe the effects of oak habitat amount and fragmentation on species richness. An additional series of papers published through a collaboration between KBO and BLM have helped to guide habitat restoration efforts with results that compare treatment approaches and identify important habitat features to retain or manage for when restoring oak woodlands (Alexander et al. 2007, Seavy et al. 2008).

KBO contributes to numerous other partnerships, such as the Rogue Forest Partners, working together for forests and communities. Through partnerships like these, KBO collaborates with teams that are applying the best available science to plan treatments that will meet desired vegetative conditions with associated ecological response. Such planning requires a multidisciplinary approach, multifaceted implementation, and effectiveness monitoring to determine whether restoration successfully reaches desired conservation outcomes (Alexander 2011).

## John Alexander

John Alexander is the cofounder and executive director of the Klamath Bird Observatory in Ashland, Oregon. He is also on the wildlife faculty at Oregon State University.

Alexander has concentrated his work on integrated management of birds and forests in the Pacific Northwest since 1992. Throughout his career, he has worked to bring together many partners on various sides of conservation issues to benefit birds and the habitats they occupy. His recent interests in the role of fire and climate change have resulted in a number of awards, including the Partners in Flight International Leadership Award, the Joint Fire Science Program's Best Scientist-Manager Partnership Award, the US Forest Service's International Wings across the Americas and the Ducks Unlimited Taking Wing Awards. John continues to serve on the following cooperative ventures: Partners in Flight, North American Bird Conservation Initiative, Avian Knowledge Network, North Pacific Landscape Conservation Cooperative, Intermountain West Joint Venture, US Department of Agriculture Federal Research Advisory Committee, North American Bird Banding Council, Western Bird Banding Association, and Ashland Rotary Community Support Foundation.

*Photo by Taylor Alexander*

## Malheur Field Station

CHRISTOPHER MATHEWS

Malheur Field Station played an important role in opening the incomparable Malheur National Wildlife Refuge to the birding public. Until recently, the nearest guest facilities were in Burns, some 35 miles from the refuge headquarters. Establishment of the field station placed inexpensive accommodations on refuge land a few short miles from headquarters.

The buildings that make up the field station were constructed in 1965 as a Job Corps training center, capable of housing 200 trainees plus resident staff. Budget cuts forced closure of the training center in 1969. Officials at colleges and universities in the Pacific Northwest recognized that the vacant facilities provided a superb opportunity for environmental education, and in 1971, several of these institutions formed a consortium with responsibility for managing and supporting what became known as Malheur Field Station. The field station's two dozen buildings occupy a 320-acre site on the refuge, which is leased on a no-cost basis from the US Fish and Wildlife Service.

Alas, in the 1980s, the cost of operating the field station became too great for its sponsors, and the field station was threatened with extinction. A group of private individuals stepped in and in 1987 formed the Great Basin Society, a nonprofit corporation whose sole function was operation of Malheur Field Station. Since then, the field station has survived on a mixture of user fees, private contributions, and grants from charitable organizations. Particularly significant were grants from the M. J. Murdock Charitable Trust, the Collins Foundation, and the Oregon Community Foundation. Several distinguished ornithologists have served on the society's board of directors, including Alan Contreras and Fred Ramsey. Today, the field station's dormitories and dining hall welcome several hundred visitors per year.

In a recent five-year period, Malheur Field Station welcomed 25 Audubon Society groups, 19 church groups, 61 K-12 school groups, 72 college and university groups, 25 Elderhostel groups, 37 workshops organized by the staff, and visits by groups such as Boy Scouts, Oregon and Idaho Native Plant Societies, Tualatin Riverkeepers, Oregon State Police, and the Oregon Department of Fish and Wildlife.

In 2016, Malheur Field Station became well known to the public when it provided housing for the Federal Bureau of Investigation (FBI) during the occupation of the refuge by a group of far-right domestic terrorists. Having survived that challenge, another threat loomed in 2018. An accident destroyed the water system that supplies the guest facilities, offices, classrooms, and dining hall. An emergency fund drive from loyal Great Basin Society members netted more than $200,000 to create a new water system and ensure the continued existence of the field station. Today, the field station is looking forward to adding additional housing facilities and staying afloat during the COVID-19 pandemic.

# Birders and Their Brethren: The Rise of Citizen Science

## SUSAN HAIG

When considering how conservation efforts are launched and sustained, there is a surprising and overlapping web of components required for a successful campaign: research, management, and advocacy are crucial components. More often than not, and even in the most sophisticated of issues, the bottom line comes down to efforts put forth by birders who have a conservation ethos and/or report bird sightings to eBird, the Breeding Bird Survey (BBS), or Christmas Bird Counts (Figure 3). Early ornithologists reported sightings so we could begin to understand Oregon's avifauna. Later, participation in these nationwide surveys helped identify distributions and migratory patterns across species ranges. In Oregon, longtime BBS compilers like Paul Adamus submit reports that form the basis of how we monitor broadscale species status (e.g., Rosenberg et al. 2019), carry out assessments for the biennial State of the Birds, and have been key to analyses of the impacts of climate change (e.g., Haig et al. 2019). There are no data sets that can replace hundreds of thousands of bird sightings from across the state, the nation, and the flyway. For example, as of October 2021, eBird reported 1.3 million check lists, 3,970 hotspots, and 23,900 observers in Oregon.

Over the past 20 years or so, the concept and practice of the "citizen scientist" have gained strength and popularity. Given how many research and conservation efforts require a great number of people to carry them out, as well as continuing budget cuts for natural resource conservation programs, many

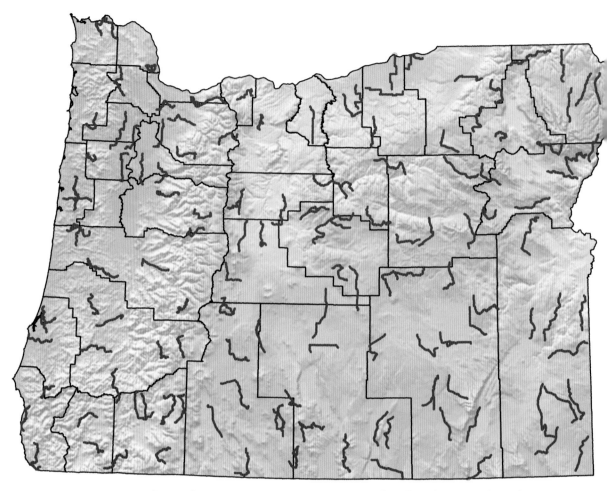

FIGURE 3. Breeding Bird Survey routes in Oregon, 2021. David Ziolkowski, US Geological Survey, Laurel, Maryland

researchers have enlisted the help of birders and other citizens to collect data. In this volume, examples can be found in the Black Oystercatcher surveys along the coast, at Joan Hagar's banding station, Frank Isaacs's state-wide Bald and Golden Eagle nest monitoring, and in Doug Robinson's Oregon 2020 project. Thus the importance of amateur ornithologists has never been greater. As this need continues to grow, the hope is to enlist the help and interest of more people across the state.

## Oregon Audubon Chapters

Oregon has 12 National Audubon Society chapters that work on conservation issues, citizen science, and environmental education programs across the state. The first and largest is the Audubon Society of Portland. It was established in 1902 (originally as the Oregon Audubon Society) to help pass the Model Bird Act and to create protections for Malheur National Wildlife Area, Three Arch Rocks, and Klamath National Wildlife Refuge. Since then, it has grown exponentially, including a diverse staff that oversees their outreach programs, wildlife rehabilitation facility, bookstore, and conservation efforts. Their role fills several niches that other professional ornithologists cannot. For example, while the USFWS cannot sponsor a program advocating for hunters to use non-lead ammunition

## Merlin "Elzy" and Elsie Eltzroth

Merlin "Elzy" and Elsie Eltzroth from Corvallis were responsible for some of the first citizen efforts for nongame bird protection, research, and laws in Oregon. The Eltzroths moved to Corvallis in 1971 and became founding members of the Audubon Society of Corvallis, the Oregon Birding Association, and Oregon Field Ornithologists. They both obtained rehabilitation permits and worked with more than 100 species of orphaned and injured birds. At the same time, Elzy enrolled in the wildlife program at OSU at age 49.

In 1985, Elzy was instrumental in the passage of Oregon laws to protect all native nongame bird species. He further established a reward program for reporting anyone harming or killing wildlife (primarily raptors) in Oregon via the Audubon Society of Corvallis. Elzy died in 2005. The Merlin S. Eltzroth Papers (housed at the OSU Library) consist of field notes compiled by Eltzroth about first and second recorded sightings of birds rarely seen in Oregon.

Elsie Eltzroth began her decades-long volunteer Western Bluebird nest box project in the central Willamette Valley in 1976 at approximately the same time that Hubert Prescott began a similar bluebird program in the northern Willamette Valley. The Western Bluebird was on the Audubon "blue list," with only six known nesting pairs in Benton County in 1977 and only seven the following year. Under the tutelage of Lloyda Thompson-Crowley, Elsie organized countless volunteers to establish a bluebird trail, learned to band bluebirds, and served as a board member of the North American Bluebird Society. Elsie died in 2019, but her leading work on the Western Bluebird is significantly responsible for recovery of the Western Bluebird in Oregon.

*Courtesy of the Audubon Society of Corvallis*

## W. Douglas Robinson

Douglas Robinson is a professor in the Department of Fisheries, Wildlife, and Conservation Sciences at Oregon State University (2002 to present) and is the Bob and Phyllis Mace Professor of Watchable Wildlife (2011–21). Robinson also serves as director of the Oregon 2020 Birds project. He directs the OSU Field and Lab Ornithology Collaboratory (the FLOCK), a student ornithology program that builds on the remarkable strengths in ornithological research at OSU to connect students with research and job opportunities. As one of the world's top contributors to eBird, he has been uniquely suited to bridge the gap between professional ornithology and citizen birders. He has also served as curator of the OSU bird collection and advised more than two dozen graduate students. In Oregon, Robinson has investigated responses of bird populations to wildfires in the Siskiyous and other mountain sage-brush communities, life history evolution, resurveys of historic bird counts in the Willamette Valley, retrospective reconstruction of historic bird populations and their changes since 1850, and development of new analytical methods to maximize the value of citizen science data. Robinson is a fellow of the American Ornithological Society.

*Photo by Tara Robinson*

## Robert Sallinger

Bob Sallinger is the director of conservation for Portland Audubon and has spent more than 30 years working on bird conservation issues in Oregon. He started at Reed College in biology and finished his education with a law degree from Lewis and Clark Law School. As conservation director, Bob develops Audubon's local, regional, and national conservation priorities, which generally focus on creating green cities and conserving native birds across Oregon. He has been involved with every major and minor bird conservation legal issue in Oregon, from Northern Spotted Owls to lead issues associated with bringing California Condors back to Oregon to the current effort to ban the bird poison Avitrol across the United States. Bob has also managed Audubon's Peregrine Project, which combined educational outreach, management, captive rearing and release, and citizen science to promote Peregrine Falcon recovery in the Portland Metropolitan Region. Today, Portland-area Peregrine eyries make up 5% of the known Peregrine nesting population in Oregon. The Audubon Program has been recognized with awards from the US Fish and Wildlife Service and the Oregon Chapter of the Wildlife Society.

*Photo courtesy of Portland Audubon*

## Jim and Karan Fairchild

Jim and Karan Fairchild have contributed to Oregon bird conservation for more than 40 years. They have each served many times in many roles for the Audubon Society of Corvallis, helping to keep the society going through thick and thin. Karan Fairchild grew up in Corvallis, the daughter of an Oregon State University faculty member. Both graduated from OSU and worked as faculty research assistants. Jim now runs an independent carpentry business, while Karan works for the US Department of Agriculture's Agricultural Research Service on the OSU campus. Avid birders, they have participated in countless Christmas Bird Counts, have a Breeding Bird Survey route that they have run for decades, and have each served as president of the Audubon Society of Corvallis, as well as in other positions and committees for ASC. Jim has served on the Oregon Audubon Council, Corvallis Watershed Advisory Commission, Marys Peak Stewardship Group, and others. Among other roles, Karan started the Birdathon for Audubon Society of Corvallis in 2000 as a fundraiser benefiting the Hesthavn Nature Center. They also have tracked birds on their own property for decades and contributed data for a number of scientific papers.

*Photo by Kate Campbell*

(to protect scavengers from lead poisoning), Portland Audubon can and does. This is a huge help in ensuring the recently released California Condors in Oregon will remain healthy (as will the hunters who also will not consume lead in the game they harvest).

Portland Audubon is the largest Audubon chapter in the country and remained the only chapter in Oregon for decades. The environmental movement of the 1960s and 1970s spurred formation of new chapters across the state, however. Salem Audubon was next to form in 1969, followed by three additional chapters in 1971. Led by OSU statistics professor Fred Ramsey, the Audubon Society of Corvallis was launched following a birding trip to Malheur National Wildlife Refuge. The Rogue Valley Audubon Society also had its beginnings in 1971 when Rev. Thomas McCamant started a bird club while serving as pastor at the local Congregational Church in Medford. One of the club's first activities was to participate in the National Christmas Bird Count. By the time the chapter held its first general meeting in September 1971, there were 131 members. Similarly, East Cascades Audubon was formed by Jim Anderson in 1971. Among their many contributions was the building and maintenance of the observation blinds at Cabin Lake.

Lane County Audubon was incorporated in the late 1970s and was headed by Dick Lamaster for many years; his wife Maeve Sowles took over in 2001 and remains to this day. The Umpqua Valley chapter was incorporated in 1977. The Kalmiopsis Audubon Society was founded in 1980 by local people interested in learning about and conserving the extraordinary values of Curry County's natural environment. They have worked aggressively to protect old-growth forests of Elk River in the Grassy Knob Wilderness and blocked dozens of damaging timber sales of old-growth forests, including ones that would have felled Oregon's last remaining redwoods. By 1983, the Endangered Species Act had been passed by Congress, and Bald Eagles were listed as endangered. Thus the Klamath Basin Audubon Society began their now decades-long tradition of the Bald Eagle Conference (now known as Winter Wings) each February. The most recent Audubon chapter to form was Lincoln City Audubon in 2006.

Addressing avian conservation challenges requires an understanding of broad policy issues as well as the ability to initiate and understand the acceleratingly complex field investigations of today's avian ecology. Over the past 50 years, conservation policy and technology have evolved at such an explosive pace that to understand how to approach one, you must keep up with the other. To ease understanding of their application throughout the rest of the book, in this section we summarize the major technological advances in the field. These range from tracing evolutionary history using genomics to exploring mating systems using experiments with hormones to tracking birds from the International Space Station. The unprecedented number of professional ornithologists in Oregon not only puts the state at the top of ornithological expertise nationally and internationally, but also provides an exciting place to witness the ever-changing ways that people study birds.

# Bird Identification

SUSAN HAIG

The foundation of bird study is being able to identify taxa at whatever level one chooses to jump in the game. The earliest of those studying birds in Oregon came prepared either to draw what they had seen or to bring someone who could. This has progressed to our current state, where online digital photography can be used to identify color-marked birds at faraway wintering sites, or molecular markers are used to do the same. But ornithology began with art, and particular aspects of bird art began with ornithology. As Corvallis's DaVinci Days motto says, "it is a celebration of art and science." Art has always been an integral tool in ornithology. The major natural history expeditions would have had so much less to report had they not been led by or included artists to describe what was seen. As an example, consider what we might have learned about Oregon's avifauna if Lewis and Clark had traveled down the Columbia River with a natural history artist. While we have rough sketches of species like the California Condor from Meriwether Lewis's notes, it could have been so much more. Even if birds were collected for museum skins, information about their habitats, behavior, and more would not have been as accurately recalled were it not for the early artists in ornithology. And we would have had no field guides until recently without ornithological artists.

## Nineteenth Century

Perhaps because of the incredible biodiversity in the state, Oregon has a strong history of phenomenal avian artists. Decorative avian art has been a part of Native culture for centuries prior to the arrival of Euro-American settlers. Of note is their use of feathers in decorating head dresses and capes for special occasions. In particular, California Condor feathers have been used extensively by Columbia River tribes in their basketry, beadwork, and sculptures (Mercer 2005, D'Elia and Haig 2013). Today also, birds play a significant role in tribal art, as evidenced by the new bronze sculpture portraying Klamath Tribe traditions outside the Favell Museum in Klamath Falls.

While decorative bird art has an important place in all cultures, accurately drawing or painting birds was key to the advent of the field of ornithology. Most noteworthy of the early Euro-American avian artists in Oregon was R. Bruce Horsfall (1869–1948). Born in Iowa, Horsfall, like his natural history artist father of the same name, largely received his art training in Europe. In the end, he was associated with the most famous natural history museums in the country, including the National Museum of Natural History (Smithsonian),

Western Snowy Plover stretching its left wing and leg simultaneously, thereby displaying the complement of colored leg bands that make this bird individually identifiable. A large proportion of this threatened population in Oregon is banded to monitor population vital rates. Photo by Roy Lowe

American Museum of Natural History in New York, and Peabody Museum (Yale University). Horsfall came to Oregon in 1914 to conduct a natural history survey of Oregon at the invitation of State Fish and Game Commissioner William Finley. By his own words, in the *Oregon Daily Journal* (Portland) in the 28 March 1921 edition,

> The survey would have been of inestimable educational value and would have proved a strong factor in the conservation of Oregon's natural resources. It would have also proved a value in attracting tourists. Unfortunately, a new Fish and Game Commission was appointed, whose members were not in sympathy with educational work, so our plans fell by the wayside.

Even so, Horsfall remained in Oregon and continued to explore and paint birds, their

habitats, and other natural history features of the state. At one point, he and Finley were the first to take a movie camera in the field near Klamath Falls to film wildlife. Among his many illustrated field guides and books was *Birds of the Pacific Coast* (1923). The Horsfall family continues to extend its Oregon roots with Bruce Horsfall Jr. and his son, who graduated from Reed College in Portland, Oregon.

## Twentieth Century

George Miksch Sutton grew up in Ashland and Eugene and went on to become one of the most noteworthy bird artists of the twentieth century. His biography reads like the history of North American ornithology. He was fascinated with birds from an early age and published his first painting (an oriole at a nest) at age 12. By age 17 (1915), he was corresponding with prominent avian artist Louis Agassiz Fuertes (a protégé of J. J. Audubon) and spent time with him in 1916. He went on to be curator of birds at the Carnegie Museum, went on Arctic expeditions with W. E. Clyde Todd, and spent a year on South Hampton Island in the Canadian Arctic. He served time in the Army Air Corps during World War II, where he was Sergeant Roger Tory Peterson's superior officer. Eventually, he went on to finish a Ph.D. at Cornell University under the eminent ornithologist Arthur Allen. Ultimately, Sutton settled into a faculty position at the University of Oklahoma, where he continued field expeditions to Mexico, the Arctic, and many other regions. His pen and ink, and often watercolor, works have been published in several of his

books, including *High Arctic*, *Iceland Summer*, *Eskimo Year*, and *To a Young Bird Artist*, which described his time working with Fuertes. Sutton's career marked the beginning of science taking more of a precedent over the art most ornithologists had previously focused on. While he painted and explored, he directed a number of graduate students in scientific disciplines who went on to be leaders in ornithology: Oregon State University (OSU) faculty member John Wiens, Lewis Oring, John Janovy, and Joel Cracraft. The University of Oklahoma built the G. M. Sutton Avian Research Center in his honor shortly after his death in 1982.

The next generation of Oregon bird artists included Hubbard, Oregon, resident Jon Janosik (Figure 4). Jon has garnered an international reputation among artists over the past

FIGURE 4. Notable contemporary Oregon avian artists (*clockwise from top left*): Julio Gallardo, Ram Papish, Larry McQueen, Shawneen Finnegan, Elva Hamerstrom Paulson, and Jon Janosik.

50 years. Training under such ornithological luminaries such as Roger Tory Peterson and Don Eckelberry, his illustrations have been featured in *Field Guide to North American Birds*, *The Audubon Society Master Guide to Birding*, *An Audubon Handbook, Western/Eastern Birds*, and *Book of North American Birds*. He painted the Yellow-breasted Chat that is the logo of the Audubon Society of Corvallis. Living in Hubbard has given Jon easy access to the mountains and the coast, which have inspired many of his works.

Larry McQueen is also a world-renowned and prolific avian artist. McQueen grew up in Pennsylvania and settled in Eugene after college and serving in the US Army. He started watching and drawing birds from a young age but did not take it up as a profession until the late 1970s. Another student of Don Eckelberry and influenced by the art of Louis Agassiz Fuertes, Larry attained early success later in life when he was invited to display his work at the prestigious Leigh Yawkey Woodson Art Museum in Wausau, Wisconsin, in 1981. His artistic career took off from there and resulted in an incredible portfolio of thousands (~2,300) of paintings in many field guides, including *The Birds of Peru* and *All the Birds of North America*. Interestingly, Larry McQueen and Jon Janosik illustrated a two-book series for Oregonian Don Kroodsma titled *The Backyard Birdsong Guide Western (and Eastern) North America: A Guide to Listening* (Kroodsma 2008a, b).

Perhaps the Oregon bird artist with the most prestigious ornithological background is Roseburg artist Elva Hamerstrom Paulson. Elva's parents were the renowned Wisconsin ornithologists Frederick and Frances Hamerstrom. Growing up on a farm in central Wisconsin "in a house with no running water that had not been painted since the Civil War and had no heat aside from a few cast-iron wood-burning stoves" did not deter the most

noteworthy of ornithologists from visiting with her parents and sitting in their blinds to watch Greater Prairie Chickens (*Tympanuchus cupido*) dance. Thus Elva was also availed of her parents' graduate advisor from the University of Wisconsin, Aldo Leopold. Elva grew up drawing and painting, which she continues to this day. Her most notable works were the drawings of all of Oregon's birds in *Birds of Oregon: A General Reference*, published by the Oregon State University Press in 2003.

## Twenty-First Century

Ram Papish is a biologist, educator, and artist from Toledo, Oregon. His work ranges from illustrating children's books to more serious works such as the *Handbook of Oregon Birds*, *California Condors in the Pacific Northwest*, *The Population Ecology and Conservation of Charadrius Plovers*, and more. But his most novel niche is preparing posters and other educational materials about birds and their habitats for agencies such as the US Fish and Wildlife Service, the Bureau of Land Management, and the National Park Service. He was awarded Artist of the Year and designed the poster for International Migratory Bird Day in 2004.

Shawneen Finnegan, from Beaverton, Oregon, had artists for parents and grew up drawing and painting. She did not get the birding bug until the age of 26, but then her bird art took off. She became the photo editor for *Birding Magazine*, a Wings tour leader, is on the advisory board for *WildBird Magazine*, and illustrated *Birds of Montana* with Jeff Marks (see the Northern Saw-whet Owls section on p. 173).

The newest ornithological artist of note in Oregon is raptor biologist Julio Gallardo, a newcomer to the faculty in OSU's Department of Fisheries, Wildlife, and Conservation Sciences. Gallardo is from Mexico and gained much of his artistic perspective by leading birding trips all over the world before going to graduate school. In addition to his research and teaching, Julio has published several field guides for local communities in Mexico and has been commissioned to prepare paintings for numerous programs, including for the annual winners of the American Ornithological Society's Conservation Award. Most recently, he was named the 2019 Donald Eckelberry Fellow from the Academy of Natural Sciences at Drexel University.

While photography can now take the place of bird art in providing a record of bird morphology and behavior for research ornithologists, bird illustrators still play a key role in the production of field guides. More importantly, bird art remains integral to our appreciation and enjoyment of birds, if not a critical gateway for people to learn about and develop a concern for the conservation and restoration of birds.

## Photography

Over the past few decades, as digital photography has advanced, photos and videos have become the primary means of depicting birds for research purposes. In the lab and field, behaviorists can use digital video cameras to film various behaviors, while in the field digital cameras ("camera traps") can be used to document the presence of various species when researchers are not present. For example, photography has long been used to track patterns of incubation and incubation recess in waterfowl and other nesting birds. More recently, with the advent of birders taking close-up bird photos with digital cameras and telephoto lenses, and posting them online, there are greater chances of seeing and identifying marked birds among the photos. Using this method, Professor David P. Craig, a behavioral ecologist at Willamette University,

has found hundreds of migratory stopovers and overwintering sites for individually marked Caspian Terns that were banded by OSU ornithologists Don Lyons, Yasuko Suzuki, and others during the breeding season.

Drones can also take digital photography to sites where it is unsafe for people studying birds to venture. For example, drones can be flown over areas where colonial nesting birds are breeding, and photos can be taken by drone-mounted cameras using remote shutter release so that colony and population sizes can be accurately estimated, sex ratios determined (in some cases), and reproductive success measured. Further, heat-sensitive photographic film can be used on drones, as we did in a study of Long-billed Curlews on the Boardman Bombing Range (Stocking et al. 2010), to fly over desert areas at night to determine the location and distribution of incubating birds, as their warm bodies showed up on the film against the cold desert ambient temperature.

# Museums and Specimen Collections in Oregon

M. RALPH BROWNING

Often, ornithologists of bygone decades spent a great deal of time shooting birds and preparing museum skins, as they had no other way of identifying birds or tracking their migrations or movements. Very little collecting is done anymore, but the information still to be collected from those skins is invaluable. Thus bird specimens should be cherished. They have long been important for identification and comparison among other individuals and species, thus providing important documentation for many aspects of ornithology (Remsen 1995). Specimens have provided documentation for distributional studies including those for American Ornithologists' Union (AOU) / American Ornithological Survey (AOS) checklists (Draheim et al. 2012) and more regional works (e.g., Gabrielson and Jewett 1940, Marshall et al. 2003, D'Elia and Haig 2013, D'Elia et al. 2016). Specimens are important for studies of morphology, modern field guides, and for evidence on molt patterns (e.g., Howell 2010, Rohwer and Rohwer 2018).

Specimens provide references for defining migratory connectivity (e.g., Novitch et al. 2015); collisions of birds with aircraft (e.g., Laybourne and Dove 1994), buildings (e.g., Kahle et al. 2016), and wind turbines (e.g., Erickson et al. 2001, Thaxter et al. 2017); and even evidence for solving crimes (e.g., Trail 2017; see also the National Fish and Wildlife Forensic Laboratory section on p. 14). Specimens also provide opportunities to discover much beyond external morphology (e.g., Holmes et al. 2016, Webster 2017), including ectoparasites and other symbionts (Lutz et al. 2017), geographic variation (Patten and Pruett 2009), and other issues concerning systematics (e.g., Zusi 2013).

Descriptive ornithology partnered with molecular studies (Rocque and Winker 2005) and tissue samples from specimens (Barker et al. 2015) now make museum collections even more important, with study skins providing vouchers for molecular studies because molecular material was obtained from the very same specimen (Dubois and Nemesio 2007, D'Elia et al. 2016).

In addition, bird skins, feathers, and eggs collected over a long period provide opportunities to study morphological and ecological changes induced by pollution, diet, and now climate change (Green and Scharlemann 2003, Rocque and Winker 2005, Kiat et al. 2019). Specimens allow detection of species difficult

to identify in the field, so-called crypto species (e.g., Barrowclough et al. 2019, Browning 2019). Besides study skins, specimens prepared as skeletons are necessary in studies of living species (e.g., Zusi 2013) and required by paleoornithologists for identification of fossil birds (James 2017).

Finally, specimens can also help measure effects of global climate change (e.g., Kiat et al. 2019). Wet specimens can be used to compare muscle anatomy (e.g., Clifton et al. 2018). Collections of nests and eggs continue to be important ornithological tools (Kiff 2005, Linck et al. 2016). Regardless of the form of preparation, bird specimens associated with as much external and internal data as available potentially will provide more uses, some of which we have not yet recognized.

Regarding the first collections of birds in Oregon, the great diversity of birds in the state drew early ornithologists from the east and beyond to observe and collect birds as a way of determining the avifauna of North America. The earliest known collectors in Oregon were Lewis and Clark (1805–6); however, no specimens remain from the expedition's time in Oregon. Scottish botanist David Douglas (1825–27, 1830) collected in Oregon, including specimens of Mountain Quail (Browning 1977a) and California Condor (Douglas 1829). The few extant specimens from Oregon that are attributed to Douglas are in the Liverpool Museum (Olson 1989), British Museum (Browning 1979), and National Museums of Scotland (Zena Timmons, pers. comm., January 2020). Otherwise, the whereabouts of specimens collected by Douglas are presently unknown and may no longer be extant.

American naturalists John Kirk Townsend and Thomas Nuttall visited Oregon from 1834 to 1836. Townsend provided valuable notes on birds observed and collected from the region

(Jobanek and Marshall 1992, Halley 2019). Specimens from Townsend and Nuttall, most of which were simply labeled "Columbia River" (Stone 1899, Hall 1934, Browning 1977a), ended up in the Academy of Natural Sciences of Philadelphia (Stone 1899) and the US National Museum (i.e., Smithsonian; Deignan 1961).

The nineteenth-century collectors were confined mostly to northwestern Oregon, with the exception of Douglas, who traveled into the Umpqua watershed (Douglas 1914). Later, T. R. Peale and others of the US Exploring Expedition—C. D. Anderson and J. S. Newberry of the Pacific Railroad Survey in 1855 (Browning 1994), and H. W. Henshaw of the Explorations and Surveys of the 100th Meridian (1881)—collected birds in Oregon, with specimens also deposited in the US National Museum (Deignan 1961, Browning 1979).

Charles Bendire is considered the father of Oregon ornithology, and unlike the haphazard collecting by his predecessors, Bendire made several systematic collections of certain species while stationed near what is now Burns (1874 to 1878) and at Fort Klamath (1882 to 1883). Specimens of study skins and eggs collected by Bendire, plus his notes (Browning 1973, 1977), are now in the US National Museum and contributed toward the well-known Bent Life Histories of North American Birds series.

None of the specimens collected by the aforementioned luminaries are known to exist in collections housed in Oregon. A. C. Shelton may have been the first major collector to donate specimens to an Oregon institution. Currently, data on bird specimens collected in Oregon are available online via VertNet, a National Science Foundation–funded collaborative project that makes biodiversity data free and available on the web (http://vertnet.org/).

## M. Ralph Browning

M. Ralph Browning grew up as an avid birder in the Rogue Valley but spent most of his professional life as an avian taxonomist for the US National Museum (Smithsonian) in Washington, DC. Browning never forgot his southern Oregon connection, and in 1975 he published *Birds of Jackson County and Surrounding Areas* as No. 70 in the North American Fauna series for the US Fish and Wildlife Service. This book was a bridge between traditional specimen-based ornithology and newer field observations. In 2019, he published a memoir, *Rogue Birder: The Making of a Modern Ornithologist*, which follows his life from childhood in southwestern Oregon, through his work at the Smithsonian, and finally returning to the Rogue River Valley, where his birding adventures began. Though retired, Browning has continued to embark upon and chronicle his birding adventures, such as a yearlong traveling expedition in 2015 that he conducted in an RV made by his wife. The Smithsonian Institution houses Browning's ornithological cataloging and research, especially focused on Oregon and the northwest United States.

*Photo by Linda Ray-Browning*

# Birdsong and Acoustic Studies

ALAN CONTRERAS

The science of understanding birdsong has advanced rapidly with ever-evolving sound-recording equipment. Technology aside, however, a person still must understand the bird to understand their communication. Don Kroodsma, perhaps the world's best-known bird sound expert and author of *The Singing Life of Birds* (2015), got his start on sound research as a doctoral student of John Wiens at OSU working on Bewick's Wrens and listening to their varied sounds. Geoff Keller was one of the region's most prolific bird recording artists. And Stewart Janes spent much of his career at Southern Oregon University studying the complex sound patterns of warblers, particularly those that live in the evergreen forests of the Pacific Northwest.

Other people have made significant contributions to the study of bird sound in Oregon. Eleanor Pugh of Wolf Creek made many recordings of Oregon bird sounds in the 1970s and 1980s, including long recordings of owls and forest birds. Dave Herr spent many years recording birds in Oregon, and Arch McCallum is best known for his work on sounds of the "Western Flycatcher complex" (*Empidonax* spp.) from Oregon. More recently, US Forest Service biologists Damon Lesmeister, Leila Duchac, and others have used acoustic monitoring to survey for Northern Spotted Owls and Barred Owls in Oregon forests (Duchac et al. 2020).

Since 1990, Stewart Janes has focused on the complex structure of song systems and functions of warblers. First, he examined the two song types of Black-throated Gray, Hermit, and Townsend's Warblers found throughout the forested regions of Oregon, but with a focus in the southern Cascades and the Klamath Mountains. The three species have complex song systems involving multiple song types, dialects, geographic variation within dialects, and songs that even change over relatively short periods, songs that in other warbler species are stable. These species appear to use song in somewhat different contexts than many other warblers. The various functions served by the different song types in *Parulid* warblers is still unsettled and more involved than was previously understood: Type 1 singing has been associated with mate attraction, but the partial convergence in Type 1 songs in local populations of Hermit and Black-throated Gray Warblers adds another layer of complexity to these relationships. Type 2 singing has been associated with territorial behavior. In these warblers, however, Type 2 singing also appears to serve to attract females for mating. The current paradigm predicts that interspecific song convergence should appear in Type 2 singing, the "territorial" song, and not Type 1 singing. Thus the mystery continues.

Janes also investigated the extensive use of call notes in the dawn chorus, a behavior not widespread within the warbler family. He investigated the song culture of Hermit and Townsend's Warblers in and adjacent to two warbler hybrid areas in the Oregon Cascades for insights into the degree to which culture and genetics are related and the role each contributes to the moving hybrid zones. He also described the song systems of Nashville and Yellow-rumped Warblers and the use of song types by Townsend's Warblers in migration.

Geoff Keller had the lofty goal of getting high-quality recordings of typical vocalizations for all avian species occurring in Oregon and throughout North America. From 1986 to 2006, he spent an inordinate amount of time making recordings of (mostly passerines) in Coos County near his homes in North Bend

and Coquille. There he found the best place to record was Eden Valley, squarely located in the Coast Range but not too far from Mt. Bolivar and the Rogue River Wilderness Area. He also focused on the South Slough Sanctuary south of Charleston.

As he ventured farther in the state, he needed pristine environmental quietness in which to make recordings. The areas he returned to time and time again included the Upper Klamath Marsh, Fort Rock, the Crooked River National Grasslands, Malheur National Wildlife Refuge, Summer Lake Wildlife Area, Crater Lake National Park, Century Drive in the Central Cascades, and the Table Rocks near Medford. As a result, he archived just under 3,000 sound recordings of about 400 species at the Macaulay Library at Cornell University, representing sites from Nome, Alaska, to the state of Nayarit, Mexico.

## Where Did YOU Get YOUR Song?

DONALD KROODSMA

In 1969, I began a graduate program at OSU under John Wiens in the Zoology Department. "Where did you get your song?" was the simple six-word question that fascinated me. No concrete answers were available for any song-bird, anywhere on the planet. Hints from laboratory-reared birds, especially White-crowned Sparrows, suggested that a young male learned his single song from his father before leaving home. Because the song dialect boundaries of this sparrow can remain highly stable over time, it was thought that those dialect boundaries might act as a barrier to dispersal. That scenario suggested that, left relatively undisturbed over evolutionary time,

each dialect could become a distinct species, and maybe that's why there are so many song-bird species. Fascinating! But in my gut, I felt it couldn't be true. That Bewick's Wren in my backyard in Corvallis had sixteen different songs in his repertoire. Sixteen! ("Take that, White-crowned Sparrows," I smiled, impressed by the wren's vocabulary.) I had recorded him on my reel-to-reel Uher recorder over several days.

The neighboring territorial wren had some of the same songs in his repertoire, but none of those songs could be found among the wrens at Finley National Wildlife Refuge. When I further learned that the wren's entire reper-toire of songs changed gradually over distance, and only individual songs in the repertoire changed at abrupt boundaries, I realized that a young male simply must learn his song repertoire after dispersing to his own territory, which he would hold for the rest of his life.

It had to be that simple. He would roam un-til he found real estate he liked, settle in, and learn the songs there, no matter how different the songs from where he had hatched. And I would bet it was the same for the White-crowned Sparrow.

In theory, the plan to test that idea was simple, and refuge manager Dick Rodgers was happy to accommodate me, as was refuge biologist Fred Zeillemaker. During the spring and summer of 1970, I would simply scour the refuge, marking as many adult male wrens and as many of their offspring as I could catch, while they were still with their parents.

Imagine my excitement when, on 19 March 1971, I found my first youngster at the north end of Cabell Marsh. The next morning, I stood in the dark, waiting for the young male to sing, along with his three neighbors, all

OPPOSITE: Nashville Warbler male defending its nesting territory with vigorous song. This spe-cies has a disjunct distribution in North America, and the song of Nashville Warblers from the population in Oregon and the West is distinct from that of Nashville Warblers from the eastern North America population. Photo by Dan Roby

## Donald Kroodsma

Donald Kroodsma received his Ph.D. from Oregon State University under Professor John Wiens in 1972. Kroodsma's dissertation addressed how a young male bird learns his songs and launched his storied career as an international expert in avian acoustics. Kroodsma spent much of his career (1980–2003) as a professor of ornithology at University of Massachusetts, Amherst. Since retirement in 2004, he has enjoyed finally having time to pursue recording, studying, and writing about birdsong to educate the general public. In addition, he and his son rode their bikes across the United States, recording birdsong as they went. Kroodsma was associate editor for the encyclopedia *Birds of North America* from 1996 to 2003 as well as for *The Auk* from 1998 to 2002. He is a fellow of the American Ornithological Society and the Animal Behaviour Society.

*Photo by Don Kroodsma*

adult wrens who were at least two years old and would have been singing when this youngster established his territory here, late summer or early fall the previous year.

Later, with recordings in hand, I made dozens of sonagrams and laid them all out on the large tables in the labs of Cordley Hall at OSU. The results were unmistakable. Even though the young male was no doubt capable of learning songs while still with his father, he rejected dad's songs in order to learn the unique songs of this little micro-dialect at the north end of Cabell Marsh.

It was the same story with the refuge's Spotted Towhees. During 1970, I recorded four males at the east entrance to the refuge, finding that each male sang seven to nine different songs. And so it had to be with the resident Song Sparrows at Finley Refuge. Young birds leaving home and dispersing just a few territories away would have to learn the unique songs there.

Oregon birdsong continues to enthrall. During 2003, bicycling from the Atlantic to the Pacific, I so well remember the magic of dipping into Hells Canyon and climbing into Oregon, out to Halfway, then Baker City. Oregon birds, special birds, Yellow-breasted Chats, Canyon Wrens, Rock Wrens, and so many more.

The best of times!

# Bird Behavior

## SUSAN HAIG

Observing the behavior of a bird or flock of birds can take many forms, and it is just as critical to theoretical studies as to applied studies. Whereas the avian ecologists of the 1950s, 1960s, and 1970s might have addressed questions about avian behavior by sitting in a blind and recording a bird's behavior every fifteen seconds for hours on end, the advent of modern tracking and molecular tools starting in the late 1980s revolutionized how we can test questions related to the cause and effect of various behaviors. Ultimately, this allows us to better understand why some individuals survive or reproduce better than others.

### Ecophysiology

During the early to middle twentieth century, Oregon ornithologists may have wondered how birds make the great migrations from Oregon to the Arctic or to South America, but they had few tools to address this question aside from shooting birds to assess fat content directly. Likewise, understanding whether one bird might produce more young compared to another came down to the unlikely task of re-sighting marked juveniles in subsequent years. Over the past twenty to thirty years, ecophysiologists have developed advanced techniques that can measure changes in hormone levels as a means of determining many things: who is ready for migration, who is ready for breeding, who might be more ready than others, and why. Starting in 1995, Oregon universities recruited several avian physiologists who have used hormone-based analyses to address these questions. Ironically, though employed at different universities, they all received their doctoral degrees from the renowned ecophysiology lab of John Wingfield and Marilyn Ramenofsky, first at the University of Washington and then at the University of California, Davis.

At the University of Portland, Katie O'Reilly studies coastal shorebirds and used fat analyses and the plasma hormone corticosterone to measure changes in stress in Western Sandpipers during their Pacific coast migration (O'Reilly and Wingfield 2003). She found that male sandpipers that were migrating more than 1,000 km had elevated stress levels relative to females. She hypothesized that perhaps birds were responding to having to continually find food to keep adequate fat levels for migration and avoiding ever-present predators. Perhaps something to consider next time you see someone's dog chasing a flock of sandpipers on the beach!

Likewise, Meta Landys in OSU's Department of Integrative Biology examines stress in birds as it relates to successful migration (Landys et al. 2006). Using Gambel's White-crowned Sparrow (*Zonotrichia leucophrys gambelii*) in a lab setting, she found one hormone, corticosterone, stimulates the hyperphagia, or super-feeding behavior, that birds go through to put on fat just before migration. Other hormones, glucocorticoids, promote the use of fat to fuel flight once the birds start migration. Distinguishing the roles for these hormones helps tease apart the various needs birds may have as they prepare for migration.

Jamie Cornelius is a relatively new faculty member at OSU in the Department of Integrative Biology and is an ecophysiologist interested in the mechanisms birds use to cope with unpredictable and extreme environmental events (Cornelius et al. 2018). Working on Red Crossbills near Fort Rock (Lake County), Cornelius has studied how the social situations that birds encounter can dictate their response to stress. She found that birds responded with higher stress levels (measured by corticosterone) whe their neighbors were stressed compared to when they were not. These results add to our understanding of the fragility of birds when they are migrating.

## Bird Flight

Our understanding of how birds fly is key to understanding how they migrate, find food, select nest sites, protect their young, and find mates. Until recently, most approaches to studying bird flight were limited by how fast a camera could record movement, the difficulty in building the wind tunnels needed to observe flying birds in the lab, and the ability to record data at an appropriate scale. These are some of the issues that Doug Warrick from OSU's Department of Integrative Biology faces as he studies the functional (ecological) morphology, aerodynamics, and evolution of avian flight. Coming to OSU in 2004, Warrick's greatest discovery to date has to do with how hummingbirds hover. Warrick led research into the hummingbird's ability to hover in flight (Warrick et al. 2009). Working with trained Rufous Hummingbirds that hovered

over a feeding syringe filled with sugar solution, Warrick and his research team employed digital particle imaging velocimetry to capture the bird's wing movements on film. This led to the discovery that the hummingbird's hovering is achieved primarily because of its wing's downstroke (which accounts for 75% of lift), rather than its upstroke (which accounts for the remaining 25% of lift). This was counter to the conventional wisdom that lift was provided 50:50 by the upstroke and downstroke. Perhaps simple in concept, the technology needed to address this question was decades in development.

# Combining Methodologies: Behavioral Ecology of Eastern Kingbirds

MICHAEL T. MURPHY

Eastern Oregon's High Desert was the last place I expected to find Eastern Kingbirds after relocating to Oregon from New York in 2000. The impetus for my study was to explore, from initial observations made in New York (Rowe et al. 2001), why female kingbirds often obtained extra-pair fertilizations. How pervasive was this behavior, and why did females seemingly seek copulations with males other than their social partner? The physical setting of Malheur National Wildlife Refuge, with nesting kingbirds being almost entirely confined to the riparian habitat of the Donner und Blitzen River, was an ideal system for detailed studies.

Our goal was to quantify the frequency of extra-pair paternity, assess the effect of nesting density or timing of breeding on the likelihood that males lost paternity, and, hopefully, determine how and why females chose particular males as extra-pair mates. We learned that females were always mothers of the eggs in the nest, but 60% of males lost paternity annually (Dolan et al. 2007, 2009). Song behavior turned out to be the most important predictor of male reproductive success; the earliest singing males were most commonly chosen as extra-pair sires (Dolan et al. 2007). Aggressiveness in nest defense did not predict extra-pair mating success (Dolan et al. 2007), nor did testosterone levels of males (Redmond et al. 2016). We did learn, however, that kingbirds often deserted areas of consistently poor nesting success to relocate to areas of high nesting success when they returned the next year (Redmond et al. 2009).

We then turned our attention to the consequences of arrival time from spring migration and found that individual variation in arrival date at Malheur varied by roughly a month. Nate Cooper showed that experienced breeders and males arrived the earliest (Cooper et al. 2009). He further showed that the earliest arriving also had the highest annual reproductive success, partially because early arriving males were commonly chosen as extra-pair sires (Cooper et al. 2011). Clearly, if you're a kingbird, don't dawdle during migration! Shortly after this, I began a collaboration with Daniel Kim to study kingbird migration using archival geolocators. We determined that kingbirds breeding at Malheur spent the winter in northwest

OPPOSITE: Red Crossbill, male. Red Crossbills are widespread in Oregon, but owing to their reliance on conifer seeds for food, they are irruptive and generally uncommon, especially when there is a failure of the cone crop. The species is highly social and can breed whenever the cone crop provides abundant food, even in winter. Photo by Dan Roby

Adult Eastern Kingbird guarding its breeding territory. In Oregon, this species is largely restricted to the eastern part of the state and is rarely spotted in the western part of the state. It is named for its habit of fearlessly harassing predatory birds many times its size that enter its territory. Photo by Dan Roby

Amazonia (southern Colombia, northern Peru, and eastern Ecuador).

Christopher Chutter tackled one of the more contentious issues in the field of extra-pair mating behavior: Do males know they have been cuckolded, and do they withhold care from broods in which few to none of the young are their offspring? Chutter filmed parental behavior at nests for which we had knowledge of extra-pair paternity and found that kingbird males did not appear to have knowledge of paternity, because their effort at feeding young had no relationship to their share of the young in the nest (Chutter et al. 2016). In this system, females appear to have the upper hand.

Anecdotal observations early in the study showed that kingbirds regularly reused old nests, usually those of American Robins (Redmond et al. 2007). Nest reuse is rare in most open-cup nesting birds, so why did Malheur kingbirds do it so frequently (10% to 15% of nests)? Sarah Cancellieri tested the alternative hypothesis that nest reuse was either a time-saving mechanism that permitted earlier breeding, or that nest reuse was attributable to a shortage of quality nest sites. Using artificial nests that we placed in trees, and that kingbirds used, we unequivocally eliminated the hypothesis that it was a time-saving mechanism and concluded that Malheur's riparian willow (*Salix* spp.) habitat lacked abundant quality nest sites (Cancellieri and Murphy 2013).

By mid-study, an ominous downward slide in kingbird numbers was emerging, and by the

## Michael T. Murphy

Professor and former chair of the Portland State University Biology Department, Michael T. Murphy has worked in Oregon since 2000. Murphy is also the curator of birds at PSU's Museum of Vertebrate Biology. Murphy's interests include population biology, behavioral ecology, and conservation biology. The latter involves work on the sustainability of wildlife populations in urban landscapes. Murphy and his students have conducted research on the population biology of forest birds in undeveloped parks in Portland, including studies of Spotted Towhees, Willow Flycatchers, Northern Pygmy-Owls, and various woodpecker species. Murphy has also done extensive work with Eastern Kingbirds, most recently at Malheur National Wildlife Refuge in eastern Oregon, where he conducted a 10-year study to assess parentage using molecular markers. His current research is focused on using geographic comparisons of kingbirds across North America to examine the relationship between climate and clutch size, egg size and composition, and nest size and insulation.

Murphy has served on the Portland International Airport Wildlife Advisory Committee since 2010. He was associate editor of *The Auk* from 2000 to 2006, and served as editor-in-chief from 2009 to 2013. He is a fellow of the American Ornithological Society.

*Photo by Karmel L. Murphy*

last year of study, in 2011, the kingbird population at our main study site (between Paige Springs and Krumbo Bridge; see Murphy et al. 2020) had halved. Our suspicion was that it was a consequence of low annual reproductive success. Was that due to us or some other agent? Numbers had not recovered four years after the end of our study (2015), suggesting that our presence was unlikely to be the driver. Population simulations confirmed that frequent nest loss was the driver of the decline, and the principal factor, we believe, was the rapid growth of the American Crow population at Malheur that began between 2004 and 2005. Littlefield described crows as uncommon in summer at Malheur, but by 2005 they were widespread regular breeders. Their apparent negative impact we suspect was not limited to kingbirds.

## Conservation Genetics

### SUSAN HAIG

Few fields of science have exploded more over the past 40–50 years than molecular genetics. In the 1970s, protein electrophoresis became the workhorse for avian taxonomists, as this was the only molecular means to evaluate differences among species. Population geneticists began using this method in the 1980s, but the technique was difficult for avian geneticists to employ because few variable proteins were ever identified for birds, and given that most birds fly, they readily mix gene pools, so differentiating populations was nearly impossible using markers with poor resolution. Furthermore, allozyme analyses or electrophoresis could be carried out only on tissue from a dead bird or from red blood cells that had been spun down and immediately frozen in liquid nitrogen. Neither was a practical method for field biologists.

In 1985, Nobel Prize winner Kary Mullis,from the University of California, Berkeley, developed a means by which deoxyribonucleic acid (DNA) could be extracted from any tissue in which it occurred (blood, skin, feathers, etc.) and replicated many times (the polymerase chain reaction, or PCR). Applications of the PCR process have revolutionized every biological field, from developing vaccines for fighting pandemics to laying the groundwork for recovering endangered species. Like other fields, ornithology grew by leaps and bounds with the application of PCR. And by using microsatellite markers via PCR, we were finally able to examine fine-scale processes such as parentage, pedigrees, and mating systems. By the turn of the twenty-first century, the emerging field of conservation biology capitalized on using the PCR process to better understand the status of threatened populations and how to rebuild them (Haig et al. 2001).

The next exponential leap in the field of molecular genetics occurred around 2015 with the development of whole-genome sequencing. Based on PCR methods, and thanks to methods devised for the Human Genome Project, it is now possible to create a genetic sequence of all the proteins in an individual's DNA. This opened a new world of questions that could be addressed with details beyond our imagination less than a decade earlier. The first bird papers are just coming out using these methods, although not yet for studies of Oregon birds. Finally, development of environmental DNA techniques is proving a useful approach for determining the presence of species in various habitats. For example, a water sample from a stream could determine whether American Dippers were present if the water sample contained some DNA from the species, such as from excreta. This coarse approach (presence/absence of species) can be a first step for delineating poorly understood

distributions of species, their habitats, or their home ranges.

Arriving at OSU and the USGS FRESC in 1994, I was the first (non-poultry) avian geneticist at both institutions. I set up a molecular lab, and we focused on issues related to recovering small populations and declining species. In our Oregon studies, we learned that the three subspecies of Spotted Owls were significantly different genetically, even though some hybridization had occurred (Haig et al. 2004a, b; Funk et al. 2008, 2010). This has helped clarify questions related to the listing (or not) of each subspecies under the federal Endangered Species Act (Haig et al. 2004a, 2006; Haig and Allendorf 2006; Haig and D'Elia 2010). Comparisons with Barred Owls provided the USFWS, National Wildlife Forensic Lab in Ashland with markers to differentiate hybrids that might have been shot "by mistake." Aside from taxonomy, our molecular markers were key in quantifying some of the genetic impacts of declining populations of Northern Spotted Owls (*S. o. caurina*; M. P. Miller et al. 2017, 2018a, b). We were able to measure levels of inbreeding across the subspecies range and learned that most occurred in the Washington Cascades (M. P. Miller et al. 2018b). Taken together with the major regional decline in Northern Spotted Owls and the invasion of Barred Owls, it will be difficult for the Washington Cascades population to recover.

We also described population and subspecies genetic differences for a number of species. In the mid-1990s, we found molecular markers were helpful in determining migration routes for various populations of Arctic-nesting shorebirds. Slogging through dense swarms of mosquitos on the North Slope of Alaska, collecting shorebird blood samples and asking other shorebird colleagues to do the same across North America, we were able to come up with definitive markers for a

American Dippers are restricted to clear, fast-flowing streams and feed primarily on aquatic insect larvae and nymphs that they capture while submerged. This foraging niche is unique among North American perching birds and makes the species an excellent indicator of stream water quality. Photo by David Leonard

number of shorebird populations, such that if a bird was caught during migration in Oregon or Kansas or wherever, we would be able to identify the general region in which the bird had nested (Haig et al. 1997).

We took it a step further with our study of Dunlin, for which we had samples from every subspecies and population of this circumpolar breeder, as well as samples from birds on migration and in winter (M. P. Miller et al. 2015). Defining population-specific markers, we were able to identify breeding origins of birds on migration or in winter sites. This was important, as there were concerns about migratory shorebirds carrying avian influenza from Asia to North America. With our markers, we would be able to catch migrating Dunlin as they entered North America (or elsewhere) and determine where they had come from and whether they had bird flu. Tracing the origins would help identify the source of the vector, and knowing the migratory routes of various populations meant we could warn

officials along the migration path whether bird flu was coming their way or not. Happily, we never had to deal with this eventuality, but the markers (and approach) remain ready if any sort of disease breaks out along their migration routes.

Our work on Snowy Plovers demonstrated that there were not subspecific differences between Oregon coast birds and Great Basin birds, but there are demographic differences (e.g., little movement among areas; Page et al. 1995), providing evidence to maintain the listing of coastal birds under the Endangered Species Act (Funk et al. 2007). Similarly, our genetics analyses of American Kestrels did not support subspecific differences across North America but did describe loss of genetic diversity in particular areas (M. P. Miller et al. 2012a). This pinpointed populations that needed the most conservation attention and concern. We worked on the genetic structure of Yellow Rail populations in Oregon and across their range, and found that despite the Oregon population

California Condor with a full crop. This adult male (204) is helping his mate raise a juvenile condor on the California coast near Big Sur. The species experienced a severe population bottleneck and loss of genetic diversity when the 20 remaining wild birds were brought into captivity for captive breeding. Careful genetic management of the captive population resulted in a current population size of a little less than 400 individuals. California Condors have now returned to Oregon skies, their success tied to reduced use of lead ammunition for hunting in the state. Photo by Tim Huntington, webnectar.com

being the only known breeding population west of the Mississippi, they also were not separated at the subspecies level (M. P. Miller et al. 2012b). Conversely, we were able to differentiate subspecies of Double-crested Cormorants along the Pacific coast of North America (Mercer et al. 2013), which provided critical information as to which subspecies was being affected by the recent culling of thousands of cormorants in the Columbia River estuary (see the Double-crested Cormorants in Oregon section on p. 95).

Our California Condor study was undertaken as part of an effort to bring these endangered birds back to Oregon (D'Elia and Haig 2013). Among other analyses, we sampled tissue from the majority of California Condors in museum collections around the world and all the available DNA from the genetic founders of the modern population. Our results demonstrated there had been an 80% loss of genetic diversity when comparing the historical population (ca. 1800s) to the contemporary population (D'Elia et al. 2016). This represents one of the greatest losses of genetic diversity ever recorded for a bird species. The results further delineated relatedness among the individuals in the current population of

## Susan Haig

Susan Haig came to Oregon in 1994 as a federal scientist for what became the US Geological Survey Forest and Rangeland Ecosystem Science Center on the Oregon State University campus. She retired from that position in 2019 as a senior scientist (ST), one of only three women to have achieved that rank in the USGS. She became the first woman wildlife faculty member in OSU's Department of Fisheries, Wildlife, and Conservation Sciences (in 1994) and continues to serve as a courtesy professor of wildlife ecology at OSU, the first woman to attain that rank (in 2000). She and her students' research interests involve efforts to recover small populations, usually of birds. She developed the USGS Conservation Genetics Lab and carried out molecular assessments and conservation planning for many Oregon species ranging from Spotted Owls to Snowy Plovers to California Condors. She further worked on waterbirds and wetland conservation efforts on the Oregon Coast, Willamette Valley, the Great Basin, and beyond. She is a former president, vice president, and board member of the Audubon Society of Corvallis. She is also a fellow of the American Ornithological Society and served as the final president of the American Ornithologists' Union. Dr. Haig was awarded the Loye and Alden Miller Award for Lifetime Achievement and the Peter Stettenheim Award for Service to the American Ornithological Society.

*Photo by Kelly Huber*

California Condors (a little more than 200 individuals) so that their pedigree could be better defined and zookeepers could ensure that captive pairings would produce young with the highest degree of genetic diversity—as it was in such an intensely inbred species.

Overall, use of these ever-evolving molecular methods have critically important conservation implications. Further, for the most part, the answers they provide cannot be found using any other methodologies. Thus they are crucial for the tool bag of conservation biologists.

# Migratory Connectivity: Tracking Bird Movement

## SUSAN HAIG

Aside from species identification, there are few other techniques or skills ornithologists have invested more effort in than individually marking birds so their identity and location can be known. Consider the difference between wondering where an unbanded bird came from versus banding birds and knowing how many birds of that species might have migrated from a particular area, which bird was its mate, whether the mated pair remained together, and how successful were they in surviving and reproducing. Not to mention who this bird's offspring might be, where they ended up, and how successful they were in surviving and reproducing. Even now, the importance of knowing the identities of individual birds has yet to be fully realized.

These possibilities for advances in ornithology changed the minute, in about 210 BC, that Roman soldier Quintus Fabius Pictor tied colored threads around the leg of a female swallow that was raising a brood and released it to fly back to its nest in a besieged garrison. There his comrades could check the

Peregrine Falcons are a highly migratory species that were formerly listed as Endangered in the conterminous United States because of ingesting DDT in prey species. With the advent of satellite telemetry techniques, much has been learned about the migration routes and seasonal habitat use of this species in Oregon and elsewhere in the Americas. Photo by Jared Hobbs

nest, capture the swallow, and know from the number of knots tied in the thread how many days before the garrison would be relieved and a sortie must be made against the enemy Ligurians (Wood 1945). Bird banding expanded in the nineteenth century when numbered metal leg bands were first used. To this day, they are used on all birds banded in Canada, the United States, and elsewhere, with reports of these marked birds going to the USGS Bird Banding Lab (BBL) in Laurel, Maryland. The

BBL then reports each re-sighting or recovery to the bander and records it for their overall files on that species. Oregonian Charles Henny was one of the first to use banding data to evaluate avian population dynamics, including populations of Ospreys (Henny and Wight 1969). His mathematical models related mortality rates (determined from band recoveries) to the reproduction and recruitment rates needed to maintain a stable population (Henny et al. 1970).

One drawback to using metal leg bands to identify individuals is that often the bird had to be dead or recaptured in order to determine where it had been banded. Once ultraviolet-stable colored bands were developed in the 1980s, a bird's identity could be determined from a distance using a telescope and potentially repeatedly over a number of years. This opened the door to answering many more questions about individual movement patterns, reproductive success, survivorship, and more.

A common method of marking birds is to set up a banding station where mist nets are erected that catch whatever happens to fly into them. Over a series of days, weeks, seasons, or years, a person gets a pretty good idea of what bird species are in the area, the timing of their migration, relative abundance, and survivorship. This is exactly what Joan Hagar has carried out at a number of locations in Oregon. Most recently, she and a group of volunteers built a banding station along the Luckiamute River in Linn County as a means of monitoring bird populations. The other traditional method for monitoring bird populations is to catch and band many individuals of the same species in several locations and see where they are re-sighted. USFWS and Oregon Department of Fish and Wildlife (ODFW) carry this out on a variety of waterfowl throughout the state, including thousands of geese and swans each year. This is part of species-wide efforts to track movements and determine hunting bag limits by the USFWS and ODFW.

In addition to individually marking birds, ornithologists have developed an untold number of techniques for tracking bird movements throughout the annual cycle. Oregon ornithologists have always been on the cutting edge of developing and implementing these tracking devices. The challenge in tracking technology

for birds is to use transmitters that are as small and light as possible, with the longest battery life, without hindering the bird's movements or causing injury or death. All this "jewelry" a bird wears must weigh less than 5% of its body weight so it does not impede flight (Gaunt and Oring 2010). This was the challenge OSU ornithologists Adam Hadley and Matt Betts faced as they addressed questions related to movements of Anna's and Rufous Hummingbirds.

Determining global movements for migratory birds was revolutionized with the advent of satellite transmitters. These transmitters send signals to orbiting satellites, which in turn identify the location of the transmitter on the earth's surface based on Doppler effects, regardless of where on earth the transmitter is located. Satellite transmitters equipped with small solar panels can transmit their location to satellites for extended periods (years) until the tag falls off the bird or is damaged. The power of these transmitters to track bird movements got even better in 2018, when the ICARUS (International Cooperation for Animal Research Using Space) project from the Max Planck Institute for Ornithology in Germany sent up a special antenna that was attached to the International Space Station. Now, anyone can watch the movement of any satellite-tagged bird by logging onto the website Movebank.org. Case in point is the monumental effort launched by Northern Spotted Owl biologists for tracking the birds' movements over their entire range (see the Northern Spotted Owl section on p. 138). This banding and movement data set is the largest of its kind in the world. Satellite transmitters also tell us about daily time budgets and flight patterns for California Condors in the ongoing effort to understand how best to reintroduce the species back to Oregon (Rivers et al. 2014, D'Elia et al. 2015).

Satellite transmitters are powerful, but they are expensive and currently too heavy to place on birds weighing less than 200 g (7 oz). Newer technologies are emerging, however, that provide for lighter and less expensive tracking devices, but there are trade-offs. Geolocators are small, light, and not nearly as expensive as satellite transmitters, and they record and store a bird's approximate location based on day length and the timing of sunrise and sunset. Geolocators can collect invaluable data on bird movements through the annual cycle, but the bird must be recaptured to retrieve the information, and this is frequently not a trivial task. Some avian ecologists have attached a standard very high-frequency (VHF) radio transmitter to each geolocator and programmed it to turn on when they expect the bird to be back in the area where it was tagged (e.g., the following spring) to make the bird easier to find and recapture. The extra transmitter adds weight to the tracking package, but it can often be the only way to find the bird and its geolocator.

Former doctoral student Hankyu Kim and Professor Matt Betts are tracking movements of Hermit Warblers in the H. J. Andrews Experimental Forest in the Cascades using geolocators. Similarly, David Johnson attached 93 geolocators to Burrowing Owls in Oregon and Washington. He recovered 27 geolocators and found that male owls go north in winter but not far north—presumably to hurry back to their nest burrows the following spring. The Klamath Bird Observatory has used geolocators to track movements of a number of passerines, from Yellow-breasted Chats to Hermit Warblers.

One of the newest means of tracking birds is to attach a small GPS (Global Positioning System) device on a bird. These tracking devices can be used to determine the precise location of a bird over a period of several hours to several days, depending on battery size. The location data obtainable from GPS tracking devices are much more precise (± a few meters) than satellite transmitters (± a few kilometers) or geolocators (± a few 100 km). The drawback of GPS tags is that they require more battery power than satellite tags, therefore weighing more than geolocators, and have a short life span. Like geolocators, GPS tags generally need to be retrieved from the bird to download the precise location data stored on the device, but GPS tags are available now from which location data can be retrieved remotely, so long as the device's battery still has power. Also, GPS tags have been developed that can communicate with cell phone towers and transmit locations (e-obs) for the bird and thereby track it as it flies by. The transmitter can record a variety of environmental factors such as ambient temperature, flight times, and so on, and sends data back to the biologist's laptop via cell signals when near a tower.

Another approach to tracking bird movements is through setting up Motus Networks across a bird's migratory route. Motus Networks are a product of Birds Canada and consist of a grid system setup whereby a bird's location is recorded as it flies through the network. This system allows the use of small transmitters that do not need to carry information-recording devices, so the bird does not have to be recaptured to retrieve information. Motus Networks are expensive and must be constructed over an extensive area to be effective; however, they can be used for multiple species or taxa fitted with the appropriate transmitters. Joan Hagar from the USGS Forest and Rangeland Ecosystem Science Center is working to bring a Motus Network to Oregon.

A breeding pair of Ospreys at their nest early in the nesting season, with the male delivering fish prey to the female on the nest site. Because of their large fish diet, Ospreys were one of the bird species that suffered the most from the widespread use of persistent organochlorine pesticides during the 1950s, 1960s, and early 1970s. Photo by Timothy Lawes

# Contaminant Studies in Oregon

CHARLES HENNY

In 1962, *Silent Spring*, by USFWS biologist Rachel Carson, woke the world to the devastation that the insecticide DDT was causing humans, animals, and ecosystems. Ten years later, I had just finished my Ph.D. at OSU in this emerging field of contaminant biology and had landed a job at USGS Patuxent Wildlife Research Center in Maryland. Several years later, I moved back to Oregon and joined my soon-to-be longtime colleague Larry Blus to begin a legacy of contaminants research, mostly on raptors, for the USFWS and the USGS.

## Osprey

My major life's work concentrated on recovering the Osprey to sustainable levels following their decline due to DDT contamination. Following their recovery, I continued to work on Osprey, as they are an ideal "sentinel species" for investigating the spread and accumulation of contaminants in the environment (Grove et al. 2009, Henny et al. 2010). Osprey have (1) a fish-eating diet atop a variety of aquatic food webs, freshwater and marine; (2) a long life span with strong nest fidelity, so one could potentially sample the same individuals over many years, monitoring changes in local environment; (3) an ability to adapt to human landscapes, often the most

Adult Osprey hunting over a coastal foraging site. Following the 1972 ban throughout the United States on the sale or use of the persistent pesticide DDT, the Osprey staged a remarkable comeback throughout most of Oregon. The Oregon Legislature declared the Osprey as Oregon's State Raptor in 2017. Photo by Dan Roby

contaminated; (4) toleration of short-term nest disturbance by researchers; (5) visible nests that are easily located; (6) nests spatially distributed at regular intervals; (7) a tendency to accumulate in its body most, if not all, lipophilic (fat-soluble) contaminants; (8) known sensitivity to many contaminants; and (9) a nearly worldwide distribution.

With Ospreys rebounding globally, I began to think about how Ospreys could be used in creative ways to evaluate newly emerging contaminants, such as the polybrominated diphenyl ethers (PBDEs), which are used as additive flame retardants in thermoplastics, textiles, polyurethane foams, and electric circuitry. PBDEs persist in the environment and bioaccumulate and biomagnify up the food web (de Wit 2002), including in birds of prey (Chen and Hale 2010). PBDE concentrations were highest in Osprey eggs from the lowest-flow rivers studied; however, the two low-flow rivers each passed through relatively large metropolitan areas. We used volumes of wastewater treatment plant discharge, a known source of PBDEs, as a measure of human activity at a location and, combined with river flow, created a novel approach (an approximate dilution index) to relate waterborne contaminants (which influence fish concentrations) to levels of these contaminants in Osprey eggs along segments of the river (Henny et al. 2011). This approach provided a useful understanding of the spatial PBDE Osprey egg patterns observed throughout the Columbia Basin.

## Peregrine Falcon

Peregrine Falcons also suffered the effects of DDT worldwide, so in 1978–79, all known historical nesting locations in Oregon were surveyed by me, Morley Nelson, and others. The known Oregon population was minimally 39 pairs in the 1930s but by 1979 was reduced to one pair and a single adult male (Henny and Nelson 1981). An unhatched egg from the peregrine nest found at Crater Lake contained 19 parts per million (ppm) DDE and an eggshell 19% thinner than normal. This finding, after decades of DDT use in Oregon, was disheartening, especially when other researchers reported that not one North American peregrine population exhibiting 15–20 ppm DDE in eggs and greater than 18% shell thinning was able to maintain a stable self-perpetuating population.

A DDT spray project (3/4 pound/acre) in Idaho, Oregon, and Washington in 1974 (two years after the nationwide ban), where 426,159 acres of forest were sprayed to control tussock moths (*Orgyia pseudotsugata*), resulted in eggshell thinning in American Kestrels from a single application. One year post-spray, kestrels laid eggs 10.4% thinner in the spray area than in the reference area (13 to 28 miles away) and 11.5% thinner than the pre-DDT norm (Henny 1977). Thus the eggshell thinning response following a single DDT spraying was almost immediate.

Once the 1972 federal ban was fully implemented, along with other countries complying with a ban, Peregrine Falcon, Osprey, Bald Eagle, and many other raptor populations have slowly recovered in Oregon and worldwide (see Table 1 on p. 6).

In the spring of 1976 and 1977, a die-off of game birds, including Canada Geese, was reported from Umatilla and Morrow Counties, Oregon. Residues of heptachlor epoxide (HE)

considered lethal were found in the brains of samples I sent to PWRC headquarters for analysis. Heptachlor was being used as a wheat seed treatment in the area to control wireworms (*Ctenicara pruinina*). This prompted two studies, one on Canada Geese nesting along the Columbia River on Umatilla National Wildlife Refuge (NWR), and a four-year American Kestrel nesting study beginning in 1978 using 217 nest boxes placed throughout Umatilla and Morrow Counties (Henny et al. 1984). If kestrels obtained heptachlor, it would indicate that residues moved through food chains (insects, mice, small birds, lizards, and amphibians), since they do not eat wheat seeds like Canada Geese. The kestrels accumulated HE in eggs that reduced productivity and caused some adult mortality. Kestrels were more sensitive to HE residues in eggs than Canada Geese; that is, reduced productivity began at >1.5 ppm in kestrels versus >10 ppm in Canada Geese. One kestrel pair nesting in 1979 was unique in two ways: (1) the 9.1 ppm HE in its egg was the highest obtained during the study, and (2) that adult female died at the nest with a lethal concentration of HE in her brain (28 ppm).

In the fall of 1979, public meetings in the area, documenting direct wildlife mortality and reduced productivity, resulted in the ban of heptachlor-treated wheat seed in irrigated lands within 10–20 km of the Columbia River. The publicity also resulted in reduced heptachlor use throughout the basin in 1979. The immediate response was decreased HE residues in 1980 kestrel eggs (geometric mean 0.44 ppm in 1979, 0.14 ppm in 1980, 0.10 ppm in 1981; Henny et al. 1984). By 1981, farmers almost completely changed their seed treatment from heptachlor to lindane. As of September 1982, production of heptachlor for use as a seed treatment in the United States was prohibited.

Adult Peregrine Falcon plucking a Bufflehead on the central Oregon coast. Owing to the widespread use of DDT and other persistent organochlorine pesticides, the Peregrine Falcon was listed as Endangered in 1970 and was close to extinction throughout the conterminous United States by 1972, when the Environmental Protection Agency banned the sale or use of DDT. Photo by Roy Lowe

Leads for studies sometimes result from unusual circumstances. For example, in 1981, a phone call from an ODFW biologist noted that his wife was rehabbing a Great Horned Owl and fed it a Black-billed Magpie that he had found dead in a field. The owl died and was analyzed at Patuxent, where they found it contained the organophosphate famphur (trade name Warbex). This prompted a secondary hazard study at Patuxent under controlled laboratory conditions with Barn Owls. The laboratory study demonstrated the potential for secondary poisoning from famphur, and a field study was initiated in October 1982. We supplied Warbex to seven ranchers (535 cattle) and assisted with the application (poured on the backs of cattle in squeeze chutes) at recommended rates. Cattle were immediately released into pastures. Carcass searches were made at one- to three-day intervals. Thirty-eight magpies were found dead at or near pastures with treated cattle (Henny et al. 1985). The first magpie died on Day 0 (treatment day), and observed mortality continued through Day 82. It peaked (58%) between Days 5 and 13. Two criteria for diagnosis of famphur mortality were met: (1) brain cholinesterase (ChE) depression of >50% in birds found dead, and (2) confirmation of famphur residues in their tissues or ingesta. Brain ChE depression in dead magpies from treatment areas ranged from 70% to 92%. Famphur was found in all 17 magpies analyzed, with the highest concentrations found in the gizzard. Gizzard contents from 13 magpies that died were pooled and analyzed for famphur and famphur oxon. All dead magpies had ingested cow hair (a trait that is poorly understood for magpies), and the contents were separated into cow hair (12%), other animal matter (37%), and plant matter (51%). Cow hair contained 4,600 ppm famphur, other animal

## Charles "Chuck" Henny

Chuck Henny has worked on avian contaminant studies in Oregon for decades. He, Eric Forsman, and Fred Ramsey are the three living ornithologists who grew up in the state and spent their professional lives addressing issues related to birds in Oregon. Chuck grew up on a farm near Salem and earned his Ph.D. in 1970 studying under Howard Wight at Oregon State University. He then moved to the US Fish and Wildlife Service Patuxent Wildlife Research Center in Maryland to conduct research on Osprey. He returned to Oregon in 1976 and established the Pacific Northwest Field Station of Patuxent in Corvallis as well as became a courtesy faculty member at OSU. Henny and his research partner, Larry Blus, carried out many studies on the effects of contaminants on waterbirds and raptors. They merged with the US Geological Survey in 1993. Chuck has had the rare privilege of growing up seeing few Osprey in Oregon owing to DDT but then through his work and others' has been able to bring the birds back to becoming common in Oregon. Chuck retired from USGS in 2008. He is a fellow of the American Ornithological Society and recipient of the Fran and Frederick Hamerstrom Award, presented by the Raptor Research Foundation.

*Photo by Susan Haig*

matter 620 ppm, and plant matter 340 ppm. Clearly, cow hair was the major source; most or all of the famphur on other material probably originated from cow hair. Furthermore, famphur persisted on the hair of live cattle for more than 90 days. It became clear that unabsorbed famphur remaining on cow hair (on the cow itself or hair shed on fences, feed bunkers, rubbing posts, etc.) caused the magpie mortality. Magpie populations in the far western states declined between 1968 and 1979 (Robbins et al. 1985), which corresponds to the widespread use of famphur as a pour-on to control warbles in cattle, although other factors may also be involved (Henny et al. 1985). Famphur-induced mortality can be eliminated by changing the application from the pour-on method. Alternatives include mixing with feed, giving in capsule form, or applying by injection.

Secondary poisoning occurs when a raptor or other carnivore eats a disabled or dead animal that has ingested poison of some sort. As anticipated, Red-tailed Hawks died during the peak of magpie mortality (Day 10). One Red-tail was found unable to fly. Brain ChE activity for the dead Red-tail was depressed 87%, while plasma ChE in the sick Red-tail was depressed 82%. The sick Red-tail was held and fed for four days and released. Crop contents of the dead Red-tail contained 21 ppm famphur. Another Red-tailed Hawk found dead near Malheur NWR contained 64 ppm famphur in the upper gastrointestinal tract (brain ChE was depressed 89%; Henny et al. 1984). During this period, the lack of organophosphate secondary poisoning reports for birds of prey in North America was likely because of limited numbers of dead raptors being analyzed for ChE depression and organophosphate residues.

## Bald Eagle

FRANK ISAACS

Bald Eagles are predators, scavengers, and pirates at the top of their food chain in North America. Fossil bones believed to be 29,000 years old collected at Fossil Lake in Christmas Valley (Lake County) indicate that they have inhabited Oregon since at least the end of the Pleistocene (Allison 1966). Consequently, the species probably witnessed the arrival of humans in the Pacific Northwest 11,000–18,000 years ago (Loy et al. 2001) and since then has been idolized, romanticized, utilized, reviled, and persecuted by Indigenous and contemporary human cultures (Stalmaster 1987). The Bald Eagle is one of the most studied birds of North America (Buehler 2000) and has received considerable attention in Oregon (Isaacs and Anthony 2011).

The first published reports on Bald Eagles in Oregon were anecdotal and indicated the species was common at the beginning of Euro-American colonization (ca. 1850) but then declined through 1935 (Gabrielson and Jewett 1940). The corresponding decline in Bald Eagles nationally resulted in the US Congress passing the Bald Eagle Protection Act of 1940 (Millar 2002; Table 1). Despite the Bald Eagle Protection Act, population declines continued into the 1970s (Stalmaster 1987). The species was listed under the US Endangered Species Act in 1973. The primary reasons for declining populations into the 1970s nationally were shooting, destruction of nests, reduced prey abundance and availability, and reproductive failure (Stalmaster 1987). Causes for the decline in Oregon were probably similar, based on anecdotal evidence.

The insecticide DDT was used extensively in North American from the mid 1940s until it was banned in 1972 (Stalmaster 1987). During that time, there was a corresponding decline in productivity of several regional Bald Eagle nesting populations (Sprunt et al. 1973). DDT and its derivatives were eventually shown to have caused those declines through its effects on reproduction. Bald Eagle reproductive success in affected areas increased after DDT was banned in 1972 (Stalmaster 1987, Buehler 2000). The effects of DDT on Bald Eagles nesting in Oregon prior to the 1972 ban are unknown because the nesting population had declined prior to the widespread use of DDT (Gabrielson and Jewett 1940), and the population was not monitored systematically statewide until after DDT had been banned for several years (Isaacs and Anthony 2011). But the lingering effects of DDT, its metabolites, and other environmental contaminants on Bald Eagle reproduction were documented in the Columbia River estuary in the 1980s (Anthony et al. 1993) and the species was still declining in the mid 1990s (Buck et al. 2005).

Marshall (1969) was the first to attempt to quantify Bald Eagle nesting in Oregon. He listed 41 known and possible nesting sites and concluded that there had been little change in distribution since the 1930s and that the size of the nesting population was unknown because there were no systematic surveys. During the early 1970s, systematic nest monitoring was initiated in central Oregon by the Deschutes National Forest, in the Klamath Basin by Weyerhaeuser Company (Anderson 1985), and statewide by USFWS (Isaacs and Anthony 2011). In 1971, Jim Anderson was the first to document Bald Eagle nest tree characteristics in Oregon and propose forest management for their protection (Anderson 1971).

In the late 1970s, a series of events led to the subsequent work on Bald Eagles in Oregon. These included the controversies surrounding proposed logging of a Bald Eagle winter roost in the Klamath Basin and a nest tree along the Columbia River, concerns about the effects

Adult Bald Eagle defending a salmon carcass from other eagles. As an apex avian predator, Bald Eagles were highly susceptible to biomagnification of persistent organochlorine pesticides and other contaminants, and suffered severe population declines throughout North America. Listed as Endangered under the US Endangered Species Act in 1973, Bald Eagles had largely recovered and were removed from the Endangered Species List in 2007. Photo by Dan Roby

of environmental contaminants on nesting Bald Eagles nationally, and the 1978 listing of the species as Threatened in Oregon under the Endangered Species Act (ESA) of 1973. Subsequently, Robert Anthony of the Oregon Cooperative Wildlife Research Unit at Oregon State University and a series of graduate students, research assistants, and cooperators studied multiple topics regarding Bald Eagles, including winter habits and energetics, communal roost characteristics, nesting home range and habitat use, nesting habits, nesting population dynamics, nest site and nest tree characteristics, food habits, environmental contaminants, effects of human activities on behavior, and causes of nesting failure (Isaacs and Anthony 2011). Results from those projects, among others, were used by land and resource managers to address tasks identified in the Pacific Bald Eagle Recovery Plan (USFWS 1986).

The Pacific Bald Eagle Recovery Plan was subsequently written by a team of experts to provide guidance for recovering the species and removing it from the Endangered Species List. The Pacific Plan covered California, Idaho, Montana, Nevada, Oregon, Washington, and Wyoming; subdivided the seven-state area into 47 management zones; and recommended four recovery goals: (1) at least 800 nesting pairs in the seven-state area, (2) at least 65% nesting success over a five-year period and an annual average of at least 1.0 fledged young per nesting pair, (3) nesting population recovery goals met in 80% of management zones with nesting potential, and (4) no persistent, long-term decline in any sizable (>100 birds) wintering aggregation (USFWS 1986). In Oregon, 30 years of annual nest surveys provided data related to goals 1–3 and results of national Midwinter Bald Eagle Surveys (coordinated in Oregon by Isaacs in 1988–2007) were used to evaluate goal 4 (Steenhof et al. 2002, Isaacs and Anthony 2008). By 1996, the recovery population goal of 265 nesting pairs for Oregon was surpassed, while goals for nesting success, productivity, and distribution still were below recovery goals as of 2007 (Isaacs and Anthony 2008). Through 2000, winter population trends nationally showed no evidence of decline, suggesting that goal 4 was achieved for the Pacific States (Steenhof et al. 2002).

The banning of DDT resulted in an exponential increase in the estimated number of

Bald Eagle nesting pairs in the contiguous United States (487 in 1963 to more than 10,000 ca. 2000). Thus the species was removed from federal ESA protection on 8 August 2007 (USFWS 2007). The minimum number of nesting pairs in Oregon also increased dramatically from 65 in 1978 to 496 in 2007 (Isaacs and Anthony 2008). ESA listing and delisting by the state of Oregon occurred in 1987 (Marshall et al. 1996) and 2012, respectively, based primarily on federal ESA status and the results of statewide nesting surveys (Oregon Department of Fish and Wildlife 2012). Even though no longer protected under state or federal ESAs, the Bald Eagle still is protected by the Bald and Golden Eagle Protection Act (Golden Eagle added in 1962; Millar 2002) and Migratory Bird Treaty Act nationally, and as a sensitive-vulnerable species on the Sensitive Species List by ODFW (2012). In addition, "active nest trees" and "identified key components" of nesting habitat have protection on state and private land under Oregon Department of Forestry Specified Resource Site Protection Rules (2021).

Federal ESA delisting included a provision for national post-delisting monitoring, which was conducted in 2009 and 2018–19 (USFWS 2020a). Results indicated that the nesting population nationally was 4.4 times larger than estimated in 2009 (USFWS 2020a). The post-delisting national survey results were not intended to provide data by state. Thus there has not been a statewide survey in Oregon since 2007 (Isaacs and Anthony 2008), and none are planned. Reports of new nesting locations and continued occupation of historical (pre-2008) nesting areas observed incidentally or during localized monitoring projects indicate that nest distribution has expanded, and population size has continued to increase in Oregon (F. B. Isaacs, pers. obs.; see the Predators and Climate Squeeze Oregon's Seabirds section on p. 81).

## Frank Isaacs

In 1979, Frank was hired by Robert Anthony from the Oregon Cooperative Wildlife Research Unit at Oregon State University to conduct six months of field work on nesting Bald Eagles in Oregon. That temporary job started Frank's career as a wildlife biologist studying raptors. For 31 years, Frank and Bob Anthony researched Bald Eagles across the state. In addition, Frank coordinated the Midwinter Bald Eagle Survey for Oregon from 1988 to 2007, coordinated Peregrine Falcon nest surveys in Oregon from 2003 to 2007, served as a member and chairman of the Bald Eagle Working Team for Oregon and Washington, and was cofounder of the Oregon Eagle Foundation (OEF) in 1986.

OEF was founded to ensure that the Bald Eagle population in Oregon was adequately monitored and that its habitats were protected. With the federal Endangered Species Act delisting of the Bald Eagle in 2007 and rapid growth of wind energy projects in Oregon, OEF shifted its focus to nesting Golden Eagles. Frank served as secretary of OEF and coordinated statewide Golden Eagle nest surveys from 2011 to 2020. Frank received several statewide awards for his work and has educated people at public events and through his work with eagle project volunteers.

*Photo by Jane Olson Frank*

# III  Avian Conservation in Oregon

# Conservation Efforts for Birds in Oregon

SUSAN HAIG

Most prominent environmental issues in Oregon have had an avian species as a key component to the discussion or as an indicator of the issue at hand: Northern Spotted Owls, Marbled Murrelets, Caspian Terns, Double-crested Cormorants, Snowy Plovers, Greater Sage-Grouse, to name a few. Addressing issues such as timber harvest, salmon restoration, coastal habitat degradation, or the impacts of contaminants, climate change, or alternative renewable energy development takes an enormous effort on the part of many agencies, nongovernment groups, researchers, and private citizens filling many roles.

Prior to the environmental movement of the 1960s and 1970s, there were few efforts to address environmental problems affecting Oregon's birds, with the exception of the Migratory Bird Treaty Act (MBTA) of 1918 (see Table 1 on p. 6). Under the MBTA, people were no longer allowed to harm, harass, or kill native birds, nor were they allowed to keep eggs, feathers, or any other parts of live or dead birds unless they had the appropriate hunting or collecting permits. Passage of the National Environmental Policy Act (NEPA, 1970), the Clean Air Act (1970), the Clean Water Act (1972), the Endangered Species Act (ESA, 1973), and the National Forest Management Act (1973), among others, provided recognition that the environment and its components could be harmed and outlined a process in which the issues could be identified and addressed (Table 1). This gave birth to a whole new world for natural resource management agencies that had not had to deal with regulations or had only dealt with hunted or "pest" species. It also initiated

programs worldwide for the conservation of "nongame" species (see Table 1 on p. 6). Agency programs grew as the new emphasis called for information and mitigation to a Pandora's box of issues.

Initially, agencies acted independently; however, as the issues became broader and more complex, interagency cooperation was needed. State wildlife commissioners formed the Association of Fish and Wildlife Agencies to promote interstate cooperation. Waterfowl biologists and managers set up migratory bird flyway councils as they considered the needs of (mostly) waterfowl throughout their annual cycle across North America (Figure 5). A crisis in waterfowl productivity in the early 1980s led waterfowl managers to draft the North American Waterfowl Management Plan (NAWMP) in 1986. NAWMP was the first international plan to incorporate non-waterfowl into a more comprehensive migratory bird management plan. Over the next several decades, a series of regional joint ventures or habitat joint ventures were formed to develop and implement conservation plans for various parts of North America (Figure 6).

Oregon is part of the Pacific Birds Habitat Joint Venture and the Intermountain West Joint Venture (Figure 6). In actuality, they overlap in eastern Oregon, and each has extensive programs that span research, management, and outreach for all avian species and their habitats. While many organizations and people have worked to make these joint ventures the success they are today, Bruce Taylor (see Figure 6) has worked from their inception to blend the needs of state, federal, and private interests in bird conservation across the state.

During this period of growth and in recognition of the need and utility for bird conservation, many people who had trained to be ornithologists ended up in administrative roles and likely accomplished far more

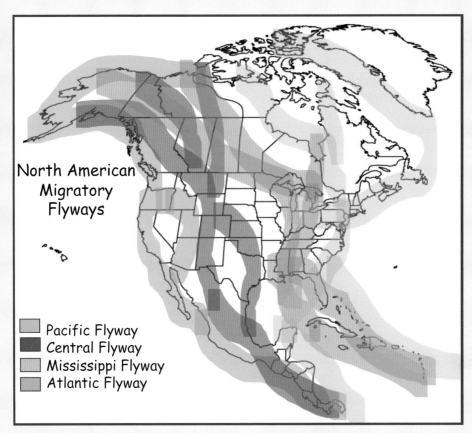

FIGURE 5. Migratory bird flyways in North America. Susan Haig, Oregon State University

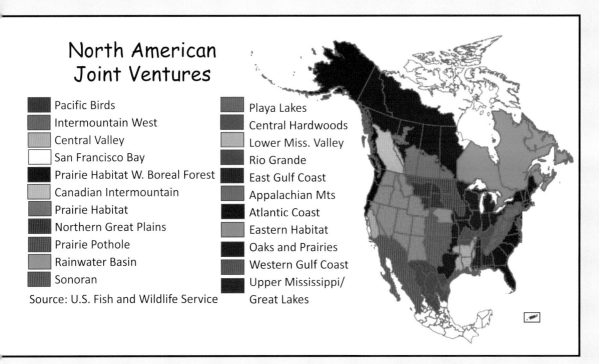

FIGURE 6. Wildlife habitat joint ventures in North America. From the US Fish and Wildlife Service

Adult Western Snowy Plover illuminated by evening light on the central Oregon coast. Western Snowy Plovers are listed as Threatened in Oregon, but their population has started to recover owing to habitat restoration and protection on Oregon's publicly owned beaches. Photo by Roy Lowe

for bird conservation by giving up their color bands and mist nets. For example, Charlie Bruce became the first nongame biologist for the Oregon Department of Fish and Wildlife (ODFW) and developed the new nongame and endangered species programs for ODFW.

Oregon State University (OSU) graduate Carol Schuler spent her forty-plus-year career working for the US Fish and Wildlife Service (USFWS) and the US Geological Survey (USGS) in Oregon by bringing bird conservation issues to the forefront, finding funding to help resolve the issues, and working with disparate groups to reach an agreed-upon plan for moving forward.

Similarly, Bruce Taylor coordinated bird conservation efforts among many groups for decades. OSU graduate Mark Stern has also led many research and conservation efforts as part of The Nature Conservancy's growth in the state over a forty-year period. In each of these cases, the concrete accomplishments that can be cited for these administrators pale in comparison to the thousands of calls, documents, and negotiations that were needed to accomplish each small step forward in avian conservation.

## Charlie Bruce

Charlie started work in 1971 for what was then the Oregon Game Commission, later renamed the Oregon Department of Fish and Wildlife. He was the first "nongame" wildlife biologist in the state and developed the first provisional rare, threatened, and endangered species list. Charlie served as secretary for the first State-Federal Interagency Endangered Species Task Force, which later focused on Northern Spotted Owls, and served on early committees developing guidelines for the Northern Spotted Owl and the Pacific States Peregrine Falcon recovery teams. He chaired the early Oregon Marbled Murrelet guidelines committee and later the coastal Snowy Plover working team. He served as the state advisor to the federal Interagency Scientific Committee for the Northern Spotted Owl and as assistant to the governor's representative on the federal Endangered Species Committee ("God Squad"). Later, he was lead state biologist associated with Columbia River Basin dams. Charlie was the lead biologist working with the Oregon Department of Forestry on state forest management plans and the first Habitat Conservation Plan for the Elliott State Forest. Charlie finished his career as the state coordinator for threatened and endangered wildlife species. He continues in "retirement" with conservation projects on Peregrine Falcons, Golden Eagles, and eastern Oregon Snowy Plovers.

*Photo by Cindy Bruce*

## Carol Schuler

Carol Schuler has had a remarkable career leading up to the highest levels of the US Department of the Interior, without leaving the Willamette Valley. Starting with her graduate work at OSU in 1983, she studied contaminant issues for birds at Kesterson National Wildlife Refuge with Bob Anthony at the Oregon Cooperative Wildlife Research Unit. Carol then worked for the US Fish and Wildlife Service for 19 years in leadership roles ranging from endangered species, ecological services, and fisheries, to refuges. She established the US Fish and Wildlife Service contaminants program for the Pacific Northwest in the 1990s, focusing on the Columbia River, Willamette River, and damage from the *New Carissa* oil spill on the south coast. She was the manager at Finley National Wildlife Refuge and then directed endangered species efforts for USFWS in Portland before becoming director of the US Geological Survey Forest and Rangeland Ecosystem Science Center in Corvallis. There she participated in research and conservation efforts for many species, including Northern Spotted Owls, Greater Sage-Grouse, and more. She also led the creation of the Landscape Conservation Cooperatives in the Pacific Northwest and the Pacific Islands. Her final position was as the science advisor to the ecosystems director of USGS in Washington, DC, which she managed from Corvallis. She was named to the Oregon State University Department of Fisheries, Wildlife, and Conservation Sciences Registry of Distinguished Graduates in 2009.

*Photo by Susan Haig*

## Mark Stern

Mark Stern recently retired as the director of Forest Conservation for The Nature Conservancy in Oregon. He has worked on wildlife projects and natural resource issues across much of Oregon, including stints in Harney County for the Burns District Bureau of Land Management and the US Fish and Wildlife Service at Malheur National Wildlife Refuge, and for nearly 40 years with The Nature Conservancy focusing on wildlife, water, forestry, and related policy issues in the Klamath Basin and around the state. Mark's wildlife work included studies of waterbirds in the Warner Valley in Lake County; Sandhill Cranes, Black Terns, and Yellow Rails at Sycan Marsh and in Lake and Klamath Counties; Snowy Plovers at Abert Lake and the Oregon coast; neotropical migrant songbirds in the Willamette Valley; the bird banding station on Sauvie Island; and many projects involving wetland restoration and waterbirds in the Klamath Basin and elsewhere. Mark served on the Management Board for the Pacific Bird Habitat Joint Venture and as a board member for the Oregon Chapter of The Wildlife Society. In 2002, Mark received the Oregon Wildlife Society Award for career achievement. Mark continues his long-term monitoring of fall Sandhill Cranes on Sauvie Island. He has an M.S. in wildlife ecology from Oregon State University.

*Photo by Ginny Stern*

## Bruce Taylor

Bruce Taylor's name is synonymous with Oregon migratory bird conservation. In early 1992, he started working for the Pacific Coast Joint Venture (now Pacific Birds), whose territory included Oregon west of the Cascades. When the Intermountain West Joint Venture (IWJV) began in 1994, Bruce joined their efforts, picking up Oregon east of the Cascades. For the next 20 years, Bruce's work with the two Joint Ventures was under the umbrella Oregon Habitat Joint Venture (OHJV). For his vision and leadership of the OHJV, Bruce received the 2015 John E. Nagel Award from the IWJV. Some of Bruce's local work involved development of the High Desert Partnership, an organization dedicated to finding common ground on natural resource and community challenges in Harney County. Bruce has also been an active collaborator in the Oregon Sage-Grouse Conservation Partnership, established in 2012 to guide Oregon state policy, leverage funding, and build conservation actions that balance natural resource conservation with local livelihoods. He has contributed to multiple local wetland conservation projects across Oregon and took the lead on Pacific Birds Oak and Prairie conservation work throughout the Pacific Northwest from 2015 until his retirement in 2020. Currently, Bruce serves as the vice chair of the Oregon Agricultural Heritage Commission.

*Photo by Jana Taylor*

# Birds That Have Come and Gone

ALAN CONTRERAS

Using current nomenclature, Gabrielson and Jewett (1940) listed 338 species as occurring in Oregon. Marshall et al. (2003) listed 486. Currently, eBird has accepted 549 species records for Oregon and the Oregon Bird Records Committee recognizes 542 (see the Appendix). Many of the additions have been seabirds (e.g., Glaucous-winged Gulls were first documented as an Oregon breeding bird in 1969; Scott 1971), and most of the rest have been vagrants or relatively rare species that do not occur often in the state. Three of the birds included in Gabrielson and Jewett (1940) are not currently recognized as being in Oregon.

Some additions have clearly expanded into Oregon since midcentury, and a few are probably new arrivals. From the southeast came the Cattle Egret, originally from Africa and probably a natural invader via Brazil and the Caribbean. Also, from a generally southern direction came Great-tailed Grackle, now a regular local breeder in Malheur and Jackson Counties and an occasional breeder elsewhere.

Anna's Hummingbird, White-tailed Kite, Red-shouldered Hawk, Black Phoebe, and Blue-gray Gnatcatcher have moved in from California. The kite, hawk, and phoebe were known by 1935 from one or two records. Virginia's Warbler and Gray-headed Junco (*Junco hyemalis caniceps*) are known to reach the edge of their breeding range, at least occasionally, in southeastern Oregon.

Not all arrivals are from the south. Least Flycatcher, Northern Waterthrush (first territorial birds in 1977), Grasshopper Sparrow, Franklin's Gull (first reported in 1948), and Barred Owl have arrived from the northeast or east (the sparrow may have been present

in some areas, as it is hard to discover). All but the flycatcher are regular annual breeders, and the owl and gull have become common. In addition, a few species that are not easy to find in breeding season have been determined to be Oregon breeders, including Horned Grebe, Red-necked Grebe, Black Swift, and, rarely, Common Goldeneye, Greater Yellowlegs, and probably Broad-tailed Hummingbird (annually) and Virginia's Warbler (irregularly).

Among the major expansions of birds that were already in the state in 1935 are European Starling, House Finch, California Scrub-Jay, Caspian Tern, White-faced Ibis, Wrentit, Acorn Woodpecker, and Great Egret. Bald Eagle, Peregrine Falcon, and Osprey have become far more common than they were in the mid-twentieth century, owing in part to changes in pesticide practices and an increased number of reservoirs. For Bald Eagles especially, the nesting population was greatly reduced before pesticides became a factor (see the Bald Eagle section on p. 56). Anecdotal evidence suggests that the increase in Bald Eagles was largely the result of reduced direct persecution (shooting, poisoning, and trapping), starting with the 1940 Federal Bald Eagle Protection Act (amended in 1962 to include Golden Eagles) and accelerating with its listing under the federal ESA in 1978. Another important factor was the shift from sheep to cattle production that occurred midcentury. Cattle ranchers were probably more tolerant of eagles than sheep farmers (F. B. Isaacs, pers. comm., 2022).

Postbreeding movements of Elegant Terns have become regular since the 1980s. Palm Warbler and Clay-colored Sparrow are now known to be regular migrants in small numbers, mainly in fall, and Tropical Kingbirds have become regular and predictable on the coast every October–November. Even Brown Boobies now occur semi-regularly in late summer and fall.

Perhaps the most dramatic change is the hundreds of thousands of Canada and Cackling Geese that now winter in the state, mainly in the Willamette Valley but also in the Columbia Basin. These birds were present in small numbers until the grass seed industry expanded after World War II (Henny and Naughton 1998). Also using the grass seed fields are increased numbers of Killdeer and Dunlin (Sanzenbacher and Haig 2001, 2002a). But recent conversion of grass seed fields to hazelnut orchards has reduced habitat for all wintering waterbirds. Trumpeter Swans were introduced as a breeding species in the 1960s and have hung on in small numbers (see the Trumpeter Swans section on p. 218). It is a regular but local wintering species in small numbers, mostly in the northern half of the Willamette Valley and in the Lower Columbia River region, expanding recently to the lake basins of south-central Oregon.

There have been withdrawals, too. Rock Sandpiper was a fairly obvious winter bird as far south as Cape Arago through the 1970s, with some farther south. Today, it is rare and local south of Tillamook County, with small numbers occasionally reaching Cape Arago. Upland Sandpipers nested in small numbers, mostly in Grant and Umatilla Counties until the early 2000s, but now are apparently extirpated as a breeder. Often the goal of Steve Herman's field trips to Bear Valley and other northeastern areas, it remains unclear why they are no longer in Oregon. The same is true of American Redstart, which bred in Umatilla, Union, and Wallowa Counties and occasionally west to the east slope of the Cascades until the 1990s, after which it has not been found as a breeder. It has also withdrawn from southeastern Washington.

The status of the Yellow-billed Cuckoo in pioneer times is not clear, though it seems to have been a fairly local bird found mainly along the Snake-Columbia River system. Today, it is seen only as a vagrant to the state and does not breed, though there are occasional reports of birds in appropriate habitat in summer.

Sharp-tailed Grouse is one of the biggest losses, as it was a common bird in parts of eastern Oregon into the early twentieth century and remained in the state until 1967. ODFW has reintroduced birds to Wallowa County, and that population seems to be holding on.

One of the most dramatic avian changes that has occurred in the past 50 years in Oregon is the rapid decline in the Spotted Owl population (see the Northern Spotted Owl section on p. 138). Based on a region-wide monitoring program in the Pacific Northwest, the Spotted Owl population has declined by >75% in Washington, >60% in Oregon, and from 30% to 60% in northern California since monitoring began in the late 1980s (Franklin et al. 2021). Initially, this decline was thought to be primarily a result of habitat loss from logging and fire. But that concern has begun to pale in light of more recent analyses that suggest that competition with the invasive Barred Owl is a much more serious threat than previously realized (Franklin et al. 2021).

A few species have lost huge numbers but have turned around and are coming back. Most obvious are the raptors (Bald Eagle, Peregrine Falcon, Osprey, etc.) that recovered following the ban of the pesticide DDT in

OPPOSITE: Adult male Anna's Hummingbird during a snowstorm in the mid-Willamette Valley. This resident species has expanded its year-round range into western Oregon in the last few decades, where it relies on hummingbird feeders to survive over winter. Photo by Timothy Lawes

1972 (see the Contaminants section on p. 50). Another conservation victory is the Western Bluebird, brought back in part by extensive nest-box programs run in the 1970s–1990s by dedicated amateurs such as Elsie Eltzroth (Corvallis; see her biography on p. 21), Hubert Prescott (Portland), and Aaron Skirvin and Al Prigge (Eugene). A species whose status was questionable until people learned how to find their nests is the Yellow Rail (see the Yellow Rails section on p. 201). Perhaps the best news about a missing Oregon bird is the recently reintroduced California Condor (see Return of the California Condor on p. 240).

## Alan Contreras

Alan Contreras is a birder extraordinaire in Oregon. A fourth-generation Oregonian born in Tillamook County, Contreras began birding at the age of 11 and created a bird newsletter, *The Meadowlark*, shortly thereafter. Contreras has since been editor of *Oregon Birds*, contributed to the founding of the Oregon Field Ornithologists (now the Oregon Birding Association), and currently operates Oregon Review Books as well as Oregon Review Special Editions, which issues specialty reprints and historical material related to the Northwest. He was the lead author of *Birds of Oregon*, *History of Oregon Ornithology*, *Birds of Lane County*, as well as numerous other volumes on natural history. Alan has been an enthusiastic mentor for many birders, young and old. Contreras is a volunteer at Malheur National Wildlife Refuge, where he assists tourists and enjoys keeping track of the many birds he spots while at the refuge. Currently, he is also president of the Friends of Malheur National Wildlife Refuge. A graduate of University of Oregon's law school, Contreras worked as an education consultant in combination with his writing career and birding activities.

*Photo by Thomas Meinzen*

# The Oregon Coast and Near-Coastal Areas

ROY LOWE

Migratory birds from around the world pass through Oregon's marine and coastal waters, and many spend the winter in coastal wetlands. Maintaining healthy ecosystems and habitats throughout the flyways is essential for the continued survival of these world travelers. Oregon's conservation efforts to conserve and protect seabirds and their habitats began when William L. Finley and Herman Bohlman traveled to Oceanside, Oregon. In 1901 and 1903, they visited Three Arch Rocks and photo-documented the seabirds breeding there (Mathewson 1986). Later, Finley traveled to Washington, DC, to meet with President Teddy Roosevelt and convinced Roosevelt that protection of this area was needed. In 1907, President Roosevelt established Three Arch Rocks as a Bird Reservation (name later changed to National Wildlife Refuge) by executive order. Because of Finley and Bohlman's efforts, Three Arch Rocks became the first National Wildlife Refuge west of the Mississippi River and, in effect, Oregon's first Marine Protected Area. But it would be another 78 years (1985) before the USFWS established a permanent position on the Oregon coast to monitor and manage wildlife, a position I was fortunate to be selected for. Oregon Islands National Wildlife Refuge (NWR) was established in 1935 when Goat Island was designated a Migratory Bird Refuge, and between 1968 and 1996, almost all the remaining coastal rocks and islands were added to the refuge. Mainland units were established for seabird

protection, including Coquille Point (1991), Crook Point (2000), and Whale Cove (2014). All of these actions ensured that more than 1 million seabirds had secure nesting sites in Oregon.

The first coastwide survey of nesting seabirds in Oregon was funded by the USFWS, conducted in 1977 (Pitman et al. 1985), and repeated by refuge staff in 1988. This led to the publication of the *Catalog of Oregon Seabird Colonies* (Naughton et al. 2007). Establishing a refuge wildlife biologist position on the Oregon coast was essential to determining the current status of seabird breeding populations, establishing long-term monitoring programs, determining current and potential future activities that affect nesting seabirds, developing cooperative programs with other agencies, and providing interpretative and education materials for the public. Chief among the early concerns were impacts from human disturbance caused by low-flying aircraft, close approach of boats, people climbing on rocks and islands, shipwreck fuel/oil spills, fisheries interactions or stock depletions, oil and gas exploration, and later concerns for development of wind- and wave-energy impacts. When Oregon's ocean planning process began in the late 1980s, the USFWS engaged in the process to ensure that seabird conservation received equal consideration in the Oregon Ocean Resources Management Plan and later the *Territorial Sea Plan*. These efforts continued when Oregon began planning for

Adult Common Murre, the most numerous seabird species that nests along the Oregon coast. Breeding colonies of this species are highly susceptible to predation and disturbance from Bald Eagles, a keystone predator along the Oregon coast. Photo by Roy Lowe

and establishing marine reserves and marine protected areas, including the Cape Perpetua Seabird Protection Area.

The loss of estuarine wetlands and impacts to migratory birds, along with overall impacts to estuarine health, productivity, and biodiversity, were major concerns (Wiens and Scott 1975). Losses of tidal wetlands in Oregon are high, including the loss of 95% of the historical tidal forest wetlands and 59% of the historical tidal marsh habitat (Brophy 2019). The bulk of these habitat losses resulted from diking and subsequent draining and clearing for agricultural purposes, but alterations and sedimentation from logging, splash dams, and

other watershed activities also greatly affected estuaries. Similar to marine resources, there are no pre-disturbance data to accurately evaluate wildlife populations prior to and during settlement of the coast by Euro-Americans. In 1985, monthly aerial winter waterfowl surveys of the Oregon coast from Nehalem Bay to the California border were expanded from a one-day to a two-day survey. The expanded surveys allowed for a more accurate estimate of the waterfowl population and use of diked lands that can serve as seasonal wetland habitat when flooded. Site-specific data were recorded for some of the diked sites to provide information for potential future restoration planning

Female and male Harlequin Ducks on the central Oregon coast. Harlequin Ducks rely on rocky intertidal and subtidal habitats for foraging during the nonbreeding season but nest along interior streams and rivers. This habitat split makes Harlequin Ducks especially vulnerable to degradation of either of the habitats they depend on. Photo by Roy Lowe

efforts. Once waterfowl surveys of the Coquille Valley were added, it was discovered that this area hosts the largest coastal concentration of wintering dabbling ducks between San Francisco Bay and the Columbia River. In recognition of the importance of the Coquille Valley, ODFW established the Coquille Valley Wildlife Area in 2013, restoring some of the important wetland functions to this area. Similarly, USFWS had established Bandon Marsh NWR (1983), Siletz Bay NWR (1991), and Nestucca Bay NWR (1991) to protect existing tidal wetlands and restore former wetlands. Wetland restoration projects have been completed at all three refuges, and others are

currently ongoing or planned. Restoration projects including long-term monitoring are accomplished through partnerships with many groups, such as other federal and state agencies, Native American tribes, local watershed councils and associations, nongovernmental organizations, universities, landowners, and many others. In 2013, a 410-acre tidal marsh restoration project was completed on the Ni-les'tun Unit of Bandon Marsh NWR. At the time, this constituted the largest tidal marsh restoration project in Oregon.

Since establishment of the first position on the Oregon coast, USFWS has expanded coastal programs, with a main office located

The Black Brant is a small goose that winters in Oregon's shallow coastal estuaries and bays, where it largely forages on eelgrass beds. The species has seen more than a 50% decline in overwintering numbers because of habitat destruction and degradation along the Oregon coast. Photo by Roy Lowe

in Newport and satellite offices at Bandon Marsh and Nestucca Bay. Long-term seabird monitoring surveys continue along with other monitoring and management programs. A successful long-term cooperative management program between the Bureau of Land Management and the USFWS at Yaquina Head resulted in a large increase in the nesting seabird populations, while visitor use of the area has also increased sharply. An ODFW/ USFWS cooperative seabird disturbance study at Three Arch Rocks NWR in 1993 led to the

establishment of an annual seasonal 500-foot boat closure zone around the refuge to protect nesting seabirds. Oil spill restoration funds from the *Nestucca*, F/V *Tenyo Maru*, and *New Carissa* spills provided funding for production of interpretative and educational materials focused on seabird conservation, management of invasive mammals at seabird colonies, purchasing occupied habitat for Marbled Murrelets, and assisted with funding tidal marsh restoration. The *Nestucca* and *Tenyo Maru* spills occurred on the Washington coast, but successful cases were made that seabird populations originating in Oregon were affected by both of the Washington spills. Thus habitat restoration of estuarine wetlands, riparian areas, and adjacent upland habitats has been going strong all along the Oregon coast for several decades now, by a variety of entities.

Looking back reminds me of how difficult it is to predict the future in natural resource management and conservation. Looking ahead, global issues such as the impacts from climate change, including sea level rise and ocean acidification, challenge our ability to comprehend what is coming. More locally, the inevitable impacts of a Cascadia Subduction Zone earthquake and the stochastic effects of oil spills add to the risks and uncertainty in the future. El Niño / El Niño–Southern Oscillation events may provide a window into the potential impacts to the marine ecosystem owing to reduced productivity, should future weather patterns result in insufficient upwelling winds. In a mere half of a human generation, we've seen many unpredicted or unexpected changes, such as Bald Eagle recovery negatively affecting seabird colonies, an order of magnitude decline in the Tufted Puffin population in Oregon, the appearance of huge nesting colonies of Double-crested Cormorants and Caspian Terns in the Columbia River estuary, continuing declines of anadromous fish and Marbled Murrelets, the 2014–16 North Pacific marine heat wave, sea star wasting disease, expansion of urchin barrens, and resulting loss of kelp. Conservation, management, research, and continued long-term monitoring will be essential for charting the future and understanding the past.

### Roy Lowe

Waldport resident Roy Lowe was the longtime biologist and refuge manager of the US Fish and Wildlife Service Oregon Coast National Wildlife Refuge Complex. When he arrived on the Oregon coast in 1985, he was the first agency employee to be stationed there. Lowe was involved in the establishment of two new national wildlife refuges (Nestucca Bay and Siletz Bay) in 1991 as well as the expansion of land on four existing refuges along the coast. He contributed to one of the largest tidal restoration projects in Oregon and the cataloging of more than 1 million nesting seabirds. For his efforts, he was named National Wildlife Refuge Manager of the Year in 2005, US Department of the Interior Federal Land Manager of the Year in 2006, and one of 50 Great Leaders in Oregon in 2006. He received the US Department of the Interior Meritorious Service Award in 2012 and the US Fish and Wildlife Service National Land Legacy Award in 2014. He served as the Oregon and Washington regional representative for the Pacific Seabird Group for 14 years. He retired in 2015 after 30 years of service on the Oregon coast. In retirement, he has been training coastal wetland managers in the People's Republic of China.

*Photo by Dave Ledig*

# Offshore Renewable Energy Development and Bird Mortality

SUSAN HAIG

Offshore renewable energy development (ORED), whether it be wave energy or wind energy, is exploding around the world and in particular off the Oregon coast. The Oregon Department of Energy says the nearshore energy development project approved in March 2021 will have the ability to power more than 28 million homes annually. The PacWave South Project is a joint venture with OSU located 6 nautical miles offshore from Newport that will spread over about 8 square miles of ocean. While the projects are phenomenal for their advancement of renewable energy development, conservation issues must be investigated and mitigated if they are to be fully successful (Marques et al. 2014).

As this technology is so new, there is a lack of understanding of its potential impact on resident or transient wildlife nearby (e.g., seabirds, shorebirds, bats, marine mammals; Grecian et al. 2010, Brabant et al. 2015, Fijn et al. 2015). Similar to onshore towers, siting of the power apparatus, whether above or below the sea surface, is the most important means so far identified for avoiding bird collisions (Suryan et al. 2012). Also, it is difficult to measure the impact of offshore technology on wildlife populations, even more so than with onshore wind turbines and transmission towers, because when animals collide with offshore facilities, they fall into the ocean and float away (Desholm et al. 2006). Thus direct measures of mortality rates (i.e., searching for carcasses) are not feasible. Early in ORED development, OSU oceanographer George Boehlert and Andrew Gill (2010) described positive and negative aspects of ORED technology related to seabird conservation. On the positive side, ORED structures can alter habitats, marine community structure, and prey distributions such that certain seabirds would have enhanced feeding opportunities and aggregate near the site. Conversely, changes to beach processes or tidal excursions may affect shorebird foraging habitat. Diving birds may be subject to entanglement, collision, or blade strike with subsurface components or devices. And there are the unknown number of fatalities from collisions with wind turbines and transmission towers. Adding to the difficulty is the nocturnal timing of many bird movements (Kunz et al. 2007). Thus it remains difficult to assess the potential impacts of ORED until accurate means of assessing damages are developed.

OSU seabird biologists, mechanical engineers, and their colleagues have been working on a sensor that would detect a bird striking an offshore tower and lower the risks to other birds in the area (Flowers et al. 2014, Suryan et al. 2016). Thus the lead bird in a migrating flock might strike a wind turbine, but that collision would shut down nearby turbines and thereby save other birds in the flock. Of course, there are cumulative effects of this sort of approach to consider. In the lab, the researchers developed a multi-sensor instrumentation package that includes vibration sensors and optical cameras networked together in an event-triggered data acquisition system. Hsu and Albertani (2021) tested this system using tennis balls to mimic bird strikes. If a turbine could recognize when it is hit by a bird, it could slow down or stop to minimize risks to other birds in the area. Not only will this help save birds like the endangered Marbled Murrelet and Short-tailed Albatross, which could be harmed by development of offshore wind power facilities off the Oregon coast, but it will also reduce some legal risks to the ORED industry associated with environmental impacts.

Pigeon Guillemot adult in breeding plumage. Pigeon Guillemots are a diving seabird in the family Alcidae (auks, murres, and puffins) that feeds close to shore on fish that are found in shallow water. They nest in crevices on coastal bluffs and cliffs along the Oregon coast and unlike most alcids can raise two-chick broods. Photo by Timothy Lawes

Short-tailed Albatrosses, adult in foreground and subadult behind. The species is listed as Endangered by the federal government and the state of Oregon, where nonbreeders are occasionally found offshore within the 200-mile Exclusive Economic Zone. This species and other albatrosses are highly vulnerable to bycatch and entanglement in longline and other fisheries. Most Short-tailed Albatrosses seen in Oregon waters are juveniles that have traveled long distances from their natal colony on islands in the western Pacific Ocean. Juvenile Short-tailed Albatrosses (inset photo) have all-dark plumage and the bubblegum pink-colored bill characteristic of adults. Photos by Dan Roby

# Predators and Climate Squeeze Oregon's Seabirds

DON LYONS

The state's first comprehensive seabird surveys were initiated by Palmer Sekora, Bob Pitman, Mike Graybill, Dan Varoujean, Roy Lowe, Jon Anderson, Daniel Matthews, and others with the USFWS in 1977 and 1988 (Naughton et al. 2007). These baseline data came at an unusual time ecologically, corresponding to historically depressed populations of species most sensitive to DDT. Although DDT use was banned in the United States in 1972, populations of severely affected species were still in the early stages of recovery at the time of these initial systematic surveys. These depressed populations included some prominent seabirds such as California Brown Pelicans (*Pelecanus occidentalis californicus*) and Double-crested Cormorants (also subject to periodic persecution owing to perceived conflicts with fisheries), and some significant seabird predators including Bald Eagles and Peregrine Falcons. The effects of the absence of these historically abundant species on other taxa was unclear in the 1970s but would become evident in the ensuing decades. Over a 30-year period beginning in 1978, Oregon's Bald Eagle breeding population grew an average of 7.3% per year (Isaacs and Anthony 2011), with growth in estuary and ocean habitats occurring at an even more rapid rate (8.2% and 9.7% per year, respectively). The impact of this growing eagle population was closely observed at two prominent mixed-species seabird colonies where monitoring was ongoing across a two-decade-long period (1998–2020): the cliffs and offshore rocks of Yaquina Head near Newport, notable for what was often the largest colony of Common Murres in the state, and the low-lying sandy habitat of East Sand Island in the Columbia River estuary near Astoria,

notable for colonies of Caspian Terns and Double-crested Cormorants that were among the largest for these species in North America and, for Caspian Terns, the entire world. Robert Suryan and Daniel Roby led studies at these respective sites in collaboration with many other partners.

At Yaquina Head near Newport, Oregon, Bald Eagles became the leading predator causing disturbance, inflicting direct mortality on adults and nest loss through egg predation (Horton 2014). Secondary nest predation by more abundant gulls, corvids, and Turkey Vultures resulted in far higher rates of nest loss, however, and in some years, complete colony failure was observed. Similar disruption was also seen at East Sand Island, with eagle disturbance increasingly leading to extensive secondary nest predation, lowered breeding success, failure of peripheral satellite sub-colonies, and occasionally complete colony failure there as well (Collar et al. 2017). Recovering populations of California Brown Pelicans, delisted under the Endangered Species Act in 2009, increasingly dispersed to the Pacific Northwest following the breeding season (Wright et al. 2007) and became a notable disturbance factor at Yaquina Head, East Sand Island, and other seabird colonies in Oregon (Suryan et al. 2012, 2014). Pelicans occasionally consume small numbers of chicks and eggs of other species but can often cause significant disturbance by landing within nesting areas, trampling nests, flushing nesting adults, remaining present for hours, and ultimately facilitating extensive secondary predation.

Patterns similar to those at Yaquina Head and East Sand Island (on the Columbia River; Figure 7) also played out elsewhere on the Oregon coast and throughout the Pacific Northwest. Eagle disturbance rates and seabird breeding failure were greater along the north coast of Oregon (Horton 2014), where the

Adult Bald Eagle preying on an adult Common Murre. As Bald Eagle populations in Oregon have recovered from the impact of persecution, DDT, and other persistent organochlorine pesticides, eagle disturbance has had an increasing impact on breeding colonies of seabirds, especially those of the Common Murre, the most numerous species of seabird nesting in Oregon. Photo by Roy Lowe

density of eagle nest territories was highest. This led to a significant redistribution of Common Murre nesting in the state. Between 1988 and 2014, the murre breeding population declined 74% on the north coast, while the central coast breeding population increased 105% (Stephensen 2020). Redistribution was not sufficient to sustain a stable murre population in Oregon as the statewide breeding population declined 23% over this period.

While the top-down effects of eagles and other predators were being well documented, long-term studies also provided for identification of bottom-up factors that regulate Oregon's seabird population and likely influence their resilience to predators. Efforts to understand relationships between Oregon's seabird populations and their forage base date back to at least the 1970s, when

in a notable first, OSU professor John Wiens and graduate student Michael Scott (1975) explored the prey demands of four prominent species (Common Murres, Brandt's Cormorants, Leach's Storm-Petrels, and Sooty Shearwaters) using a bioenergetics modeling framework. Just a few years later, the severe 1982–83 El Niño became a watershed event in the advancement of our understanding of Oregon's coastal oceanography and the potential effects of hemispheric-scale climate fluctuations on seabirds through changes in their marine prey base (e.g., Pearcy et al. 1985). Increased adult mortality and decreased breeding success were documented among several species, including Common Murres, Pelagic Cormorants, Brandt's Cormorants, and Pigeon Guillemots (Hodder and Graybill 1985, Bayer 1986).

A Western Gull kleptoparasitizes (steals) a salmon smolt from a Caspian Tern that has just delivered the fish to its mate on the tern colony. Western Gulls are major predators on seabird eggs and chicks along the Oregon coast, and readily take advantage of eagle disturbances to depredate seabird nests. Photo by Dan Roby

Since this seminal reckoning, a sequence of climate events have affected seabirds, including a regime shift in the Pacific Decadal Oscillation in 1998, a historic drought in the Columbia River Basin during the winter of 2000–2001, an anomalously late onset of upwelling in 2005, and the severe marine heatwave in the northeast Pacific Ocean (the "Blob") during 2014–16 (Piatt et al. 2020, among others). Several of these events have been associated with significant predator impacts and complete nesting failure of Oregon seabird colonies. Widespread, persistent ocean warming and the ectothermic vise it creates for forage fish populations have eclipsed persistent organochlorine pesticides in providing bottom-up constraints on Oregon's seabirds (Scott et al. 1975). And while Bald Eagles are a major top-down factor limiting nesting success, far more murres, the most numerous seabird species breeding in Oregon, are killed during die-offs like the one during the marine heat wave ("Blob") than by eagle predation. In years of normal ocean conditions and presumably good prey availability, murres at the Yaquina Head colony have successfully raised young despite predator disturbance (Thompson et al. 2019). Unfortunately, climate predictions for the near future suggest persistent ocean warming may be the norm rather than the exception, raising serious concern for the future of seabird populations and other components of marine ecosystems.

At East Sand Island, Caspian Terns have bred more successfully in years of reduced Columbia River flows and greater prey availability for themselves and their predators (Collar et al. 2017). Additionally, behavioral adaptations are also allowing murres and terns to reduce disturbance rates: murres are more successfully nesting on vertical (cliff) surfaces that are more difficult for predators to access than on the flat tops of offshore rocks previously relied on, and terns are more frequently shifting their colony sites and more often using artificial sites in close proximity to people, where predators like eagles are less common (e.g., barges, fenced mainland areas, rooftops).

Consistent with the reintroduction or recovery of other keystone predators, the return of eagles and other species is significantly restructuring seabird populations in Oregon and elsewhere in the Pacific Northwest. Our perception of baseline levels of some Oregon seabird populations is likely skewed owing to the absence of these predators in initial surveys or when large colonies developed.

Climate fluctuations also play a significant role in mediating these predator-prey interactions. Warming oceans dramatically impact the distribution and abundance of seabird prey, often to the detriment of nesting birds tied to a specific colony site, but in favorable climate and prey base conditions, seabirds can be resilient and successfully produce young. It remains to be seen whether the frequency of unfavorable climate events, expected to increase under warming climate projections, will allow some of Oregon's most iconic seabird species sufficient nesting opportunities to sustain their populations.

## Don Lyons

Don Lyons's first career was as an engineer working for Hewlett-Packard in Corvallis. But in 1998, he changed careers, became a seabird biologist in Dan Roby's lab, and went back to graduate school. He earned his M.S. in wildlife science at Oregon State University in 2004 and a Ph.D. in wildlife science there in 2010. He served as a postdoctoral researcher for OSU's Oregon Cooperative Fish and Wildlife Research Unit from 2010 to 2014. There he assisted with conservation efforts involving piscivorous waterbirds and fish species on the Oregon coast. Since 2014, Lyons has held the position of assistant professor in OSU's Department of Fisheries, Wildlife, and Conservation Sciences, where he investigates factors that limit seabird populations and works to restore species of critical conservation concern. Lyons is also the director of conservation science with the National Audubon Society's Seabird Institute (also known as Project Puffin) and has provided leadership on several national and international conservation projects. In the Pacific Northwest, Lyons has most notably assisted resource managers with assessing and reducing conflicts between Caspian Terns and threatened salmon populations.

*Photo by Susan Schubel*

# Marbled Murrelet Conservation in Oregon

S. KIM NELSON

The Marbled Murrelet is a small, nonmigratory diving seabird (family Alcidae) that occurs along the Pacific coast of North America (Nelson 1997, 2020). They forage in small to large groups for schooling fish or invertebrates in nearshore, sheltered marine waters, generally within 5 km of shore. They are secretive alcids that breed at low densities in older-aged coastal forests from Alaska to central California, but also nest on the ground and on rock ledges in parts of Alaska and British Columbia (one ground nest known from Washington; Bradley and Cooke 2001, Carter and Sealy 2005, Bloxton and Raphael 2008, Nelson et al. 2010). Murrelet nests are notoriously difficult to locate because adults are difficult to detect as they enter and leave their nests, which are high up in trees or cliffs in remote, rugged terrain (Bradley et al. 2004, Peery et al. 2004, Baker et al. 2006, Barbaree et al. 2014). Murrelets do not build a nest but lay their single egg on moss or other substrate on a large tree limb, on the ground, or on a cliff face.

Historically, Marbled Murrelet populations have declined over much of their range, primarily due to loss and fragmentation of older-aged forest breeding habitat and mortality from oil spills and gill nets at sea (see Table 2; Marshall 1988a; Ralph et al. 1995; Nelson 1997; USFWS 1992, 1997, 2009; McShane et al. 2004; Piatt et al. 2007). Despite being listed as Threatened in Canada in 1990 (Rodway 1990); Threatened in Washington, Oregon, and California in 1992 (USFWS 1992); and classified as Endangered by the International Union for Conservation of Nature (IUCN) in 2010 (BirdLife International 2018), populations have continued to decline in many areas (Piatt et

al. 2007, S. L. Miller et al. 2012, Environment Canada 2014, Lorenz et al. 2021). Current issues at sea, such as changes in prey distribution and climate change, are also affecting murrelet populations (Betts et al. 2020). Therefore the current primary reason for continued declines includes sustained low recruitment from (1) the continued loss of quality nesting sites (contiguous, unfragmented forests); (2) increases in predation, primarily by corvids (ravens and jays), at nest sites in fragmented or edge habitat; and (3) lack of adequate forage fish close to nesting areas (McShane et al. 2004; USFWS 2009, 2012a, 2019a).

In Oregon, as elsewhere in the listed range, murrelets have been at the center of controversies concerning older forest management since they were listed in 1992 and especially with the creation of the Northwest Forest Plan in 1994 (see Table 1 on p. 6 and Table 2 on p. 88; US Forest Service and Bureau of Land Management 1994a, Thomas et al. 2006, DellaSala et al. 2015). They have been featured in newspapers, court proceedings, and television programs, yet murrelets have taken a back seat to the Northern Spotted Owl in terms of research funding and management plans based on the assumption that if the owl was protected, the murrelet would be as well.

## Inland Nesting

Seabirds have evolved a habitat-split strategy in which their nesting and foraging areas are often spatially distant from one another. Because they depend on more than one habitat, and especially within a single season, seabirds are believed to be more sensitive to anthropogenic pressures than those that survive in one habitat year-round (Becker et al. 2007, Betts et al. 2020).

Marbled Murrelets are affectionately known as "Enigmas of the Pacific," fog larks, and buzz bombs, among other nicknames (Guiguet 1956, Ruth 2005). Their distinctive "keer" calls can

be heard as they fly and circle over the forest before sunrise. But they are rarely seen as they speed (>100 km/hr) and dart through the dawn darkness in the dense forest canopy or among thick banks of fog. Prior to 1974, there were no known nests of Marbled Murrelets, despite the fact that a few naturalists and ornithologists had searched determinedly and unsuccessfully in Oregon and elsewhere to discover the first nest of this elusive species (e.g., Grinnell 1897, Bent 1919, Taylor 1921, Brooks 1926). Oologists (scientists who study bird eggs) surmised that murrelets nested near the sea, in the forests, in the mountains, in cavities in trees, in burrows, under tree roots, on cliffs, or on offshore islands (Brooks 1926, 1928; Guiguet 1956). Native Americans believed they nested in the forest, either in cavities in trees, or in burrows in the soil or under tree roots (Grinnell 1897, Dawson and Bowles 1909). All indications were that they nested inland, but where? In what was believed to be the first reward in American field ornithology, the National Audubon Society offered $100 for the first person to find a murrelet nest (Arbib 1970). The reward went to Hoyt Foster of the Davey Tree Company, who was climbing an old-growth Douglas-fir tree (*Pseudotsuga menziesii*) in a campground at Big Basin State Park, California, on 7 August 1974 to cut off a branch that had been deemed unsafe (Binford et al. 1975). He accidentally found a chick near fledging age on a branch 45 m above the ground. Finally,

the nest of the last bird species on the North American continent (north of Mexico) had been found! Through later research, however, it turns out that unknown nests and indications of nesting had been documented well before 1974 (Carter and Sealy 2005).

In Oregon, hints about their breeding habits began to be pieced together as more discoveries of chicks and fledglings were recorded. The first inland chick was found by A. B. Johnson on a road in an old-growth forest on the North Fork of the Siuslaw River, 10 km inland from the Heceta Head Lighthouse near Minerva, Oregon (Lane County), on 8 September 1918 (Jewett 1930). The chick was dead and appeared to have been near fledging age, given the loss of most of its down feathers. This inland encounter was followed by the discovery of a downy chick on the ground in a logged-off area adjacent to an old-growth forest 1.6 km inland near Devil's Lake, Oregon (Lincoln County), on 4 September 1933 by S. Jewett Jr. and L. Scott, and a fledgling on the ground at the base of a cliff 40 km inland along the Coos River, Oregon, on 22 July 1940 by O. Barber (Jewett 1934, Gabrielson and Jewett 1940, Barber 1941, Carter and Sealy 1987, Nelson et al. 1992). These three discoveries proved conclusively that murrelets nested inland some distance from the coast in or near older-aged forests in Oregon.

The first known inland observations of adult murrelets in Oregon occurred in 1958, when A. Thoresen detected 4–10 birds flying up

OPPOSITE, TOP: Adult Marbled Murrelet incubating its single egg on a mossy platform high in an Oregon old-growth forest. Because Marbled Murrelets are seabirds that forage in the nearshore ocean but nest in late successional old-growth forest, they are a split-habitat species, susceptible to habitat degradation in two distinctly different habitats. Marbled Murrelets are listed as Endangered by the state of Oregon and Threatened by the federal government. Photo by Brett Lovelace

OPPOSITE, BOTTOM: Adult Marbled Murrelet in breeding plumage taking off from the ocean surface at a nearshore foraging site during the nesting season. Marbled Murrelets in Oregon are affected by the availability of old-growth forest as nesting habitat and ocean conditions as they affect food availability in the nearshore. Photo by Dan Cushing and Kim Nelson

**TABLE 2. History of Marbled Murrelets in Oregon**

| Year | Event |
|---|---|
| 1902 | First known mention of murrelet at-sea abundance off Yaquina Bay, Oregon (Woodcock 1902). |
| 1918 | First evidence of inland nesting in Oregon with the discovery of a dead chick on the ground in an old-growth forest on the north fork of the Siuslaw River about 10 km from Minerva, Oregon, on 8 September 1918 (Jewett 1930). |
| 1933 | Discovery of a fledgling on the ground at Devil's Lake, Oregon, on 4 September 1933 (Jewett 1934). |
| 1940 | Discovery of a fledgling on the ground at Coos River, Oregon, on 22 July 1940 (Barber 1941). |
| 1958 | First known inland detections of adult murrelets in Oregon (Nelson et al. 1992). |
| 1982 | The Pacific Seabird Group (1982) identified the Marbled Murrelet as a species of concern and wrote a resolution on considerations for old-growth forest management. |
| 1985 | S. K. Nelson first heard Marbled Murrelets on the east side of Marys Peak (Benton County), Oregon, while conducting surveys for forest birds and woodpeckers. She eventually heard them in seven of her old-growth and mature research sites, up to 47 km inland between 1985 and 1986. |
| 1986 | The Pacific Seabird Group (1986) again identified the Marbled Murrelet as a species of concern and created the Marbled Murrelet Technical Committee to work with agencies and others on research and conservation of murrelets. |
| 1988 | US Fish and Wildlife Service publishes the first detailed summary of the status of the murrelet in California, Oregon, and Washington (Marshall 1988b).<br><br>First inland survey protocol developed for surveying murrelets at dawn in the forest (Paton et al. 1988). This was followed by updates in 1990, 1992, 1993, 1994, and 2004. |
| 1990 | First Marbled Murrelet tree nest found in Oregon by S. K. Nelson and her crew on 19 May 1990 in the Siuslaw National Forest. |
| 1992 | Marbled Murrelets are listed as a Threatened species in California, Oregon, and Washington under the Federal Endangered Species Act (USFWS 1992). |
| 1994 | Northwest Forest Plan created to protect older-forest habitat for Marbled Murrelets, Northern Spotted Owls, salmon, and other species (US Forest Service and Bureau of Land Management 1994a, b). |
| 1995 | Marbled Murrelet listed as Threatened under the Oregon Endangered Species Act (OAR 635-100-0125). |
| 1996 | Critical habitat for murrelets designated in California, Oregon, and Washington (61 FR 26256). |
| 1999 | Effectiveness Monitoring Plan for the Marbled Murrelet, Northwest Forest Plan, developed for monitoring murrelet populations at sea (standardized surveys) and monitoring inland habitat distribution and abundance (Madsen et al. 1999). |
| 2002 | The National Audubon Society, in cooperation with BirdLife International, designated Important Bird Areas in Oregon and elsewhere to protect important foraging and nesting habitats for birds. The largest IBA is globally significant and encompasses more than 32,000 ha in Lincoln and Lane Counties along Cape Perpetua (Marbled Murrelet IBA). |
| 2009 | Five marine reserves and nine marine protected areas are created along the Oregon coast to protect forage fish, salmon, and seabirds. The Cape Perpetua marine reserve and marine protected area are directly adjacent to the Marbled Murrelet IBA. |
| 2016 | Oregon Wildlife Commission petitioned by six environmental groups to up-list the murrelet to Endangered in Oregon under the Oregon Endangered Species Act. |
| 2021 | Oregon Wildlife Commission up-lists the murrelet to Endangered under the Oregon Endangered Species Act. |

the Chetco River Valley (Curry County) on a regular basis (Nelson et al. 1992). I first heard Marbled Murrelets call at dawn on the east side of Marys Peak (Benton County), 48 km inland, in the spring of 1985 while conducting variable circular plot surveys for forest birds and woodpeckers (Nelson 1986). During 1985 and 1986, I eventually heard them in 7 of my 49 old-growth and mature forest study sites. It was these observations and the realization that no nests were known in Oregon that spurred my interest in pursuing research on Marbled Murrelets. In 1988, I began compiling all the historical and recent at-sea and inland records (from 1899 to 1987; Nelson et al. 1992), and in 1989–90, I conducted systematic surveys in forests of all ages throughout the Coast Range to determine habitat preferences (Nelson 1990).

The first actual tree nest in Oregon was found by my crew on 19 May 1990, 25 km inland on the Siuslaw National Forest (Lincoln County). We found it while monitoring an occupied site we had located in 1989, where murrelets were seen flying below the forest canopy. Between 1991 and 1992, we went on to find 10 more murrelet nests on the central Oregon Coast (1–40 km inland, Lincoln and Lane Counties) using dawn surveys (Nelson and Peck 1995) and 50 more nests from 1993 to 1999 using dawn surveys and tree-climbing in randomly selected plots in known occupied sites (7–30 km inland in Tillamook, Lincoln, Douglas, and Coos Counties; Nelson and Wilson 2002, Hamer et al. 2021, S. K. Nelson, unpublished data). We developed the first guidelines for using tree-climbing to locate old and active murrelet nests (Peck et al. 1994). Other nests were located opportunistically between 1993 and 2013 by agencies and individuals during dawn surveys, tree-climbing, or by finding eggshell fragments on the forest floor ($n$ = 16 nests, 2–48 km inland in five of the seven coastal counties). During 2018 and

2019, OSU's Oregon Marbled Murrelet Project (College of Forestry, Departments of Forest Ecosystems and Society and Forest Engineering, Resources, and Management, and College of Agricultural Sciences, Department of Fisheries, Wildlife, and Conservation Sciences) located 21 additional nests on the Siuslaw National Forest, 4–40 km inland (Lincoln, Lane, Benton, and Douglas Counties) using radio telemetry (J. W. Rivers, S. K. Nelson et al., unpublished data). As of August 2021, 98 tree nests had been located in Oregon up to 51 km inland. Hundreds of occupied sites, where murrelets have been seen flying through the canopy but nests have yet to be found, are also known up to 76 km inland (Nelson 2003, unpublished data).

Most nests were found in coniferous trees of a variety of species in mature and old-growth forests. Some nests, however, were found in younger stands that included western hemlock (*Tsuga heterophylla*) trees with dwarf mistletoe (*Arceuthobium tsugense*) infestations, which make branches larger in size (Nelson and Wilson 2002). The broadened definition of murrelet habitat now included coniferous forest stands >60 years in age with remnant or older-aged trees with one or more platforms (>10 m in height and >10 cm in diameter; Evans Mack et al. 2003). The first nest in a deciduous tree, a bigleaf maple (*Acer macrophyllum*), was found in Oregon in 2018 in the Siuslaw National Forest (J. W. Rivers, S. K. Nelson et al., unpublished data). Only two other nests in deciduous trees are known (one bigleaf maple and one red alder [*Alnus rubra*]), both from British Columbia (Bradley and Cooke 2001, Ryder et al. 2012). At present, coniferous trees are still considered preferred nesting habitat for Marbled Murrelets.

Advances in technology would help in locating nests for this elusive and cryptic species. The Oregon Marbled Murrelet Project has been testing the efficacy of using drones and platform terminal transmitter (PTT) tags on murrelets to increase nest discoveries. Unfortunately, drone battery life limited nest searches, and adult murrelets were slightly disturbed by the sounds of the small-sized drones needed in dense forests (J. W. Rivers, S. K. Nelson et al., unpublished data). PTT and satellite tags that are small enough to attach to small seabirds that cannot be readily recaptured, like murrelets, are currently not available (murrelet average weight = 220 g; maximum tag size = 2 g). The PTT tags the Oregon Marbled Murrelet Project tested (5 g) were bulky and did not include Global Positioning System (GPS) technology, limiting our ability to pinpoint murrelet locations or find nests (Northrup et al. 2018).

Low fecundity levels across California, Oregon, and Washington as measured by nest success indicate a population that cannot currently maintain itself (McShane et al. 2004, Beissinger and Peery 2007, USFWS 2019a). Most murrelet nests (>70%) have been unsuccessful primarily because of predation by corvids (ravens and jays) and birds of prey (owls, accipiters, and hawks; Nelson and Hamer 1995a, USFWS 2012a). Nest predation is affected by forest fragmentation (amount of edge and stand size) and proximity to human developments (Raphael et al. 2002; McShane et al. 2004). Forests in coastal Oregon have been intensively harvested since the early 1900s. The landscape today is a mosaic of younger forests and clear-cuts, with some older-aged stands remaining mostly on federal lands or in state parks. Climate change is predicted to alter the terrestrial environment within the range of the murrelet by changing precipitation (the amount, type, and when) and temperatures, extending the fire season, increasing fire severity, and increasing the prevalence of disease. Although it is uncertain to what degree climate change will influence high-intensity, stand-replacing fires within Oregon, warmer, drier summers are likely to produce more frequent and extensive fires, thus reducing the extent and connectivity of older-aged forests and potentially resulting in severe consequences for the murrelet (Sheehan et al. 2015, Wan et al. 2019).

Large, contiguous blocks of suitable nesting habitat need to be created and preserved throughout the coast ranges of Oregon to diminish the threat of nest predation, increase murrelet reproduction, and minimize catastrophic wildfires. Creating buffers to existing murrelet habitat and minimizing the effects of disturbance will also be important for reversing negative fecundity trends and helping to minimize wildfire risk.

### At Sea

Historically, murrelets were considered common along the entire Oregon coast (Woodcock 1902, Taylor 1921, Gabrielson and Jewett 1940, Bayer and Ferris 1987), with the central coast (Tillamook, Lincoln, and Lane Counties) being the center of abundance (Woodcock 1902, Gabrielson and Jewett 1940, Nelson et al. 1992). Today, the central Oregon coast continues to be the center of abundance for murrelets; however, populations have significantly declined, and they are no longer considered common along the entire coast (McShane et al. 2004, S. L. Miller et al. 2012, McIver et al. 2020, Strong 2020). See details of historical at-sea counts and numbers of beached birds in Nelson et al. (1992).

Since 2000, murrelet populations have been monitored with standardized surveys every other year by zone in California, Oregon, and Washington (Raphael et al. 2007). At present, there are thought to be about 10,300 murrelets (confidence interval [CI] = 7,100–13,600) off the Oregon coast (McIver et al. 2020).

Threats at sea in Oregon include gill-net fishing, oil spills and pollution, and changes in prey distribution and availability related to changes in climate, currents, and winds (Nelson et al. 1992, Garcia-Reyes et al. 2015). Gill-net fishing was a threat at one time but is no longer an issue on the Oregon coast as it was outlawed before 1942 and currently only occurs in the Columbia River. Hundreds of murrelets were estimated to have been killed in two oil spills off Oregon since 1999 (*New Carissa* and a mystery spill at the mouth of Columbia River; M. Szumski, pers. comm.); three other oil spills occurred off Oregon between 1978 and 1984, but effects on murrelets are unknown (Nelson et al. 1992). The mitigation for these spills has included purchasing habitat for the murrelet and producing signage and educational materials for all species affected. To date, more than 1,620 ha of private land have been purchased to protect occupied sites or stands adjacent to known occupied sites in Lincoln County.

A further threat involves changes in ocean currents related to climate. The Northern California Current, which extends from southern British Columbia to southern California, has undergone changes in the past several decades related to shifts in basin-scale climate variables such as the El Niño–Southern Oscillation (Wang et al. 2019). And most alarmingly, between 2013 and 2016, the Northern California Current experienced an extensive warming event when a reduction of storms in the Gulf of Alaska created the "Blob," where warm surface water shifted south and prevented sustained upwelling (Bond et al. 2015). This was followed by one of the warmest El Niño events on record (Jacox et al. 2016) and was then followed by another blob and warm water event in 2018–19 (Harvey et al. 2020). This extended warming period is thought to have had extensive impacts on the distribution and abundance of forage fish populations. It also appears likely that these types of anomalous events will increase in the future with projected climate change. The effects of these events on

murrelets in Oregon are likely significant. In fact, Betts et al. (2020) found that during poor ocean years, murrelets often forgo nesting and do not readily fly inland. In 2017 the Oregon Marbled Murrelet Project found no tagged murrelets (n = 61 with VHF radio tags) flew inland to nest, and many left Oregon for better foraging conditions in northern California (as far south as San Francisco Bay; J. W. Rivers, S. K. Nelson et al., unpublished data). Of 129 marked birds in 2018 and 2019, only 16 flew inland to nest: 12 nests and two occupied sites were found; both the male and female at two nests were tagged.

On a positive note, in cooperation with a variety of agencies including Oregon State University and the Department of Energy, there is a proposal for a series of Motus towers to be constructed along the entire Oregon coast and at sea to help with research on seabirds, shorebirds, and marine mammals (R. Orben, pers. comm.; see also Part II of this volume). Should these towers be constructed, it may help in tracking murrelet movements in the nearshore and remove the need to fly in fixed-wing aircraft to track murrelets with VHF tags.

## Management and Conservation

In 1982 and 1986, the Pacific Seabird Group identified the Marbled Murrelet as a species of concern because of the significant loss of old-growth forest nesting habitat throughout their range (Pacific Seabird Group 1982, 1986). In 1986 the Pacific Seabird Group created the Marbled Murrelet Technical Committee to (1) act as a technical authority on the status, distribution, and life history of the murrelet; (2) identify, encourage, and facilitate research; (3) address conservation and management problems; (4) act as a liaison between research and management; and (5) develop a science-based inland dawn survey protocol for management and research that provides for the conservation of the murrelet over the long term. The first inland survey protocol was developed in 1988 (Paton et al. 1988), followed by revisions in 1990 (Paton et al. 1990), 1992 (Ralph and Nelson 1992), 1993 (Ralph et al. 1993), 1994 (Ralph et al. 1994), and 2003 (Evans Mack et al. 2003). A new revised protocol will be available for the 2023 survey season.

It was the early efforts of the Pacific Seabird Group that spurred the USFWS to write a biological report on the status of the Marbled Murrelet in 1988 (Marshall 1988b). In that same year, a variety of conservation groups, including National Audubon and the Oregon Natural Resources Council, submitted petitions to the USFWS to list the species as Threatened in California, Oregon, and Washington, and to the Oregon Wildlife Commission for listing as Threatened under the Oregon Endangered Species Act. The key threats outlined in Marshall (1988a, b) and the listing petitions were the extensive loss and fragmentation of old-growth forest nesting habitat, mortality from oil spills and gill nets, and lack of existing regulatory mechanisms to protect the murrelet. After several lawsuits the species was finally listed at the federal level in 1992 (USFWS 1992), followed by listing in Oregon in 1995 (OAR 635-100-0125). The murrelet was recently (2021) up-listed to Endangered in Oregon (Oregon Department of Fish and Wildlife 2017).

From 1987 to 1994, efforts were made to update the science on the species. First, PSG held two symposia on Marbled Murrelets in 1987 and 1993 (Carter and Morrison 1992, Nelson and Sealy 1995) to share the latest discoveries, and the US Forest Service in northern California spearheaded the writing of a Marbled Murrelet Conservation Assessment that summarized all known data on the species throughout its range (Ralph et al. 1995). Along with these efforts came a list of research priorities developed by the Marbled Murrelet Technical Committee.

In 1994, in response to concerns over the loss of old-growth forests and recent information on the decline of the Northern Spotted Owl, the Northwest Forest Plan was created to preserve large blocks of older-forest habitat on federal lands for Marbled Murrelets, Northern Spotted Owls, and other species associated with older forests in northern California, Oregon, and Washington (US Forest Service and Bureau of Land Management 1994a, b; Madsen et al. 1999; Raphael 2006; Raphael et al. 2018). In concert with this action, in May 1996, the USFWS designated 1,573,340 ha of critical habitat for the Marbled Murrelet in the listed range (61 FR 26256). The thought was that the combination of the Threatened listing and these habitat management designations would help to recover murrelet populations. After 25 years, however, murrelet populations are still declining in some areas, especially in Washington (McIver et al. 2020), and few improvements have been made in the abundance and distribution of older forest habitat (Raphael et al. 2018, Lorenz et al. 2021). Given that (1) it takes >200 years to create murrelet habitat from the ground up, (2) the landscape is still largely fragmented because of historical and current logging and ownership patterns, and (3) thinning in younger stands is now occurring over large areas within the Northwest Forest Plan reserves and on state lands, which is likely to increase predation risk in nearby occupied sites, forest conditions for murrelets are not expected to improve in the short term.

In 2002 the National Audubon Society, in partnership with Birdlife International, began designating Important Bird Areas around the world. These areas are meant to help protect habitat for imperiled bird species and protect important stopover or foraging areas for migrating and resident species alike. In Oregon, inland areas were designated in Clatsop, Tillamook, Lincoln, Lane, and Coos counties, along with numerous nearshore and offshore habitats to help benefit the murrelet. The largest Important Bird Area is globally significant and encompasses more than 32,000 ha in Lincoln and Lane counties near Cape Perpetua. In an attempt to improve conditions for forage fish, salmon, and seabirds in the nearshore off Oregon, five marine reserves and nine marine protected areas were designated from Cape Falcon to Redfish Rocks in 2009. The idea was that the marine reserves would act as nurseries for forage fish and over time help create abundant populations of high-quality forage fish close to shore, and for murrelets, abundant prey adjacent to inland nesting sites. The Marbled Murrelet Important Bird Area is directly adjacent to the Cape Perpetua marine reserve (37 sq km), marine protected area (49 sq km), and seabird protection area making it one of the largest contiguous inland and at sea "reserves" for Marbled Murrelets anywhere. At this location Portland Audubon conducts an annual "citizen science" survey in July to track murrelet presence inland and at sea, and helps spread the word about the plight of the murrelet in Oregon.

Numerous lawsuits have been filed on behalf of or against murrelet conservation in Oregon and within the listed range between 1991 and 2022. To date, attempts to prevent listing or to delist the species at the federal and state levels have failed. A suit against the Oregon Department of Forestry in 2016 led to better protections being established on state lands in Oregon. It also led to a proposal

to sell the Elliott State Forest. This sale was subsequently dropped by the Oregon State Land Board, which then put in motion a plan to sell the property to Oregon State University on the condition that management priorities would include forest research as well as conservation of murrelets, owls, salmon, and other sensitive species. Also in 2016, the Oregon Wildlife Commission was petitioned by six environmental groups to up-list the murrelet to Endangered under the Oregon Endangered Species Act. The decision to up-list the species to Endangered was finally made in 2021.

Controversies and uncertainty surrounding the Marbled Murrelet will undoubtedly continue. Currently, populations are still declining in some areas, habitat loss continues, the forests remain fragmented, predator numbers and predation risk continue to increase, climate change is increasing the chances of catastrophic wildfires, and ocean conditions, especially in the nearshore, may never return to historical states (ODFW 2017). Given that we are dealing with a relatively uncommon seabird that is facing a plethora of challenges, things do not bode well for the long-term persistence of the Marbled Murrelet in Oregon.

## S. Kim Nelson

Kim Nelson has been an avian ecologist and senior faculty research assistant for Oregon State University's Department of Fisheries, Wildlife, and Conservation Sciences since 1989. She is also a co-principal investigator on the OSU College of Forestry's Oregon Marbled Murrelet Project. Her research has focused on the ecology and habitat associations of seabirds. She has primarily studied the nest-site characteristics and stand and landscape associations of the threatened Marbled Murrelet. She and her crews located the first Oregon Marbled Murrelet nest in 1990 and have since found more than 60 nests in the state. She has also studied the nesting behavior of other seabirds of the Pacific, including Japanese Murrelets, Long-billed Murrelets, Caspian Terns, and a variety of alcids at mixed colonies in the Bering Sea.

Kim is the North American representative on the board of the World Seabird Union and past chair of the Pacific Seabird Group. She has also served the Oregon Chapter of The Wildlife Society as an executive board member. She received the Oregon State University Agricultural Research Foundation Faculty Research Assistant Award, the Homer Campbell Award from the Audubon Society of Corvallis, and the Pacific Seabird Group's Special Achievement Award.

*Photo by Keith Larson*

## James Rivers

Jim Rivers is an associate professor in Oregon State University's Department of Forest Engineering, Resources, and Management. Since coming to OSU as a postdoc in 2009, Jim has worked on a variety of projects, but overall he studies bird success as a result of various forest management practices. He is the principal investigator on the Oregon Marbled Murrelet Project, which examines predictors of space use and reproductive success in murrelets. He also focuses much of his time on cavity nesters and pollinators.

Rivers was one of the founders of the Annual OSU Willamette Valley Bird Symposium, which began in 2014. He is a fellow of the American Ornithological Society and was the recipient of its Ned K. Johnson Young Investigator Award in 2012. He is an associate editor for *Ornithology* and *Frontiers in Ecology and Evolution*. In 2020, he was presented with the Research Award by the Oregon Society of American Foresters.

*Photo by Luke Pangle*

# Double-crested Cormorants in Oregon

DAN ROBY

Double-crested Cormorants (*Nannopterum auritum*) are a piscivorous colonial waterbird native to Oregon and a widespread species throughout the state and much of North America (Dorr et al. 2014). Double-crested Cormorants have proven to be adaptable and opportunistic, equally at home in Oregon's inland lakes, rivers, and reservoirs as well as coastal marine and estuarine habitats (Matthews et al. 2003). The species nests colonially, and nesting habitat preferences are similarly eclectic, including on the ground on rocky and sparsely vegetated islands; in trees on islands, peninsulas, and headlands; on cliffs and rocky promontories; and on anthropogenic structures over water, such as bridges, navigational aids, and powerline towers. Their food habits are also catholic, and include a wide variety of freshwater, estuarine, and marine forage fishes from a few centimeters to as much as 50 cm in length (Lawes et al. 2021b). Four subspecies of Double-crested Cormorants have been recognized, and the one found in Oregon (*N. a. albociliatus*) nests from southern British Columbia to northwestern Mexico, and from the Continental Divide to the Pacific coast (Adkins et al. 2014). Recent analysis of the genetic structure of Double-crested Cormorants in North America (Mercer et al. 2013) indicates, however, that the western subspecies is not genetically distinct enough to be considered a different subspecies than the more widely distributed and abundant subspecies in eastern and central North America (*N. a. auritum*). Nevertheless, the Pacific Flyway population of the species warrants consideration as a separate management unit from the population east of the Continental Divide because of differences in population status and biology (Anderson et al. 2004, Wires and Cuthbert 2006, Mercer et al. 2013, Wires 2014).

Double-crested Cormorants in particular, and cormorants generally, have become infamous among fisheries proponents as gluttonous consumers of many fishes that are valued as subsistence, commercial, or recreational resources (King 2013, Wires 2014). From the outset, European settlers in North America transferred the antipathy and prejudice toward cormorants in the Old World to the New World, and Double-crested Cormorants were soon subjected to the persecution and harassment that the Great Cormorant (*Phalacrocorax carbo*) was subjected to in Europe (King 2013). Cormorant numbers in Oregon plummeted following Euro-American settlement during the late nineteenth and early twentieth centuries owing to a combination of aquatic habitat destruction, disturbance, and persecution (Wires 2014). The Migratory Bird Treaty Act of 1918, which provided much-needed protections for most migratory bird species in North America, did not include protection for Double-crested Cormorants, which were considered a pest species like crows and blackbirds. Double-crested Cormorants were frequently the subject of lethal persecution in Oregon during the first half of the twentieth century, as exemplified by a documented take of about 1,300 cormorants in 1938 at Crane Prairie Reservoir in Deschutes County at the behest of the director of the Oregon Game Commission (Bayer 1989). Until 1958, it was legal in Oregon to kill all species of cormorants, plus Common Mergansers, Hooded Mergansers, and Belted Kingfishers, as fish-eating pests. Then, in the mid-twentieth century, persistent organochlorine pesticides, especially DDT, contaminated the food supply of cormorants and other piscivorous waterbirds (Scott et al. 1975) and caused widespread reproductive failure.

The Double-crested Cormorant population in the western United States began to increase in the mid-1970s, following a ban on the use

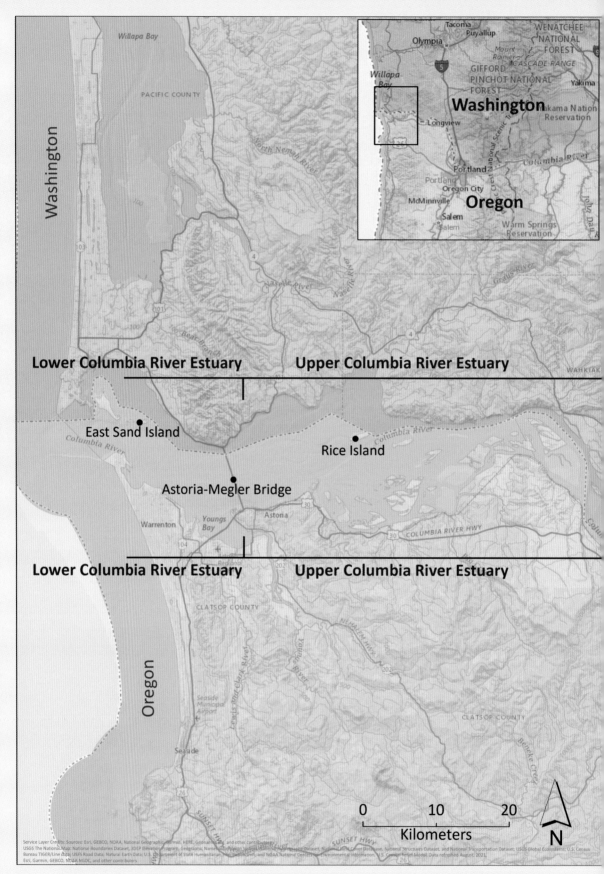

FIGURE 7. The Columbia River estuary, with sites labeled where large colonies of Double-crested Cormorants and Caspian Terns have formed in Oregon. Tim Lawes, Oregon State University

Double-crested Cormorant in breeding plumage at the nesting colony on East Sand Island in the Columbia River estuary. Many Double-crested Cormorants in the Pacific Flyway population have white crest feathers early in the nesting season. The breeding population of the species in the Pacific Flyway is currently in decline following the abandonment of the large breeding colony on East Sand Island in Oregon. Photo by Dan Roby

of DDT and protection of the species for the first time under the Migratory Bird Treaty Act in 1972. Recent population expansion in Oregon can also be attributed, in part, to the species' efficiency in adapting to foraging and nesting in anthropogenic habitats, such as dams, dredged material disposal islands, hatcheries, and aquaculture facilities (Nettleship and Duffy 1995). Because Double-crested Cormorants had been rare or absent from large portions of their former range in North America for nearly a century, the population recovery was perceived by some publics as an invasion by a non-native species (Wires and Cuthbert 2006, Wires 2014).

The breeding distribution of the Pacific Flyway population of Double-crested

Cormorants has changed dramatically over the past 50 years, flyway-wide and in Oregon. Population size increased by about 3% per year during 1990–2009 through the growth of some known colonies and the formation of many new colonies (Carter et al. 1995, Wires and Cuthbert 2006, Adkins et al. 2014). Nearly all of the increase in the Pacific Flyway population during 1990–2009 was attributable to increases in the number of breeding pairs in the Columbia River estuary, in particular the breeding colony on East Sand Island in the Oregon portion of the estuary (Figure 7). By 2013 the East Sand Island colony had grown to nearly 15,000 breeding pairs and accounted for more than 40% of all breeding pairs in the Pacific Flyway population (Adkins et al. 2014, Lawes et al. 2021b). The East Sand Island colony had grown to be the largest known breeding colony of Double-crested Cormorants anywhere (Adkins et al. 2014). Studies on the genetic structure of the Pacific Flyway population (Mercer et al. 2013) and the postbreeding dispersal of Double-crested Cormorants nesting in the Columbia River estuary (Courtot et al. 2012, Peck-Richardson 2017) suggested a high degree of population connectivity from southeastern California to southern British Columbia. Fisheries managers raised concerns over the impact of predation by Double-crested Cormorants from this large breeding assemblage on survival of out-migrating juvenile salmonids (*Oncorhynchus* spp.) from throughout the Columbia River Basin, especially those populations of salmonids that are listed under the US Endangered Species Act (National Marine Fisheries Service 2010, 2019a; see the Birds and Salmon in Oregon section on p. 102).

The large size of the Double-crested Cormorant colony on East Sand Island was made possible by anthropogenic changes to stabilize the island, an abundance of forage fish near the mouth of the Columbia River, and an absence of terrestrial predators and human disturbance on the island. Further, the large size of the colony apparently provided safety in numbers from Bald Eagle predation, a limiting factor implicated in the declines of cormorant nesting colonies in the Salish Sea region to the north (Chatwin et al. 2002), and likely contributed to the initial emigration to the East Sand Island colony. Bald Eagle disturbance and predation became an important factor influencing nesting success of Double-crested Cormorants at the East Sand Island colony after 2000 (Isaacs and Anthony 2011; see the Predators and Climate Squeeze Oregon's Seabirds section on p. 81). In addition to direct mortality of cormorants from predation by

eagles, eagle disturbances at the colony were associated with heightened predation on cormorant eggs and nestlings by Glaucous-winged / Western Gulls (*Larus glaucescens* x *L. occidentalis*).

The diet of Double-crested Cormorants nesting on East Sand Island consisted of a wide variety of marine, estuarine, and freshwater forage fishes, including juvenile salmonids listed under the Endangered Species Act (ESA). Starting in 2015, the US Army Corps of Engineers (USACE), the USFWS, and US Department of Agriculture (USDA)–Wildlife Services implemented a management plan with the goal of substantially reducing predation on ESA-listed juvenile salmonids by Double-crested Cormorants in the Columbia River. The objective of the *Double-crested Cormorant Management Plan to Reduce Predation on Juvenile Salmonids in the Columbia River Estuary* (hereafter *Cormorant Management Plan*; USACE 2015) was to reduce the size of the East Sand Island colony to no more than 5,380–5,939 breeding pairs, or at most 42% of the average pre-management colony size. The outcome of implementation of the *Cormorant Management Plan* is described in the Birds and Salmon in Oregon section on page 102.

The estimated size of the Pacific Flyway population of Double-crested Cormorants was 22,890 breeding pairs in 2019 (95% CI = 15,925–29,855 pairs; USFWS 2020b), the most recent year when the Flyway Council's monitoring strategy (Pacific Flyway Council 2013) was implemented. This represents about a 43% decline (95% CI = 10% to 66%) in the estimated number of breeding pairs in the Pacific Flyway population since 2014, just prior to implementation of the *Cormorant Management Plan* (USFWS 2016a). Such a large decline in the number of breeding pairs in the Pacific Flyway population following implementation of the *Cormorant Management Plan* was,

Double-crested Cormorant nesting on East Sand Island on the Oregon side of the Columbia River estuary. Double-crested Cormorants can lay up to seven eggs in a clutch and nest on the ground, in trees, on cliffs, or on artificial structures over water, such as bridges, navigational aids, and power transmission towers. Photo by Dan Roby

however, predicted by the USFWS in the Final Environmental Impact Statement (EIS) for the *Cormorant Management Plan* (USACE 2015). In that EIS, the USFWS anticipated that by the fifth year of implementing the *Cormorant Management Plan* (2019), the Pacific Flyway population of Double-crested Cormorants would have declined to about 19,182 breeding pairs (USACE 2015), considerably less than the estimate from the monitoring strategy in 2019 (22,890 breeding pairs; USFWS 2020b).

The USFWS has taken the position that the Pacific Flyway population of Double-crested Cormorants is resilient, and that even if the population were reduced below 20,000 breeding pairs during implementation of the *Cormorant Management Plan*, the population would stabilize and then slowly recover, given that the population has increased from a population size much less than 20,000 breeding pairs in the past (USACE 2015). The future trajectory of the Pacific Flyway population is highly uncertain, however, based on the apparent complete abandonment of the formerly productive colony on East Sand Island. The most recent status assessment of the Pacific Flyway population of Double-crested Cormorants (Adkins and Roby 2010, Adkins et al. 2014) identifies the central role of the East Sand Island colony in sustaining the flyway-wide population. Growth of the East Sand Island colony during the 1990s and early 2000s was the sole reason that the Pacific Flyway population increased during that period; total numbers of breeding Double-crested Cormorants elsewhere in the Pacific Flyway were stable (Adkins et al. 2014). Not only was the East Sand Island colony by far the largest breeding colony for Double-crested Cormorants in the Pacific Flyway, it also supported consistently high nesting success in a region where nesting failure due to disturbance by Bald Eagles and humans was increasingly frequent (Carter et al. 1995, Chatwin et al. 2002, Hipfner et al. 2012, Adkins et al. 2014). Further, nesting success at the East Sand Island colony may have compensated for the loss in 2013 of the second largest colony of Double-crested Cormorants in the Pacific Flyway (Mullet Island in Salton Sea, California, due to drought) and declines at other large inland colonies (e.g., Upper Klamath Lake, Oregon, and Potholes Reservoir, Washington; USFWS 2020b).

The single recent development that suggests that the trajectory of the Pacific Flyway population of Double-crested Cormorants could stabilize and then slowly increase, as predicted by the USFWS in the EIS (USACE 2015), is the dramatic increase in colony size on the Astoria-Megler Bridge, 15 km (9 miles) upstream of East Sand Island (Figure 7). As of 2020, the cormorant nesting colony on the Astoria-Megler Bridge had supplanted the East Sand Island colony as the largest

Double-crested Cormorant breeding colony in the Pacific Flyway at more than 5,000 breeding pairs. If left unmanaged, the Double-crested Cormorant colony on the bridge is expected to continue to grow, and it is estimated that there is enough space to support a colony of as many as 10,000 breeding pairs (M. J. Lawonn, ODFW, pers. comm.). The growth of this colony, however, is largely an unintended and undesirable consequence of management of the East Sand Island colony. Given the apparent impact of the Astoria-Megler Bridge cormorant colony on survival of ESA-listed salmonid smolts in the Columbia River estuary (Cramer et al. 2021a; Banet et al., in review) and the impact of guano from thousands of nesting cormorants on the structural integrity of the bridge, plans are currently being developed to prevent cormorants from nesting on the Astoria-Megler Bridge starting in the 2022 nesting season. Therefore the future trajectory of the Pacific Flyway population of Double-crested Cormorants largely depends on whether a large breeding colony is allowed to persist in the Columbia River estuary.

A federal rule for management of Double-crested Cormorants issued in the waning days of the Trump administration further enhances uncertainty over the future population trajectory of the Pacific Flyway population (50 CFR 21.48; USFWS 2003, 2020c). The rule established a new system for issuing special permits to states and tribes for lethal take of Double-crested Cormorants and applies not just to regions of the United States formerly included under the Public Resource Depredation Order for Double-crested Cormorants, but includes the contiguous 48 states (USFWS 2020c). The rule sets a maximum allowable take of 4,539 Double-crested Cormorants per year within the Pacific Flyway (USFWS 2020c), or about 10% of the ~45,780 breeding individuals enumerated during the most recent count of the

Pacific Flyway population, based on the Pacific Flyway Monitoring Strategy (Pacific Flyway Council 2013, USFWS 2020b). While it is not clear what level of annual take will occur in the Pacific Flyway under the new rule, if the take level approaches the maximum allowable level, stabilization and recovery of the Pacific Flyway population in the aftermath of the *Cormorant Management Plan* is unlikely.

For the above reasons, it is important to sustain the USFWS's Double-crested Cormorant Monitoring Strategy for the Pacific Flyway (Pacific Flyway Council 2013) to ensure that the Pacific Flyway population does not decline to a level where it becomes of conservation concern. The monitoring strategy, which is based on a dual-frame sampling approach, is currently conducted every three years and has as its objective to detect a 5% change/year in the size of the Pacific Flyway population of Double-crested Cormorants with 80% power and 10% Type I error rate (Pacific Flyway Council 2013). Because of the transitory nature of many Double-crested Cormorant breeding colonies and the high interannual variability in colony size among extant colonies, the Cormorant Monitoring Strategy has struggled to achieve its monitoring objectives (USFWS 2020b). The dual-frame sampling approach for monitoring the Pacific Flyway population of Double-crested Cormorants can achieve its objectives for accuracy and precision if the strategy is modified and updated after each implementation and if surveys are conducted throughout the flyway to identify new and incipient colonies as they form. Periodically, perhaps every 10–12 years, a flyway-wide census of Double-crested Cormorant breeding colonies may be required to ensure that the dual-frame sampling approach is achieving its objectives and that significant changes in population size are detected. The conservation of Double-crested Cormorants in Oregon will depend on continued, accurate monitoring of the flyway-wide population by the USFWS and its partners for the foreseeable future.

# Birds and Salmon in Oregon

DAN ROBY

Anadromous salmonids (salmon and steelhead; *Oncorhynchus* spp.) are an invaluable natural resource in Oregon and throughout the Pacific Northwest that are prized for their cultural, subsistence, commercial, and recreational values. Salmonids are also an icon of the quality of life for many in the Pacific Northwest, and the intrinsic value of salmonids rivals the utilitarian values for a wide variety of stakeholders. Salmonids, including adults, eggs, fry, and smolts, are also a key resource for a wide variety of wildlife in Oregon, including birds (Cederholm et al. 2001, Hilderbrand et al. 2004).

During the past 150 years, however, populations of salmonids in the Pacific Northwest have experienced dramatic declines that in many cases have imperiled their existence. Currently, 13 of 20 populations of anadromous salmonids that spawn in the Columbia River Basin are listed under the US Endangered Species Act (National Marine Fisheries Service 2019a). Four anthropogenic factors, known as the "4 H's," have played an oversized role in threatening salmonid populations: *h*arvest, *h*abitat degradation, *h*ydrosystem development, and *h*atchery production (Lichatowich 1999). Many billions of dollars have been spent by federal, state, and tribal agencies to halt these declines and restore salmonids to just a fraction of their former abundance (Lackey et al. 2006).

Beginning in the late 1990s, research on piscivorous colonial waterbirds nesting in the Columbia River estuary revealed an additional source of mortality for ESA-listed juvenile salmonids from throughout the Columbia River Basin (Schreck et al. 2006). Research on the impact of predation by Caspian Terns on survival of juvenile salmonids in the Columbia River estuary during the 1997–98 breeding seasons revealed that terns from the large breeding colony on Rice Island, an artificial dredged material disposal island 29 km (18 miles) upriver from the mouth of the Columbia River (Figure 7), consumed millions of salmonid smolts annually (Roby et al. 2003a). Data on the diet composition of Caspian Terns nesting on Rice Island indicated that the vast majority of prey items were young salmon and steelhead (*O. mykiss*; Collis et al. 2002). The colony consisted of more than 8,000 breeding pairs, by far the largest known Caspian Tern colony anywhere in the world (Cuthbert and Wires 1999, Wires and Cuthbert 2000, Roby et al. 2021a). Whereas the first record of Caspian Terns nesting in the Columbia River estuary was in 1984, by 1999 about two-thirds of all Caspian Terns in the Pacific Flyway were nesting on Rice Island (Gill and Mewaldt 1983, Roby et al. 2003b, Suryan et al. 2004, Naughton et al. 2007). Data on colony size and diet composition were entered into a bioenergetics model (Furness 1978) to estimate the number of juvenile salmonids consumed by Caspian Terns nesting at the Rice Island colony; the model indicated that about 12.4 million young salmon and steelhead were being consumed annually (Roby et al. 2003a), or about 10% of all juvenile salmonids that survived out-migration to the Columbia River estuary.

This magnitude of mortality from avian predation for out-migrating juvenile salmonids, many of which were listed under the ESA and were within a few miles of the Pacific Ocean, was unacceptable to regional managers tasked with restoring threatened and endangered salmonids in the Columbia River. Even though managers recognized that Caspian Terns and other fish-eating waterbirds were not primarily responsible for the heightened conservation concern over Columbia Basin salmonids, avian predation on young salmon and steelhead had become an apparent impediment to restoration of some ESA-listed stocks (Roby et al.

A Caspian Tern flying over the large breeding colony on East Sand Island in the Columbia River estuary, Oregon. The pictured tern is transporting a salmon smolt back to the colony to feed to its mate or its young. The Caspian Tern nesting colony on East Sand Island was formerly the largest of its kind in the world, but management has reduced colony size to less than a third of its former size to protect salmon smolts. Photo by Dan Roby

2021b). Steelhead smolts were particularly susceptible to predation from Caspian Terns, and there was no indication that steelhead smolts raised in hatcheries were more susceptible to tern predation than their wild counterparts (Evans et al. 2012, 2019). Even though the abundance of hatchery-raised smolts in the Columbia River estuary early in the nesting season likely helped attract fish-eating birds to nest there, there was little evidence that avian predators preferred hatchery smolts over wild ones (Hostetter et al. 2021).

In 1999, regional managers led by the USFWS, USACE, and NMFS started to implement a management plan intended to reduce the numbers of juvenile salmonids consumed by Caspian Terns in the Columbia River (Roby et al. 2002). The first step was to try to relocate the Rice Island tern colony to East Sand Island, 21 km (13 miles) closer to the Pacific Ocean (Figure 7), where it was hoped that fewer juvenile salmonids would be consumed (Lyons et al. 2005) without negatively affecting the size or nesting success of the tern colony. Nesting habitat for Caspian Terns was enhanced on East Sand Island, social attraction (tern decoys and audio playback of tern vocalizations) was deployed (Kress 1983, 1998), and the restored colony site was carefully monitored to identify any factors that might limit tern colony formation and nesting success. Predation on tern nest contents by Glaucous-winged / Western Gulls was identified early on as a limiting factor for the formation of the incipient tern colony, so limited gull control was implemented at the tern colony on East Sand Island during 1999 and 2000 (Roby et al. 2002). Concurrently, habitat modification techniques (plantings and erecting rows of silt fencing) were used to make the former tern colony site on Rice Island less attractive to prospecting Caspian Terns.

A Caspian Tern colony quickly formed on East Sand Island in 1999, and by 2001, all Caspian Terns nesting in the Columbia River estuary were on East Sand Island, and the Rice Island colony had been completely abandoned. Consumption of juvenile salmonids by Caspian Terns nesting on East Sand Island was less than half what it had been when the colony was located on Rice Island because the bulk of the diet now consisted of marine and estuarine forage fishes instead of salmonids (Roby et al. 2002). Tern nesting success on East Sand Island was more than twice what it had been on Rice Island, and the colony on East Sand Island increased to nearly 10,000 breeding pairs by 2002. Despite the successful relocation of the tern colony from Rice Island to East Sand Island, however, the terns nesting on East Sand Island consumed about 5 million young salmon and steelhead annually, and predation rates on steelhead ranged from 15% to 25% of the smolts that survived to the Columbia River estuary (Roby et al. 2021a).

Regional managers sought to further reduce the impact of Caspian Tern predation on survival of juvenile salmonids, especially steelhead, in the Columbia River estuary by reducing the size of the breeding colony on East Sand Island, while preventing Caspian Terns

from nesting on Rice Island or other dredged material disposal sites in the Columbia River estuary (USFWS 2005a). The strategy for accomplishing this reduction in colony size was to reduce the area of tern nesting habitat on East Sand Island by about 80% and to force about 60% of the terns nesting on East Sand Island to disperse and nest outside the Columbia River estuary, thereby reducing tern predation rates on juvenile salmonids by a similar percentage (Roby et al. 2021a). Concurrent with the managed reduction in tern nesting habitat in the Columbia River estuary, nesting habitat would be restored or created outside the Columbia Basin to attract nesting Caspian Terns to areas where fish species of conservation concern were not vulnerable to tern predation (USACE 2006, USFWS 2006).

*Caspian Tern Management to Reduce Predation of Juvenile Salmonids in the Columbia River Estuary* (hereafter the *Estuary Tern Management Plan*) was jointly developed by the USFWS, USACE, and NMFS, and first implemented in 2008. Its objective was to relocate most of the Caspian Terns nesting on East Sand Island to alternative colony sites outside the Columbia Basin and thereby increase survival of ESA-listed juvenile salmonids. The area of tern nesting habitat on East Sand Island was gradually reduced from 2 ha (5 acres) to 0.4 ha (1 acre) during 2008–15 and remained at 0.4 ha during 2016–20 (Roby et al. 2021a). Also, during 2008–15 the USACE built or enhanced 14 islands totaling 4.1 ha (10.25 acres) of prospective Caspian Tern nesting habitat outside the Columbia River Basin and attempted to attract nesting Caspian Terns to these alternative colony sites (Lawes et al. 2021a). Most of these new and restored colony sites were in southern Oregon and northeastern California, the so-called SONEC region, but one island was built in the southern Willamette Valley, and five small islets were enhanced in southern

San Francisco Bay (Hartman et al. 2019).

In response to the large reduction in tern nesting habitat on East Sand Island, however, Caspian Terns nested at higher-than-expected densities on the 1-acre colony site, attempted to establish satellite colony sites on East Sand Island near the designated colony site, and prospected for nest sites in large numbers at the former colony site on Rice Island (Roby et al. 2021a). Nesting success at the Caspian Tern colony on East Sand Island also declined during this period, mostly because of high disturbance rates by Bald Eagles, high nest predation rates by Glaucous-winged / Western Gulls, and low availability of marine forage fish owing to poor ocean conditions and high river discharge (Collar et al. 2017; Peterson et al. 2017, 2019; Bailey 2018). By the 2020 nesting season, the targeted reduction in size of the Caspian Tern colony on East Sand Island (3,125–4,375 breeding pairs) had been achieved, and with it an average reduction in predation rates on steelhead smolts by East Sand Island terns of about 63%. The future of the Caspian Tern colony on East Sand Island, however, formerly the largest known colony for the species in the world and supporting the majority of breeding Caspian Terns in western North America, has become highly uncertain (Roby et al. 2021a). During the 2021 nesting season, only about 2,050 breeding pairs attempted to nest at the East Sand Island colony, and for the third time in five years, no young terns were raised at the colony.

In 2018, the most recent year during which all of the alternative tern colony sites constructed by the USACE were monitored, about 1,440 breeding pairs of Caspian Terns nested on the islands (Lawes et al. 2021a). Most Caspian Terns that colonized these alternative nesting sites in the SONEC region were from the East Sand Island colony, as indicated by re-sightings of marked terns that were banded at the East Sand Island colony. Although 1,440 breeding pairs is far less than the approximately 5,600 breeding pairs that were displaced from East Sand Island as part of the *Estuary Tern Management Plan*, these alternative colony sites nevertheless provided a significant boost in numbers of nesting Caspian Terns in parts of the Pacific Flyway where the species was in decline (Lawes et al. 2021a). Artificial islands that were built in Oregon and support Caspian Tern breeding colonies in most years include a 0.4-ha (1-acre) island in Malheur Lake, Harney County; a 0.4-ha island in Crump Lake, Warner Valley; and a 0.2-ha (½-acre) island in Summer Lake Wildlife Area, Lake County. Each of these islands provided nesting habitat for Caspian Terns in areas of the state with a history of nesting by Caspian Terns, but where no Caspian Terns were nesting when the islands were built (Lawes et al. 2021a).

Concurrent with implementation of the *Estuary Tern Management Plan*, the USACE and its partners implemented the *Inland Avian Predation Management Plan* (IAPMP) in the Columbia Plateau region of eastern Washington State and northeastern Oregon (USACE 2014). The focus of the IAPMP was to eliminate the two largest Caspian Tern breeding colonies in the Columbia Plateau region: on Crescent Island in McNary Reservoir near Pasco, Washington, and on Goose Island in Potholes Reservoir near Moses Lake, Washington. These two Caspian Tern colonies were small in comparison to the large colony on East Sand Island (~450 and ~350 breeding pairs, respectively), but predation on smolts from some populations of ESA-listed salmonids (e.g., Upper Columbia River steelhead) by terns from these two colonies exceeded 18% of all available smolts in some years (Collis et al. 2021). Based on the magnitude of predation on some ESA-listed salmonids, the IAPMP sought to reduce all Caspian Tern colonies that were within foraging distance of the mainstem Columbia and Snake Rivers to no more than 40 breeding pairs, and the entire breeding population in the Columbia Plateau region to no more than 200 breeding pairs (USACE 2014). The management strategy was to eliminate all Caspian Tern nesting habitat on Crescent Island and Goose Island using passive and active dissuasion techniques, and thereby force Caspian Terns to nest elsewhere (Collis et al. 2021). To compensate for the reduction in Caspian Tern nesting habitat and breeding population size in the Columbia Plateau region, the USACE enhanced five small islands in Don Edwards National Wildlife Refuge, southern San Francisco Bay (about 1,000 km distant), as alternative tern nesting habitat (Hartman et al. 2019).

The IAPMP was implemented during 2014–18 by the USACE and the Bureau of Reclamation. By 2019 the colonies on Crescent Island and Goose Island had been abandoned, and the breeding population of Caspian Terns in the Columbia Plateau region had declined by about half, from a pre-management average of 875 breeding pairs to 445 breeding pairs (Collis et al. 2021). Some of the emigrants from the two managed tern colonies recruited to smaller tern colonies in the Columbia Plateau region, which prevented the regional population of Caspian Terns from reaching the target size of no more than 200 breeding pairs. Two of the five small islands in Don Edwards NWR that were enhanced as alternative nesting habitat for Caspian Terns in south San Francisco Bay were colonized by Caspian Terns, and an average of about 450 breeding pairs nested on those two small islands during 2015–18 (Lawes et al. 2021a).

In a study of Caspian Tern predation on passive integrated transponder (PIT)–tagged Upper Columbia River and Snake River steelhead smolts, results indicated that decreases in tern predation rates on steelhead were associated with statistically significant increases in steelhead survival rates in all evaluated years and salmonid life stages (i.e., smolt out-migration and smolt-to-adult returns; Payton et al. 2021). The results also indicated that, in the absence of Caspian Tern predation, significantly more steelhead smolts would have survived out-migration, and more importantly from a conservation perspective, more adult steelhead would have returned to the Columbia River to spawn (Payton et al. 2020). These results provide strong evidence that predation by Caspian Terns is not primarily a compensatory source of mortality for steelhead, and instead can be a largely additive source of mortality during the smolt life stage and a partially additive source of mortality for the smolt-to-adult life stage (Payton et al. 2021).

Following implementation of the *Estuary Tern Management Plan* and the *Inland Avian Predation Management Plan*, a demographic model was developed for the Pacific Flyway population of Caspian Terns, one of the most sophisticated population models available to date for colonial waterbirds, in order to project the Pacific Flyway population of Caspian Terns under 16 different scenarios of management and environmental conditions for breeding (Suzuki et al. 2018). The model incorporated life history complexity (e.g., delayed maturity, age-structured survival, and breeding propensity) and spatial specificity (e.g., natal and breeding dispersal, region-specific demographic rates across five distinct regions of the breeding range) and was based on empirical data collected throughout the flyway over more than a decade. Population simulations from the model indicated resiliency of the Pacific Flyway population of Caspian Terns under the vital rates and proposed tern management regimes in 2018. Model population trajectories depended, however, on the quantity and quality of nesting habitat available for Caspian Terns in the Pacific Flyway, the predominant limiting factor for this population. Also, certain combinations of management actions coupled with realistic changes in breeding conditions that follow recent trends (e.g., simultaneous reductions in nesting habitat in the Columbia River estuary coupled with recent unfavorable environmental conditions for reproduction on East Sand Island) would likely result in long-term declines in the Pacific Flyway population, even when assumptions are made for the continued availability of nesting habitat outside of the Columbia River estuary at the maximum levels observed to date. In particular, circumstances that substantially reduce the unique capacity for fledgling production at the East Sand Island colony in the Columbia River estuary, if not mitigated by a similar high capacity for fledgling production elsewhere in the Pacific Flyway, are likely to result in a population of substantially reduced size, and potentially a long-term downward trend that could put the Pacific Flyway population of Caspian Terns at risk of greatly reduced resiliency (Suzuki et al. 2018). Avoiding such a long-term downward trajectory will depend on maintaining suitable Caspian Tern nesting habitat on East Sand Island, as stipulated in the *Estuary Tern Management Plan*, or creating an alternative coastal nesting site in the Pacific Flyway capable of supporting a large and productive Caspian Tern breeding colony (Roby et al. 2021a).

As the *Estuary Tern Management Plan* was being implemented, salmonid managers turned their attention to a different species of piscivorous colonial waterbird that nested in large numbers in the Columbia River estuary: Double-crested Cormorants. The breeding colony of Double-crested Cormorants on East

Sand Island had grown from just 91 breeding pairs in 1989 to nearly 15,000 breeding pairs in 2013, making it the largest known breeding colony for the species anywhere (Lawes et al. 2021b; see the Double-crested Cormorants in Oregon section on p. 95). Although the diet of Double-crested Cormorants nesting on East Sand Island consisted of a wide variety of marine, estuarine, and freshwater forage fishes, juvenile salmonids belonging to 11 of the 13 populations of anadromous salmonids from the Columbia River basin that are listed under the ESA averaged 12.5% of the diet (percentage of total prey biomass). Despite the relatively low proportion of juvenile salmonids in the diet of Double-crested Cormorants nesting at East Sand Island, cormorants from this large colony annually consumed an estimated average of 10.9 million juvenile salmonids during 1998–2013 (Gremillet et al. 2003, Lyons 2010, Lawes et al. 2021b), more than twice the average annual consumption of juvenile salmonids by Caspian Terns nesting on East Sand Island. Estimates of the number of juvenile salmonids consumed annually ranged from a low of about 2.9 million smolts in 2005 to a high of about 20.9 million smolts in 2011, two years when the size of the Double-crested Cormorant colony on East Sand Island differed by just 6%. Millions of salmonid smolts were consumed annually by Double-crested Cormorants nesting at East Sand Island during 1998–2013, and smolt losses likely represented 5% to 15% of all smolts that survived to the Columbia River estuary (Lyons et al. 2014).

An investigation of predation rates on juvenile salmonids (percentage of available fish consumed) by Double-crested Cormorants nesting on East Sand Island indicated that impacts were highly variable depending on the salmonid population and the year. Estimated predation rates by East Sand Island cormorants were consistently the highest on

Part of the large breeding colony (nearly 15,000 breeding pairs when this photo was taken) of Double-crested Cormorants on East Sand Island on the Oregon side of the Columbia River estuary. The colony on East Sand Island was the largest known colony for the species in the first decade and a half of this century but has since been abandoned. The structures in the photo are observation blinds and aboveground tunnels connecting blinds so that the colony can be studied without disrupting nesting cormorants. Photo by Dan Roby

ESA-listed Lower Columbia River Chinook salmon (*O. tshawytscha*), with an average annual predation rate of 27.5% (Sebring et al. 2013, Lawes et al. 2021b). At the other extreme, the estimate of average annual predation rate on Upper Willamette River spring Chinook salmon smolts during 2003–14 was just 1.8%. As with Caspian Tern predation, there was no evidence that cormorant predation rates differed based on the rear-type (hatchery vs. wild) or out-migration history (in-river vs. transported by truck or barge) of smolts, suggesting that East Sand Island cormorants indiscriminately preyed on smolts within a population, despite high variation in predation rates among populations.

Pilot studies to test the feasibility of nonlethal methods to reduce the size of the Double-crested Cormorant colony on East Sand Island, and thereby its impact on survival of juvenile salmonids, without causing colony abandonment, were conducted during 2007–13. A combination of human hazing of prospecting cormorants and destruction of partially built cormorant nests on one side of a large privacy fence transecting the colony proved highly effective at limiting the area of nesting habitat available on East Sand Island for cormorant nesting (Lawes et al. 2021b). Social attraction techniques (decoys and audio playback of cormorant vocalizations) and nesting substrate enhancement were also demonstrated to be effective in encouraging Double-crested Cormorants to nest in previously unused habitats on East Sand Island and on other islands in the Columbia River estuary with previous histories of cormorant nesting attempts (Suzuki et al. 2015). Double-crested Cormorants, however, were slow to respond to social attraction and nesting substrate enhancement at sites with no previous history of cormorant nesting attempts, such as the artificial islands constructed by the USACE

as alternative colony sites for Caspian Terns (Lawes et al. 2021a). Nevertheless, the artificial tern islands constructed by the USACE were eventually colonized by Double-crested Cormorants within four to eight years of Caspian Terns and gulls (*Larus* spp.) initiating nesting on the islands.

Starting in 2015, the USACE, USFWS, and USDA–Wildlife Services implemented a management plan with the goal of substantially reducing predation on ESA-listed juvenile salmonids by Double-crested Cormorants in the Columbia River estuary. The *Cormorant Management Plan* (USACE 2015) consisted of two phases. Phase I was primarily lethal in approach and consisted of culling up to 10,912 adult Double-crested Cormorants and oiling the eggs in up to 15,184 cormorant nests on the East Sand Island colony over a four-year period, with the objective of reducing the size of the colony to no more than 5,380–5,939 breeding pairs, or at most 42% of the average pre-management colony size. Phase II was primarily nonlethal in approach and consisted of reducing the available cormorant nesting habitat on East Sand Island to an area capable of supporting no more than 5,380–5,939 breeding pairs of Double-crested Cormorants in the long term (USACE 2015).

During Phase I (2015–17), USDA–Wildlife Services reported culling 5,576 adult Double-crested Cormorants near the East Sand Island colony and oiling the eggs in 6,181 cormorant nests on the colony. The reported take was far less than intended because, beginning in May 2016, the cormorant colony on East Sand Island was repeatedly abandoned, and most nesting cormorants dispersed from the colony site (Lawes et al. 2021b). The Double-crested Cormorant colony on East Sand Island had fewer than 5,380 active nests during the 2017 breeding season, and adaptive management provisions of the *Cormorant*

*Management Plan* triggered the initiation of Phase II in 2018 (USACE 2015). During Phase II (2018–19), most of the former habitat used by nesting cormorants on East Sand Island was converted to intertidal mudflats, and a privacy fence was erected to separate the designated cormorant colony area from parts of the island where cormorant nesting would be dissuaded by hazing (Lawes et al. 2021b). In 2018, a Double-crested Cormorant colony of about 3,670 breeding pairs formed in the designated colony area, and many pairs nested successfully. During 2019, after the conversion of much of the former cormorant colony area to intertidal mudflats, only a few hundred pairs of cormorants attempted to nest on East Sand Island; the colony experienced repeated partial and total abandonment events during the 2019 nesting season, and no nesting attempts were successful. In 2020, there were no sustained nesting attempts by Double-crested Cormorants on East Sand Island, and the colony was apparently completely abandoned (Lawes et al. 2021b).

The factors responsible for the repeated and unprecedented colony abandonment events by cormorants attempting to nest on East Sand Island during 2016–17 are not completely understood, but the primary factor appears to be the cumulative effects of disturbance and stress on cormorants from Bald Eagles and the activities of USDA–Wildlife Services technicians engaged in culling and egg-oiling (Duerr et al. 2009, Lawes et al. 2021b). It is likely that these two factors acted synergistically to dissuade cormorants from nesting on East Sand Island during Phase I of the *Cormorant Management Plan*. The conversion of much of the former cormorant colony area on East Sand Island to intertidal mudflats and other modifications on the west end of East Sand Island during the winter of 2018–19 as part of Phase II of the plan apparently resulted in

nearly a complete dispersal of cormorants from the East Sand Island colony. Concurrent with these repeated abandonment and dispersal events at East Sand Island, the formerly small breeding colony (fewer than 400 breeding pairs until 2015) of Double-crested Cormorants on the Astoria-Megler Bridge, 15 km (9 miles) upriver from East Sand Island (Figure 7), became a heavily used roosting site for nonbreeding cormorants, and the number of Double-crested Cormorants nesting on the bridge increased to more than 5,000 breeding pairs by the 2020 nesting season (Lawonn, in press).

To evaluate the efficacy of the *Cormorant Management Plan* for reducing predation rates on ESA-listed salmonid populations by Double-crested Cormorants nesting on East Sand Island to the levels specified by NMFS (USACE 2015), average annual estimates of predation rates were compared prior to and following implementation of the plan. Results indicated that average annual predation rates by East Sand Island cormorants on all salmonid populations evaluated were significantly lower during Phase II (2018), with average annual predation rates decreasing by 70% to 90%, depending on the salmonid population (Lawes et al. 2021b). During 2016–19, however, attendance by Double-crested Cormorants at the East Sand Island colony was intermittent, with multiple episodes of dispersal from the colony and partial or complete colony abandonment, resulting in the rapid expansion of the nesting colony on the Astoria-Megler Bridge. Predation rates on juvenile salmonids by Double-crested Cormorants nesting on the Astoria-Megler Bridge were measured in 2021 and were found to be two to four times greater on a per capita basis because the bridge is located in or near the mixing zone of the estuary (Figure 7), where juvenile salmonids are relatively more prevalent compared to

California Brown Pelican in breeding plumage. The species was federally listed as Endangered until 2009, when it was delisted, but is still listed as Endangered by the state of Oregon. The largest postbreeding roost site for the species is on East Sand Island in the Columbia River estuary, Oregon, where a few nesting attempts have been documented recently, the first for the state. Climate change is apparently responsible for the species' northward range expansion. Unlike Caspian Terns and Double-crested Cormorant, there is no evidence that California Brown Pelicans consume significant numbers of salmonid smolts. Photo by Dan Roby

marine forage fish (Cramer et al. 2021a; Banet et al., in review).

The future of the East Sand Island cormorant colony, like that of the East Sand Island Caspian Tern colony, is uncertain. In 2020 and 2021, no persistent cormorant nesting attempts were detected on the East Sand Island colony. Concurrently, the largest Double-crested Cormorant nesting colony in the Pacific Flyway was the colony on the Astoria-Megler Bridge, 15 km upriver from the East Sand Island colony (Lawes et al. 2021b). With the growth of the Double-crested Cormorant colony on the bridge concurrent with the abandonment of the colony at East Sand Island, it is likely that cormorant predation on ESA-listed juvenile salmonids throughout the estuary has not changed, and certainly has not been reduced to the level intended by the *Cormorant Management Plan* (NMFS 2008, 2019b). As such, the objectives of the *Cormorant Management Plan* could potentially be accomplished if the breeding colony on the Astoria-Megler Bridge were relocated back to East Sand Island. Colony relocation could be achieved by dissuasion and hazing

## Dan Roby

Dan Roby is a physiological ecologist who studies piscivorous birds. He joined the Oregon State University faculty in 1995 and became leader of the US Geological Survey Oregon Cooperative Wildlife Research Unit on campus. Dan's lab has focused on research questions related to the ecology, conservation, and reproductive energetics of seabirds and colonial waterbirds in Oregon and internationally. In Oregon, he has led a long-term research project investigating the ecology of piscivorous colonial waterbirds and their impacts on the survival of juvenile salmonids that are listed as Threatened or Endangered in the Columbia River Basin. Much of this work has focused on Caspian Terns and Double-crested Cormorants, two species of piscivorous colonial waterbirds that have suffered from persecution, overkill, and pesticide poisoning that had driven them to very low levels by the mid-twentieth century. Dan and his lab have worked on developing nonlethal approaches for resolving fisheries and fish-eating bird conflicts by relocating colonies of avian predators to alternative sites. Dan also serves as a co-principal investigator on the Oregon Marbled Murrelet Project. Dr. Roby was chair of the Pacific Seabird Group in 2005 and was presented with their lifetime achievement award in 2022. He is an elected fellow of the American Ornithological Society, which awarded Dan its Ralph W. Schreiber Conservation Award in 2017.

*Photo by Rob Suryan*

of cormorants nesting on the Astoria-Megler Bridge combined with nesting substrate enhancement and social attraction at the former colony site on East Sand Island (Suzuki et al. 2015). Colony relocation would not only reduce impacts on survival of ESA-listed juvenile salmonids to levels intended by the *Cormorant Management Plan*, but it would also minimize damage to the bridge from cormorant guano deposition and help ensure the long-term conservation of the Double-crested Cormorant population in the Pacific Flyway (Lawes et al. 2021b).

Following the implementation of management actions to reduce the impact of avian predation on survival of ESA-listed juvenile salmonids in the Columbia River Basin (i.e., the *Estuary Tern Management Plan*, the *Inland Avian Predation Management Plan*, and the *Cormorant Management Plan*), the Pacific Flyway populations of Caspian Terns and Double-crested Cormorants have declined. The estimated size of the Pacific Flyway population of Caspian Terns in 2021 (~7,660 breeding pairs) has declined by about 59% since the population size peaked in 2009 (~18,870 breeding pairs), shortly after the implementation of the *Estuary Tern Management Plan* (Lawes et al. 2022). The estimated size of the Pacific Flyway population of Double-crested Cormorants in 2019 (22,890 breeding pairs, 95% CI = 15,925–29,855 pairs; USFWS 2020b) was about 43% lower (95% CI = 10% to 66%) than in 2015, just prior to implementation of the *Cormorant Management Plan* (USFWS 2016a). The large, recent declines in the flyway-wide populations of these two piscivorous colonial waterbirds are clearly related to the management actions that have been implemented to date to help restore ESA-listed populations of salmonids in the Columbia River Basin. These three management plans directed at reducing avian predation on Columbia Basin

Caspian Tern adults in breeding plumage at a loafing site on the Oregon side of the Columbia River estuary. The breeding population of Caspian Terns in the Pacific Flyway increased dramatically during the last few decades of the twentieth century, but the breeding population in the flyway is in decline and is now less than half the size it was in the first decade of the twenty-first century. Photo by Timothy Lawes

salmonids have been highly controversial, in part because many ornithologists and avian conservationists have perceived management as scapegoating birds for salmon declines, while many fisheries managers are frustrated by any constraints on controlling predators that are taking threatened fish of such high resource value. Whether the flyway-wide declines in Caspian Terns and Double-crested Cormorants persist will depend on future management actions, and whether stakeholders and other publics advocate for more balanced approaches to managing piscivorous colonial waterbirds to benefit salmonid species of conservation concern.

Caspian Terns and Double-crested Cormorants have been the focus of nearly all management efforts to reduce the impact of avian predation on salmonids of conservation concern in Oregon. But other piscivorous waterbirds have also attracted the attention of fisheries managers tasked with restoring depleted stocks of salmonids. California Brown Pelicans, American White Pelicans, Western

Gulls, Glaucous-winged Gulls, California Gulls (Cramer et al. 2021b), Ring-billed Gulls (Ruggerone 1986), Great Blue Herons (Sherker et al. 2021), Ospreys, Common Mergansers (Wood 1987), and Common Murres (Wells et al. 2017) have all been implicated or suspected of causing significant losses of juvenile salmonids. All of these species are indigenous to the state of Oregon (Gabrielson and Jewett 1940) and have co-evolved with juvenile salmonids as prey. But that was before the dramatic increase in human population within the region and the cumulative impacts of dams, logging, wildfire, and anthropogenic climate change on both freshwater and marine habitats for salmonids.

When piscivorous waterbirds cause sufficient losses of salmonids to warrant management concern, there is generally an anthropogenic environmental change that has led to the conflict. These changes can include artificially enhancing forage fish availability (e.g., hatchery releases) and thereby eliciting a functional or numerical response from avian predators, habitat alterations that make juvenile salmonids more susceptible to avian predation (e.g., hydroelectric dams), creation of new waterbird nesting habitat (e.g., dredged material disposal islands), or depleted stocks of salmonids that lack resilience to natural predation rates (Osterback et al. 2013). Restoring salmonid populations in Oregon and throughout the Pacific Northwest while conserving native species of piscivorous waterbirds will require creative approaches and collaborative management initiatives that recognize the extrinsic and intrinsic values of salmonids and their avian predators. A general approach to reducing fisheries–piscivorous waterbird conflict is through control of where breeding colonies of piscivorous waterbirds are allowed to become established and persist; colonies at some sites lead to large smolt losses, while others have little or no impact. Such a management approach avoids expensive and unsustainable culling programs that seek to control wildlife populations through widespread lethal take, an approach that is distasteful to a large segment of the public, obsolete in practice (Wires 2014), and the antithesis of ecosystem-based management (Tallis et al. 2009). When warranted, management of piscivorous waterbirds to benefit beleaguered salmon populations must recognize the intrinsic value and protected status of the birds themselves.

## Oregon Coastal Beaches

### CHARLIE BRUCE

The Oregon coast is approximately 365 miles long, of which about 250 miles are beach interspersed with rocky shoreline, estuaries, and jetties (ICF International 2010, Hapke et al. 2014). Oregon beaches are home to more than 15 species of shorebirds for nesting, wintering, and migration (Marshall et al. 2003). The beaches are currently backed by a mix of geomorphology that includes dunes (39%), bluffs fronted by sandy beaches (33%), and cliff/rocky shore (25%), and 6% are a mix of sand and gravel or cobble (J. Allan, pers. comm., 2021). Prior to about 1930, sand beaches lacked the large grass-covered foredunes that now extend along much of the Pacific coast (Wiedemann and Pickart 1996). Sand moved freely inland to the native tree line or cliffs. Non-native grasses, especially *Ammophila arenaria* and *A. brevigulata*, began to be planted experimentally in the 1930s to help stabilize drifting sand as coastal communities and roads were developed. Major plantings were made up and down the coast in the 1950s and 1960s by multiple agencies and local governments (Morgan 1952, Green 1965, Wiedemann 1987), resulting in the tall foredunes that exist today.

Back in 1952, the last bridge crossing on Oregon Highway 101 was completed over the Coquille River. The entire coast was open to greater leisure travel, commerce, and development, all of which took off. The population of major coastal cities more than doubled between 1950 and 2020 to 114,000, and the state population almost tripled to 4,268,000. Human population growth along the entire coastal zone of Oregon has resulted in many and continuous planning efforts and impacts to all beach and rocky shore areas.

Stepping back briefly, the Oregon Highway Commission, established in 1913, did much

Sanderlings foraging on an Oregon beach for sand crabs. Sanderlings breed in the High Arctic and some spend the winter on Oregon's beaches, although others continue migrating to the Pacific coast of South America to spend the winter. This highly migratory species is dependent on pristine beaches without contaminants in Oregon to complete the annual cycle. Photo by Dan Roby

more than construct the coast highway. Then-governor Oswald West, a champion of public access to Oregon beaches, which were used for early coastal travel, approved legislation that declared beaches a state highway. Included in state law, this was changed in 1947 to designate all coastal beaches as a state recreation area, setting the groundwork for public access and protection as we know it today.

> 390.615 Ownership of Pacific shore; declaration as state recreation area. Ownership of the shore of the Pacific Ocean between ordinary high tide and extreme low tide, and from the Oregon and Washington state line on the north to the Oregon and California state line on the south, excepting such portions as may have been disposed of by the state prior to July 5, 1947, is vested in the State of Oregon, and is declared to be a state recreation area.

The early Oregon Highway Commission also began purchasing scenic lands adjacent to the highway for public enjoyment and later for park lands. This has continued to the present, where we now have beach and shoreline access at some 80 state parks and scenic waysides along 143 miles of coastline (i.e., 37%). But protection and access to Oregon beaches was not guaranteed until passage in 1967 of what is known as the "Beach Bill." The bill (HB 1601) legally established a recreational easement to all beaches west of the vegetation line or a statutory line established along the coast in 1969 (Oregon Parks and Recreation Branch 1977). The Oregon Parks and Recreation Department, originally a division of the Highway Commission, administers implementation of the Beach Bill.

In response to uncontrolled growth statewide and impacts on natural resources, the 1970s saw major environmental laws and planning programs initiated that affected coastal areas, statewide and nationally. In 1971 the legislature passed bills to establish the Oregon Coastal Conservation and Development Commission to address natural resource protection given increased and unplanned development along the coast (Department of Land Conservation and Development 2021a, b). Legislative action that year also gave jurisdiction over "nongame" wildlife species to the Oregon Game Commission (now the Oregon Department of Fish and Wildlife), and early survey efforts in the 1970s focused on declining coastal Western Snowy Plovers (*Charadrius nivosus nivosus*; Oregon Department of Fish and Wildlife 1994).

The 32,000-acre Oregon Dunes National Recreation Area (DNRA) was established by Congress in 1972 (Pub.L., 92-260) between Florence and Coos Bay and is managed as part of the Siuslaw National Forest (USFS 1977). The DNRA includes 38 miles of beach and dunes, parts of which are key to Snowy Plover recovery and are habitat for migrating/wintering shorebirds. To the south of the DNRA, Coos Bay Bureau of Land Management manages about 12 additional miles of beaches and dunes critical to ongoing Snowy Plover recovery efforts as well as other shorebirds. About 25 miles of these beaches are now designated critical habitat for Western Snowy Plovers and

are key to recovery efforts in the state (USFWS 2012b).

The volunteer work of the Oregon Coastal Conservation and Development Commission in the early 1970s became the foundation for what later evolved into state coastal planning goals that included conserving and protecting beaches and dunes by cities and counties. With the leadership of then-governor Tom McCall, Senate Bill 100 was passed in 1973 that established the current program for statewide planning administered under the Department of Land Conservation and Development (2021a). Over the next three years, the Department of Land Conservation and Development adopted 19 statewide planning goals and associated guidelines that counties and cities were required to address in local comprehensive plans. Importantly, four goals adopted in 1977 were developed for the coastal zone: Goal 16, Estuarine Resources; Goal 17, Coastal Shorelines; Goal 18, Beaches and Dunes; and Goal 19, Ocean Resources. In addition, Goal 5 addresses a broad category of natural resources, including significant fish and wildlife resources, such as key shorebird habitats. Although initial coastal county plans made some reference to beaches as wildlife habitat, little inventory information was available at the time to identify important shorebird habitat, and most plans have not been properly updated.

Complementing Oregon's coastal management programs is the federal Coastal Zone Management Program established in 1972. This volunteer program, administered by the National Oceanic and Atmospheric Administration (NOAA), currently assists 34 states through cooperative funding with marine resource protection. NOAA approved the Oregon Coastal Management Program in 1977 that is now administered by Department of Land Conservation and Development (2021b).

Important elements of the program include addressing non-point-source pollution, sea level rise, and ocean energy development, all of which already have impacts on coastal beaches and shorelines.

In response to concerns about federal proposals for leasing offshore areas for oil and mineral exploration, the 1991 Oregon Legislature established the Ocean Policy Advisory Council to develop a plan that coordinates management of the resources and activities in the state's territorial sea from the high tide line out to the 3-mile limit. The first Oregon *Territorial Sea Plan* was completed in 1994 and has been updated many times since (Department of Land Conservation of Development 2021c). A key section in the plan addresses protecting rocky shores and the ecological functions of marine habitats. Rocky shores cover about 25% of the Oregon coastline interspersed with beaches and are important to resident Black Oystercatchers, migrant and wintering shorebirds, marine mammals, and seabirds (Oregon Department of Fish and Wildlife 2016a). ODFW has developed a specific nearshore strategy as part of a statewide conservation strategy for Oregon fish, wildlife, and their habitats (Oregon Department of Fish and Wildlife 2016b).

Guidelines for implementing coastal management by planning cities and counties as part of Goal 18 attempt to address protection of significant wildlife habitats in the face of continued growth and development adjacent to beaches. A key element is that permits for beachfront protective structures (e.g., riprap, sea walls), sand removal, and grading will only be issued for adjacent lands developed prior to 1 January 1977, with some exceptions (Department of Land Conservation and Development 2021b). Permits to allow installation and maintenance of protective structures are administered by Oregon Parks

and Recreation Department. As of 2015, about 23 miles of beach coastline have some kind of shoreline protective structures for private properties, and 43 miles are eligible for armoring under current law (Gardner 2015). For context, of the approximately 9,050 individual oceanfront properties along the coast, about 3,500 could legally apply to install shoreline armoring under Goal 18 (Beasley and Dundas 2021). Over 90% of the current beachfront armoring has taken place in Lincoln, Tillamook, and Clatsop Counties and is expected to increase owing to ongoing and future impacts of climate change.

With the federal listing of coastal populations of the Western Snowy Plover in 1993, extensive conservation efforts have increased to this day (see the Western Snowy Plovers section on p. 118). Federal critical habitat was designated in 2005 and revised in 2012 on about 34 miles of coastal beaches, primarily from the Florence area south (USFWS 2012b). Following completion of the federal recovery plan in 2007, the Oregon Parks and Recreation Department (OPRD) developed a habitat conservation plan to cover beaches along the entire Oregon coast (ICF International 2010). Among other tasks, OPRD committed to developing Snowy Plover site plans for five state parks in occupied and unoccupied areas, primarily along the north coast where the species had not nested in most areas since the early 1980s (Oregon Parks and Recreation Department 2005).

Ongoing global climate change is already having impacts on coastal areas and is expected to increase flooding and erosion. The warming ocean is contributing to increased sea level rise, increased wave height, and greater-intensity storms, leading to continual beach and shoreline erosion and loss. Expected sea level rise by 2100 is predicted to be 3–8 feet (Mote et al. 2019, Oregon Department of Fish and Wildlife 2020). These and other climate impacts together with decades of anthropogenic shoreline development continue to reduce and negatively impact intertidal beaches and rocky shorelines.

## Western Snowy Plovers
### DAVID LAUTEN

Western Snowy Plovers (*C. n. nivosus*) were once a common breeding shorebird in coastal Oregon, inhabiting open sandy beaches, spits, sand and mudflats, and associated dunes (Gabrielson and Jewett 1940). ODFW began annual breeding season surveys for plovers on the coast in the early 1970s. Data indicated a steadily declining population due to loss of habitat, predation pressures, and disturbance (ODFW, unpublished data). By the late 1980s, plovers were absent from all historical breeding sites north of Lane County, and only several nesting sites remained occupied in Coos and Curry Counties.

In the late 1980s, Mark Stern from the Oregon Natural Heritage Program at The Nature Conservancy, with the cooperation of the Oregon Parks and Recreation Department, the ODFW, the US Bureau of Land Management, and the USFWS, began monitoring and studying plovers in conjunction with Gary Page at Point Reyes Bird Observatory based in Marin County, California (now Point Blue Conservation Science). In 1988, Ruth Wilson Jacobs completed a study of Snowy Plovers at New River, Coos County, that indicated low plover numbers and poor reproductive success (Wilson-Jacobs and Meslow 1984, Wilson-Jacobs and Dorsey 1985). In 1990, Mark Stern began the first systematic surveys of plovers and nests at Bandon Beach, New River, and Floras Lake as well as Coos Bay North Spit on the southern Oregon coast. After the first season, the coastal plover population was estimated to be 35–50 individuals.

Decline in plover numbers on the coast was primarily attributed to introduction of European beachgrass (*Ammophila arenaria*) that has been used to stabilize drifting dunes and sand. The growth of beachgrass eliminated nesting habitat and introduced cover for predators that plovers had not evolved to avoid or were new to plovers. Owing to very low nest success, in 1991 metal cages called exclosures were used to protect nests from predators, particularly Common Ravens, which were the main cause of nest failure.

By 1993, nest success improved from about 10% prior to exclosure use to nearly 70% after exclosure use, which resulted in a growing plover population. Additional nesting areas were identified on the US Forest Service's Oregon Dunes National Recreation Area and the Siuslaw National Forest in the 1990s. In 1993 the Pacific Coast population of Western Snowy Plovers was listed as threatened under the federal Endangered Species Act (USFWS 1993, 2005b), and ODFW listed the Western Snowy Plover as threatened throughout the state (Oregon Department of Fish and Wildlife 1994). Recovery goals for Recovery Unit 1 (Washington and Oregon) were determined to be 200 breeding individuals in Oregon and 50 breeding plovers in Washington.

By 1995, there were approximately 100 plovers breeding on the Oregon coast, with nesting plovers at Baker/Sutton Beach and Siltcoos in Lane County, Tahkenitch Creek in Douglas County, Tenmile Creek, Coos Bay North Spit, Bandon Beach / New River, and Floras Lake. In the late 1990s the plover population leveled off and hovered around 100 individuals. Research indicated that predators remained the main cause of nest and brood failure. Exclosures improved survival of nests, but since chicks leave the nest shortly after hatching, predators

Adult Western Snowy Plover brooding a newly hatched chick on the central Oregon coast. Nesting success for Western Snowy Plovers on Oregon beaches is generally low without conservation actions to enhance nest survival. This distinct population segment of Snowy Plover is listed as Threatened by the state of Oregon and the federal government. Photo by Roy Lowe

were suppressing chick survival and causing the population to remain stable.

In February 1999, the freighter *New Carissa* grounded on the beach at Coos Bay North Spit, spilling thick bunker fuel and contaminating plovers and other birds and wildlife. Emergency funding from the wreck resulted in the first efforts to reduce predation pressure by removing non-native red foxes (*Vulpes vulpes*) from the Bandon Beach / New River area. Red foxes had become locally common following their release from farms after the collapse of the fur market. By 2002, a predator management program to reduce predation pressure on plover nests and broods was implemented at the Bandon Beach / New River

area and at Coos Bay North Spit. By 2004, predator management was implemented at all known plover nesting areas from Baker/Sutton Beach to Floras Lake.

A five-year review of the predator management program indicated that reduction in predator numbers resulted in improved reproductive parameters, and by 2005, Snowy Plover numbers had increased to approximately 150 breeding individuals. Plover numbers continued to increase, and by 2009 there were an estimated 200 individuals on the Oregon coast.

The population continued to grow as a result of the effectiveness of the management program, including habitat management through the use of heavy machinery to eliminate non-native beachgrass and open habitat for the plovers, recreation management restricting public access to sections of beach where plovers nest, and predator management to reduce predation pressure on plover nests and broods. By 2011, reproductive output increased to more than 170 fledglings per year, and by 2015, more than 300 fledglings were produced. The adult population continued to increase, and with increasing numbers, more nests were found, and more fledglings were produced.

By 2018, Snowy Plovers reoccupied sites on the north coast, including Clatsop Spit in Clatsop County, Nehalem Spit, Bayocean Spit, Sitka Sedge State Natural Area in Tillamook County, and South Beach State Park, Bayshore Spit, and Patterson State Park in Lincoln County. In 2018 and 2019, Snowy Plovers were confirmed nesting in all coastal counties, likely the first time all coastal counties had nesting plovers since the early 1970s. The Oregon Parks and Recreation Department now monitors the northern coast nesting area in conjunction with many volunteers and the USFWS.

As of 2021, the Oregon Biodiversity Information Center, formerly the Oregon Natural Heritage Program and now part of the Institute for Natural Resources at Portland State University, has conducted 32 years of monitoring and managing Snowy Plovers on the coast of Oregon from Lane County to Curry County in conjunction with USFWS, BLM, USFS, USACE, Oregon Parks and Recreation Department, and ODFW. The five-member field crew monitors more than 500 plovers and 500 nests a year from Baker/Sutton Beach in Lane County to Floras Lake in Curry County.

Years of monitoring banded plovers has resulted in a broad understanding of plover

## David Lauten

David is a faculty research assistant at Portland State University's Institute of Natural Resources based in Bandon, Oregon. David has a biology degree from Rutgers University in New Jersey and a master's degree in wildlife ecology from the University of Wisconsin–Madison. Since 1997 he has worked with the Oregon Biodiversity Information Center on monitoring the threatened Western Snowy Plover on the Oregon coast. David is part of the Western Snowy Plover Working Team and has been involved in several other projects that have impacts on plovers, including the *New Carissa* removal project, the formerly proposed liquefied natural gas plant in Coos Bay, the proposed container facility in Coos Bay, the Oregon Parks and Recreation Department's Habitat Conservation Plan for the Snowy Plover, and plover projects in Washington and California. David is a master bander and has also completed research on Ruffed Grouse, Canada Geese, songbirds in the Cascades, shorebirds, pine martens, fishers, and bobcats. He has traveled extensively, chasing birds from the tundra of Canada to the jungles of South America.

*Photo by Kathleen A. Castelein*

movements along the coasts of California, Oregon, and Washington. The success of the Oregon plover project has provided Washington State and northern California counties with growing plover numbers as Oregon birds disperse to new territories. Recovery Unit 1 (Washington and Oregon) is well above recovery goals and continues to have a growing plover population, while Recovery Unit 2 (Mendocino, Humboldt, and Del Norte Counties in northern California) has not achieved its recovery goal, although Oregon plovers greatly supplement the population in these counties.

Snowy Plovers banded in Oregon winter as far north as Washington, and on 23 May 2019, two Snowy Plovers banded in Oregon were seen at Wickaninnish Beach, Pacific Rim National Park Reserve, British Columbia. More often, they are seen wintering at northern coastal sites in Oregon, including Clatsop Spit, Nehalem Spit, Bayocean Spit, South Beach State Park, and the Bayshore spit area. Oregon banded plovers also winter in Humboldt County, Mendocino County, and Sonoma County, and have been reported from numerous other California counties, including Marin, San Mateo, Monterey, San Luis Obispo, Los Angeles, San Diego, the Channel Islands, and even in Baja California.

Recovery efforts for the Western Snowy Plover have been exceptionally successful in Oregon and will hopefully be continued in the future. This recovery effort is an important success story in wildlife conservation and is a real feather in the cap of the biologists, volunteers, and local, state, and federal agencies that have made it happen.

## Black Oystercatchers

ELISE ELLIOTT-SMITH
AND JOE LIEBEZEIT

Despite being a readily visible and vocal resident along rocky portions of the Oregon coast, Black Oystercatchers occur at only low densities throughout their range. The population is currently considered stable or declining slowly, yet there are fewer than 700 birds along the Oregon coast. About 311 oystercatchers were estimated in Oregon during 2012 (Lyons et al. 2012), about 629 in 2015, and about 506 in 2016 (Liebezeit et al. 2020). These small numbers and the species' vulnerability to increasing human disturbance and climate change render them a Bird of Conservation Concern for the US Fish and Wildlife Service, and they are also on the "watch list" of the North American Bird Conservation Initiative (NABCI) as a species of high conservation concern (NABCI 2016). In Oregon, they are recognized as a species of conservation concern or "strategy species" by the Oregon Department of Fish and Wildlife (2016b) in its *Oregon Conservation Strategy*.

Surprisingly, there has not been much research or management focused on this charismatic species in Oregon or elsewhere (see Tessler et al. 2014 for a summary). Incidental data have been collected on the species going back to the 1970s, but the first coastwide Oregon population estimate (~350 individuals) was obtained in 1988 as part of a seabird survey conducted from watercraft by the USFWS Oregon Islands National Wildlife Refuge. Most of our knowledge of the species in Oregon comes from volunteer surveys, first organized by ODFW and then taken over by the USFWS in 1997 (Naughton et al. 2007). The volunteer survey was originally limited to the central coast, but USGS joined USFWS in coordinating the volunteer survey effort in 2005, expanding the survey coastwide and later taking the lead in coordination. The most recent coastwide abundance surveys for oystercatchers were coordinated by Portland Audubon from 2015 to 2019 (Lyons et al. 2012, Liebezeit et al. 2020).

Although most surveys have been conducted during the breeding season, winter monitoring and research on Black Oystercatchers in the Pacific Northwest suggests that individuals remain at many of the same locations where they nest. A study of oystercatchers tracked with VHF and satellite transmitters for almost two years found that birds breeding on Vancouver Island, British Columbia, and Kodiak, Alaska, remained within 15 km of those sites during the winter (M. Johnson et al. 2010). But that same study documented birds from three Alaska breeding sites migrated as far south as Vancouver Island, British Columbia, in winter (M. Johnson et al. 2010). It remains unknown whether birds that nest in more southerly parts of Alaska and in British Columbia migrate to Oregon for the winter, and likewise whether all oystercatchers breeding in Oregon remain in the state year-round.

Another volunteer effort includes oystercatcher nest and brood monitoring, initiated by the USGS in 2006 and now coordinated by Portland Audubon since 2015. While estimating population size is critical for understanding a species' status, understanding the underlying factors that affect their breeding success or lack thereof is important as well. This is particularly true in a long-lived species like the Black Oystercatcher, where a single year of low productivity may not cause a population decline yet multiple years of poor productivity could lead to a major population decline, and recovery from such a decline would be challenging. In addition, identifying factors that limit reproductive success (e.g., high egg predation, low rates of chicks fledging, increasing nest disturbance rates from human activities, and their interactions) helps managers understand where to place efforts to recover a declining population or species.

Adult female Black Oystercatcher opening a California surf mussel. Black Oystercatchers are a resident shorebird species characteristic of Oregon's rocky shorelines whose populations are in decline. The species is an important bio-indicator of the health of intertidal habitats in Oregon. Photo by Dan Roby

More detailed studies of Black Oystercatcher breeding biology are needed to better understand the conservation status of the species and potential causes of nest failure. In Oregon, oystercatchers nest along the rocky coastline as well as on offshore islands (Naughton et al. 2007). Previous studies have suggested that birds nesting on offshore rocks and islands have higher nest survival owing to relatively low nest predation rates compared with sites on or connected to the mainland (Morse et al. 2006). But one mink swimming to an island can destroy all nests in that breeding season. On the mainland, nests are vulnerable to a variety of mammalian predators, such as raccoons, skunks, foxes, coyotes, mink, and river otters (Spiegel et al. 2012).

Further, oystercatcher nests are more vulnerable on mainland sites and nearshore rocks if human activity distracts adults from guarding their eggs and chicks, thereby exposing them to extreme temperatures, predators, or other hazards (Morse et al. 2006, Spiegel et al. 2012). Coastal visitation by tourists, particularly during the summer, has increased over the past several decades (6.8% annually since 2003; Dean Runyan Assoc., unpublished data, 2020) and continues to increase, suggesting an increasing threat to oystercatcher reproductive success that perhaps needs mitigation in order to enhance oystercatcher population recovery.

Once land managers can identify key sources of nest failure and whether Black Oystercatcher nest survival differs between onshore and offshore sites,

Adult male Black Oystercatcher with a newly hatched chick. Oystercatcher nesting success along the Oregon coast is largely limited by nest predation, disturbance from humans and their pets, coastal development, and oil spills. Black Oystercatchers are considered a species of high conservation concern in Oregon. Photo by Roy Lowe

they will be able to prioritize habitat conservation, site stewardship/outreach, and mitigation efforts accordingly (see Michel et al. 2021). Oregon's new management strategy for rocky coastline habitats (chapter 3 of the *Territorial Sea Plan*) is stressing the importance of regulatory and nonregulatory management to protect key rocky habitat/intertidal resources. Stewardship of sites by volunteers, friends groups, and NGOs with agency support and guidance is an important component of this plan.

Climate change is another factor that must be seriously and urgently taken into consideration in management of Black Oystercatchers. The 2010 *State of the Birds* report evaluated vulnerability to climate change for every avian species in North America (North American Bird Conservation Initiative 2010). Among all Oregon bird species, nine were given the highest rating for vulnerability, and all nine were coastal species. Two of these species were Black Oystercatchers and Pigeon Guillemots, which nest along the Oregon coast, while the other seven were species that migrate through or winter on the Oregon coast: Surfbird, Wandering Tattler, Yellow-billed Loon, Black Turnstone, Western Sandpiper, Rock Sandpiper, and Short-billed Dowitcher.

Black Oystercatchers are particularly important indicators of coastal and marine climate change for several reasons. First, climate predictions for Oregon and the Pacific Northwest are for warmer, wetter winters and hotter, drier summers (Intergovernmental Panel on Climate Change 2020), plus rising sea levels may inundate or fragment low-lying habitats such as the rocky intertidal areas where Black Oystercatchers and other shorebirds forage.

Topographic data for the Oregon coast are not available at fine enough resolution to estimate statewide changes in foraging habitat owing to sea level rise, and this is further complicated by tectonic processes. But Hollenbeck et al. (2014) demonstrated this assessment could be undertaken for a portion of the Oregon coast using LiDAR technology. Their models predicted a reduction in intertidal habitat due to climate change, further reducing foraging and nesting opportunities for Black Oystercatchers and other bird species.

Increasing ocean temperatures, changes in upwelling timing and strength, increased storm intensity, and ocean acidification are further by-products of our changing climate that may also impact intertidal prey taken by oystercatchers, and this, in addition to the adaptive capacity of oystercatchers, is an area of concern and for further study (Chan et al. 2017). Finally, the increasing number and severity of winter storms potentially threatens overwinter survival and juvenile recruitment of Black Oystercatchers in Oregon (Baron et al. 2015). Some baseline data on winter movement patterns and survival estimates for Oregon oystercatchers would assist in assessing oystercatcher vulnerability and identifying sites that provide refuge during storms.

For these reasons, Black Oystercatchers may be the "canary in the coal mine" or an indicator of intertidal community health as climate change dictates our future. For example, it may be difficult to remedy the chemical changes in seawater affecting intertidal prey species through local management actions. But protecting current oystercatcher nesting areas from disturbance by humans and other species, and taking a species-wide approach to assessing the best nesting areas and areas least affected by adverse ocean conditions due to climate change, will give us a strong foothold in saving an icon of the Oregon coast.

## Elise Elliott-Smith

Elise has worked for almost 20 years coordinating shorebird monitoring and research projects across Oregon and beyond. After working with passerines and raptors throughout the western United States, she completed her M.S. in 2003, studying shorebirds at Southern Illinois University, advised by Bruce Dugger in Oregon State University's Department of Fisheries, Wildlife, and Conservation Sciences. That same year, Elise joined Susan Haig's lab at the US Geological Survey Forest and Rangeland Ecosystem Science Center (FRESC). Her subsequent work in coastal Oregon has included research on Snowy Plovers and coordination of a coast-wide volunteer monitoring program for Black Oystercatchers. She has worked along the Columbia and in Great Basin wetlands, on research and monitoring projects with Long-billed Curlews and Snowy Plovers. Outside of Oregon, she has coordinated the International Piping Plover Breeding and Winter Census for many years. She currently works for USGS FRESC and the Northwest Climate Adaptation Science Center.

*Photo by Antonio Becerra*

# Oregon's Forested Ecosystems

BRENDA McCOMB

Forests in Oregon are highly variable in species composition, structure, and rates of vegetation growth, ranging from the tall, dense, old-growth forests of the Coast Range and west slope of the Cascades to the diverse forests of the Siskiyou Mountains and the ponderosa pine (*Pinus ponderosa*) and western juniper (*Juniperus occidentalis*) forests of eastern Oregon (Franklin and Dyrness 1988). Forests of eastern Oregon in the Blue Mountains, the Wallowas, and the far northeastern parts of the state are more similar to forests of the Rocky Mountains than those of western Oregon. Consequently, the avian communities that we find in the forests of Oregon are also highly variable, with some species found in all of these forest types, and some found highly associated with specific forest types (e.g., White-headed Woodpeckers in ponderosa pine forest, and Marbled Murrelets in coastal Douglas-fir [*Pseudotsuga menziesii*] forests). But it is not just forest types with which some bird species are associated, but also forest structure. It is tempting to think of the forests of Oregon prior to Euro-American colonization and genocide of Native Americans to be uniformly old growth with huge trees, multi-layered and stable through time in structure and composition. Certainly, there are parts of every landscape in the state that had these types of structures, but there were huge areas that did not, and the structural characteristics of forests changed over space and time because all forests in Oregon are and have

been the product of disturbances (Franklin et al. 2002).

Historically, wildfire was the dominant natural disturbance in Oregon forests, many of which would have been set by Native peoples, and many by lightning, especially in central and eastern Oregon. In coastal forests, fires tended to be large and severe, occurring every 100 years or less on the valley margin to more than 200 years near the coast (Impara 1997). Native Americans set fires along the Willamette Valley margin in the oak (*Quercus garryana*) woodlands, and sometimes these would expand into the Coast Range and the Cascades (Spies et al. 2007). Burning by Native peoples was also common in central and eastern Oregon, but fire return intervals were frequent in dry forests (every 5–20 years in ponderosa pine) and less frequent in higher-elevation moist forests. Wind and landslides have also been significant sources of disturbance in these forests, especially those close to the coast (May 2002, Harcombe et al. 2004). Forests of the past were quite different in structure and pattern from the forests of today, and while old-growth forests were represented on Oregon landscapes much more than they are today, the distribution of old forests across the state has waxed and waned with changes in frequency, intensity, and size of disturbances. It seems reasonable to conclude that avian communities also shifted in composition and pattern across Oregon landscapes in response to these changes from

Varied Thrush male, a bird species characteristic of Oregon's forested habitats. Varied Thrushes nest and are resident in higher-elevation moist coniferous forests of western and northeastern Oregon, and winter nearly throughout the state where there is damp, dense woodland. Populations of Varied Thrushes in Oregon have declined with the decline of late successional / old-growth forests. Photo by Dan Roby

disturbances; we certainly see that in contemporary landscapes affected by these same disturbances in addition to timber harvest (McGarigal and McComb 1995).

Timber harvest in Oregon forests was likely localized and modest among Native peoples. But during the mid-1800s, clearing of forests for homesteads and agriculture began in earnest, and cutting trees, especially large and old trees over vast areas, became possible with the introduction of large numbers of loggers equipped with crosscut saws, steam donkeys, and railroads. Development of the gasoline-powered chainsaw in the early 1900s further accelerated the process. The structure and pattern of forests across the state changed quickly, with the dominant disturbances being logging that replaced disturbance by fire over large areas (fires such as the 1933–45 Tillamook burns being notable exceptions, because some were due to logging, or careless or deliberate burning by the Euro-Americans who replaced the Natives). Following World War II and the Korean War, we began to see forests managed as crops, with the result being forests that were less diverse within stands and a more complex patchwork of plantation age classes across landscapes. Old forests were liquidated on most state, federal, and private lands because they were seen as decadent and not economically as valuable as rapidly growing young forests. Further, fire was seen as a threat to humans and their property, including managed forests, so fire was actively suppressed, resulting in forests that became much denser and with greater representation of shade-tolerant tree and shrub species than had been present previously (Table 3). There is now a recognition that fire should be reintroduced into Oregon forests and that forests of today may need to be reset over large areas to provide the structure and composition that can support species of animals found in these forests historically

(Hessburg et al. 2016). There is also a recognition that old forests and diverse young forests typical of natural disturbances are now underrepresented across nearly all forest types in the state (Franklin et al. 2007).

I provide this background on Oregon forests as context because it sets the stage for ornithological research that has occurred in these forests for the past 70 years. In general (and there are many exceptions), ornithological research in Oregon forests has been in reaction to the dominant forest policies in place at the time (Table 3) and evolved from primarily species-specific studies of game and high-interest species to species identified as being at risk from the forest practices of the period to characterizing bird communities at stand and landscape scales. We are now on the forefront of another shift in ornithological research in forested systems to consider the impacts of climate change on species and communities across migratory pathways (Betts et al. 2019). The following sections provide simple examples of the many studies that have been conducted over this chronology of forest bird research and are not meant to be exhaustive.

*1950–70:* Ornithological research during the 1950s, 1960s, and 1970s tended to be descriptive, providing species lists or reports on individual species, and often focusing on either game species or birds as predators of tree seeds used to regenerate forests following logging or fire (Gashwiler and Ward 1966). One notable exception was when Hagar (1960) began examining the effects of the logging practices of the time on bird communities in northern California, long before other researchers. Following the publication of *Silent Spring* by Rachel Carson in 1960, we began to see additional research on pollutants, culminating in research results dealing with spraying insecticides in forest systems (Richmond et al. 1979) and more recently on

**TABLE 3.** Areas of Oregon bird research linked to changes
in forest management policy, 1950–2020

| Forest Policy | Decade | Bird Research Focus |
|---|---|---|
| Fire control; manage forests on a rotation to create staggered-setting landscapes of various forest ages through rotation age; reestablish seedlings | 1950–60 | Descriptions of bird-forest associations and initiation of work on effects of logging; birds as effects on forest regeneration; game species |
| Multiple Use Sustained Yield Act 1960; Wilderness Act 1964; Wild and Scenic Rivers Act 1968 | 1960–70 | Effects of fire and forest chemicals on birds |
| National Environmental Policy Act 1970; Oregon Forest Practices Act 1972; Endangered Species Act (ESA) 1974; Forest and Rangeland Renewable Resource Planning Act 1974; Federal Land Management and Policy Act 1976; National Forest Management Act 1976; Clean Water Act 1977; Cooperative Forest Assistance Act 1978 | 1970–80 | Effects of forest management on birds, especially loss of old-growth trees |
| Farm Bill Forest Stewardship Assistance Act 1990 | 1980–90 | Bird–habitat relationships in managed forests and beginning of bird associations with complex forest landscapes |
| Northern Spotted Owl listed under ESA in 1990; Marbled Murrelet listed in 1992; Forest Ecosystem Management Assessment Team produced in 1993; NWFP in 1994 | 1990–2000 | Bird–habitat relationships at landscape scales; managing stands to provide habitat for birds; document effects of old-growth loss; habitat requirements of ESA-listed species |
| Northwest Forest Plan Survey and Manage on federal lands | 2000–2010 | Landscape- and biome-scale assessments of forest policies and practices on birds and other species are assessed |
| Intergovernmental Panel on Climate Change reports receive attention | 2010–22 | Climate change effects on habitat and bird phenology; synergistic effects of multiple stressors and disturbances |

effects of herbicides on bird communities (Morrison and Meslow 1984).

**1970–90:** The 1970s saw major changes in environmental policies that ignited a series of research directions, especially around ecological relationships between birds and forest structure and composition. While research continued on the effects of birds on forest regeneration (Pank 1976), the National Forest Management Act and the Endangered Species Act brought more attention to species such as Northern Spotted Owls in western Oregon

(Forsman et al. 1977) and dead-wood-associated species in forests across the state (e.g., Bull and Meslow 1977). Thomas's (1979) compendium *Wildlife Habitats in Managed Forests of the Blue Mountains of Oregon and Washington* made it clear that under these policies we needed to consider how to manage forests while providing habitat for vertebrates. It is at this time that we see continued research on the ecological interactions of birds with forests in various parts of the state (Anderson 1970a, Wiens and Nussbaum

1975) and improvements in quantifying bird abundance in a variety of habitats, especially in forests where bird detection can be imperfect (Reynolds et al. 1980). The variable circular plot technique proposed by Reynolds et al. (1980) became a standard for sampling birds in Oregon forests for several decades.

Brown's (1985) compendium for western Oregon and Washington mirrored Thomas's (1979) eastern Oregon work and brought focus to a suite of bird species that could be affected by forest management in western Oregon. It was at this time that I moved from the University of Kentucky to Oregon State University and was tasked with understanding the effects of managing forests on wildlife. Larry Harris had just published his book, *Fragmented Forest* (Harris 1984), and the Coastal Oregon Productivity Enhancement Program (a cooperative research effort between OSU and the US Forest Service Pacific Northwest Research Station) began funding research focused on birds across complex landscapes (Hansen et al. 1995, McGarigal and McComb 1995). Given the attention being paid to federal and state policies on salmonids through streamside protection rules (e.g., the Oregon Forest Practices Act in 1972), Anthony et al. (1996) and others also began examining what impact riparian management areas might have in protecting habitat for various bird and mammal species. Research into effects of forest management was also occurring in forests of northeastern (Mannan and Meslow 1984) and western Oregon (Morrison and Meslow 1983). Studies by Bull and Meslow (1977), Goggans (1985), Bull et al. (1987), and others focused on species believed to be adversely affected by the forest practices at that time. Attention was also being paid to the role that birds have in controlling (or not) insects in forests in eastern Oregon (Torgersen et al. 1984), where western spruce budworm (*Choristoneura freemani*) and Douglas-fir tussock moth (*Orgyia*

*pseudotsugata*) irruptions had led to significant tree mortality and subsequent spraying of insecticides.

***1990–2010:*** When the Northern Spotted Owl was listed as Threatened under the Endangered Species Act in 1990 and a compendium of studies was published detailing relationships between unmanaged Douglas-fir forests and many organisms, including birds (Ruggiero et al. 1991), greater concern was raised regarding conservation of old forests and the species that inhabited them. The Northwest Forest Plan was finalized in 1994 and was designed to provide protection for all old-forest-associated species on federal lands in northern California, Oregon, and Washington. This was a period in which forest bird research expanded tremendously, focused on individual species, such as Marbled Murrelets (Nelson and Hamer 1995b), as well as conservation of ecological communities and biodiversity. Work was beginning that attempted to assess whether alternative silvicultural practices and policies might allow conservation of late-successional bird species while also allowing timber extraction (Schreiber and DeCalesta 1992; Hagar et al. 1996, 2004; Chambers et al. 1999; Weikel and Hayes 1999; Sallabanks et al. 2000; McComb et al. 2007). Compendia such as those that were developed during the previous period were expanded to synthesize research that could be put into application by managers. DecAID (the decayed wood advisor for managing snags, partially dead trees, and down wood in forests of Oregon and Washington) is an example of such a compendium that synthesized information on the role of dead wood in forests and provided guidance on managing levels of dead wood to meet the needs of a wide range of bird species as well as mammals and herpetofauna (Mellen-McLean et al. 2017). Although work by Kim Mellen-McLean,

Male Black-backed Woodpecker on the charred trunk of a beetle-infested snag. This species is a sensitive Oregon Conservation Strategy species and generally rare where it occurs in recently burned, higher-elevation forest throughout the state. Fire suppression and salvage logging negatively affect habitat availability for this species. Photo by Scott Carpenter

Bruce Marcot, Janet Ohmann, and others began on DecAID during this period, it continues to be updated with new information. As climate began to change and fires became larger and more intense (Hessburg et al. 2016), the interface between forest management, fire, and subsequent salvage logging of dead wood was getting increased attention (Cahall and Hayes 2009, Fontaine et al. 2009). While much of the forest bird research of this time was applied in nature, we continued to see research focused on more basic forest bird relationships in conifer forests in eastern and western Oregon (Sallabanks et al. 2002, Cushman and McGarigal 2004, Betts et al. 2010), and in oak woodlands (Hagar and Stern 2001).

*2010–20 and beyond:* Research is continuing to examine how to balance production from forests for economic gain while protecting biodiversity. A large project is underway to tag Marbled Murrelets with radio transmitters at sea to find their nests on land and is already beginning to provide insights into the behavior and ecology of this species (Valente et al. 2021). Research also continues on Spotted Owl declines, especially with regard to competition from Barred Owls, a recent addition to Oregon's avifauna (Wiens et al. 2014). But on the horizon of Oregon forest bird research is the need to understand the effects of climate change on bird abundance, behavior, and habitat availability (Betts et al. 2017), and for migratory birds, it is increasingly important to understand climate effects across breeding areas, wintering areas (often in the Neotropics), and migratory pathways in between. Associated with climate impacts are emerging efforts to view forests as a means to sequester and store carbon, which could have significant implications for forest management and the species of birds associated with old forests (Kline et al. 2016). We are beginning to see research in Oregon focused on not only the effects of climate change, but also the synergistic effects of climate change, invasive species, forest management, fire, and more (Betts et al. 2019, Northrup et al. 2019, McComb and Cushman 2020). These stressors and disturbances interact with shifting political agendas to create uncertainty in habitat availability and bird population responses. I believe that it is these and other interacting socially and politically driven biophysical synergies that will need to be more fully understood if Oregon is to do its part in reversing the 3-billion-bird decline in the United States (Rosenberg et al. 2019).

## Brenda McComb

Brenda McComb is a retired Oregon State University ornithologist who has served many roles, including vice provost for academic affairs, a member of the OSU board of trustees, graduate dean, department head, and professor of forestry in the Department of Forest Ecosystems and Society. McComb taught and conducted research in the Department of Forest Sciences and the Department of Fisheries and Wildlife from 1987 to 1996. She then became head of the Department of Natural Resources Conservation at the University of Massachusetts, where she continued to work with Oregon colleagues and transitioned from Bill to Brenda. McComb returned to Oregon in 2009 to lead the Department of Forest Ecosystems and Society at OSU. McComb's research has focused on passerines in fragmented forests of the Coast Range and Cascades, as well as issues related to conservation, landscape ecology, and silviculture. She authored the book *Wildlife Habitat Management: Concepts and Applications in Forestry*. McComb retired from OSU in 2016, then worked as associate vice provost and interim dean of students at Stanford. She continues to teach part-time at OSU and remains deeply involved in public service as a member of the Oregon Board of Forestry and the Oregon Watershed Enhancement Board.

*Photo by Susan Simoni Burk*

# Cascade and Coast Range Forests

*The Fate of Forest Bird Populations under the Northwest Forest Plan, 1993–2019*

MATTHEW BETTS

The early 1990s marked the culmination of a decades-long war between environmental groups and the logging industry that was focused on conservation of old-growth forests, particularly Spotted Owl habitat. In 1993 the Northwest Forest Plan was adopted at the behest of President Clinton following court injunctions that halted logging on federal forests. This new plan effectively shifted the primary management objective on 10 million ha (25 million acres) of federal land to sustaining and restoring older forests and their species (Thomas et al. 2006). Thus was born one of the most ambitious forest conservation and management policies in the world. On the federal land covered by the Northwest Forest Plan, 41% was set aside for reserves in addition to the 36% already protected in wilderness areas and national parks. Timber harvest was permitted to occur on ~20% of the remaining unreserved area, with a focus on learning by doing (or "adaptive management"). Things did not go as planned by those who drafted the legislation, however; most clear-cutting on federal land ceased as a result of further court injunctions, and timber harvest shifted predominantly to large private holdings.

Monitoring was intended to be a cornerstone of the adaptive management framework of the Northwest Forest Plan. Indeed, the Northwest Forest Plan was conceived as a 100-year plan that would adapt to changing ecological and social conditions and knowledge. Remote sensing technology and existing Breeding Bird Survey (BBS) data enabled tracking of the consequences of the plan for the forest itself and for associated forest bird populations (Phalan et al. 2019).

How have bird populations fared under the Northwest Forest Plan? With so many federal forest acres now under some form of protection, one would predict that declines in mature-forest-associated species would slow or even halt. Prior to the plan—in the 1960s through 1980s—there had been an extensive pulse of old-growth logging, but as noted above, this largely ceased on federal lands after the Northwest Forest Plan. Similarly, one might also predict that populations of bird species associated with young forests ("early seral," e.g., Orange-crowned Warblers, Wilson's Warblers, White-crowned Sparrows) would have declined over the past three decades owing to the reduction in new habitat generated by clear-cutting. Whether or not such declines have happened on private land has been an open question; recent research has shown that although industrial tree plantations do serve as habitat for early seral bird species (Rivers et al. 2019), intensive management practices can be detrimental when broadleaf forest is reduced by herbicide applications (Betts et al. 2010, 2013) and when the Douglas-fir canopy closes as stands age (Harris and Betts 2021).

Interestingly, even though the management changes brought by the Northwest Forest Plan were successful at slowing the loss of mature and old-growth forest (Phalan et al. 2019), these forest types continue to decline across the Pacific Northwest. These declines correlate with declining populations of bird species associated with this age class (e.g., Hermit Warbler, Varied Thrush, Chestnut-backed Chickadee) (Northrup et al. 2019). One possibility is that these population declines are partly the result of an "extinction debt" (Tilman et al. 1994) from previous timber harvesting in earlier decades, but a

Western Tanager male on red-flowering currant. Western Tanagers are a beautiful Neotropical migrant and common summer resident in coniferous and mixed coniferous forests throughout Oregon. Photo by Dan Roby

surprising amount of old forest has been lost to fire on federal lands and harvest on private lands since the adoption of the Northwest Forest Plan.

Early seral species have also continued to decline on federal and private lands (Northrup et al. 2019, Phalan et al. 2019). Undoubtedly, the lack of clear-cutting on federal lands is responsible for some early seral habitat loss. Management intensification on private industrial forests (i.e., herbicide use and tree plantations) is likely the cause of early seral bird species declines on private land. Nevertheless, it is still a matter of debate whether declines

of this early seral species group should be of conservation concern. Historically, only a relatively small proportion of Oregon's wet, west-side forests (in the Coast Range and Cascades) would have been severely disturbed by fire (Nonaka and Spies 2005). It is possible that population declines of early seral species observed in the Breeding Bird Survey data (from 1966 to 2019) reflect a return to more historical population abundances prior to the pulse of habitat creation that occurred during the timber harvesting boom of the 1960s through 1990s (Kroll et al. 2020). Additionally, an argument could be made that early seral

Hermit Warbler male, with a bill-load of arthropods to feed nestlings in coastal coniferous forest. Hermit Warblers are widespread and common in late successional / old-growth Douglas-fir and true fir forests throughout western Oregon but are frequently overlooked because they usually remain in the forest canopy. Like Varied Thrushes, this species has declined with the decline of older forests in the state. Photo by David Leonard

forest is likely to become more common again as the frequency and extent of fires increases with climate change (Halofsky et al. 2020). In contrast, old-growth forest is rare in comparison to its historical distribution and becoming rarer (Spies et al. 2007). Further, the large trees, high biomass, and dead and downed wood structures of this forest type are challenging to create using restoration silviculture in the current era, although various experiments have attempted to speed up the onset of old-growth structures (Cahall et al. 2013, Yegorova et al. 2013).

What is the prognosis for forest birds in the coming decades? We do know that western forest birds are one of the species groups exhibiting the greatest declines in North America (Rosenberg et al. 2019). Although the causes of these declines remain somewhat cryptic, recent work has shown that habitat loss and climate change on the breeding grounds are major driving forces

(Northrup et al. 2019). Conservation emphasis should likely be placed on mature-forest associates, particularly species that are already threatened, such as the Northern Spotted Owl (Dugger et al. 2016) and the Marbled Murrelet (Betts et al. 2020). Further, forests with complex structures appear to buffer species against the potential negative effects of global warming (Frey et al. 2016). But the same forests that serve as habitat for these species are also the most in demand by humans for wood products. As the global human population continues to increase, and along with it demand for wood, we can expect mature and old-growth forests of the Pacific Northwest to come under increasing pressure. Indeed, one federal agency, the Bureau of Land Management, is already increasing harvest intensity (via variable retention harvests) to boost wood supply and create early seral habitat (Bureau of Land Management 2016). A group of rural counties in Oregon recently sued the State of Oregon Department of Forestry, arguing that protection of old forests for Spotted Owls and other wildlife violated historical agreements to maximize timber production on state lands (Sickinger 2019). The jury ruled in favor of the plaintiffs, and the issue is still unresolved.

It may well be that concentrating timber production onto existing tree plantations and establishing permanent reserves is the solution to conserving old forests and associated bird populations in the face of ever-expanding timber demand. Some bird species require disturbance for habitat creation, however, so such a dual strategy may fall short. Quite likely, a balance of forest practices ranging from intensive management (focused on wood production), ecological forestry (to maintain disturbance-associated species), and reserves (to conserve remaining old growth and species) may be the best avenue to sustaining forest bird populations and meeting timber demand in the Pacific Northwest. But rigorous tests of this hypothesis are required (Seymour and Hunter 1999, Betts et al. 2021). Of course, consumption habits in the United States are global drivers of biodiversity impacts (Moran and Kanemoto 2017), including to Neotropical birds that migrate well beyond Oregon for most of the year. Three cornerstones of bird conservation will therefore be not only to conserve breeding habitat, but also to reduce our consumption habits, and to engage in international policy and collaborations that encourage conservation of wintering habitat in the tropics.

## Matthew Betts

Matthew Betts is an avian ecologist who joined the faculty of Oregon State University's Department of Forest Ecosystems and Society in 2007. He and his students conduct research in the areas of forest, wildlife, and landscape ecology as well as integrated social and ecological systems. Much of the Betts Forest Landscape Ecology Lab's work is focused on management and conservation, but they have also focused on pollination mutualisms in tropical hummingbird systems and the mechanisms driving habitat selection in birds. Betts carries out research in the Cascades at the National Science Foundation Long Term Ecological Research Network's H. J. Andrews Experimental Forest, where recent NSF funding will ensure continued long-term ecological research on climate change and biodiversity in old-growth forests.

Professor Betts also co-leads the Oregon Marbled Murrelet Project, is director of the Forest Biodiversity Research Network, and is the lead investigator for the Oregon Intensive Forest Management project. Betts has been honored with the Institute for Working Forest Landscapes Research Professorship and is a fellow of the American Ornithological Society.

*Photo by Greg Davis*

## The Northern Spotted Owl: The Conservation History of an Iconic Oregon Species

KATIE DUGGER

The Northern Spotted Owl (*Strix occidentalis caurina*) is a medium-sized resident of coniferous forests in the Pacific Northwest. It is nonmigratory and highly territorial (Forsman et al. 1984), with a diet focused on arboreal and scansorial small mammals (Forsman et al. 1984, 2004), and a preference for mature and old-growth forest for successful survival and reproduction (Franklin et al. 2000; Olson et al. 2004; Dugger et al. 2005, 2016). Spotted Owls have a long lifespan (~20–25 years) and low rates of reproduction, as they do not breed every year (Gutiérrez et al. 2020), and when they do, they typically fledge only one or two young, rarely three (Forsman et al. 1984). This species is well known for its lack of concern for the presence of humans and can be approached closely in the wild.

Early research identified a close association between Northern Spotted Owls and old-growth and late-successional forest during all life stages (Forsman et al. 1984, Thomas et al. 1990), and since that time the positive association between increased amounts of late-successional forest and Spotted Owl survival, reproduction, and occupancy has been well documented (e.g., Franklin et al. 2000; Olson et al. 2004; Dugger et al. 2005, 2016). Because of the subspecies' dependence on old forests, the Northern Spotted Owl has been at the center of land management controversy since even before the Endangered Species Act was first implemented in 1973 (Meslow 1993; Table 4). Concerns about Northern Spotted Owl population declines because of the harvest of old-growth forest during the 1970s and 1980s eventually resulted in the listing of the species as Threatened under the Endangered

Species Act in 1990 (USFWS 1990). Intensive and sometimes violent conflict between the timber industry, local communities, and environmental groups over the management of forests on federal lands within the range of the owl put federal land management agencies like the USFS and the BLM in the unenviable position of playing arbiter between the different sides. It was a particularly difficult time for many employees within the federal agencies, where previous performance standards had been based primarily on lumber production as opposed to worrying about the needs of forest animals like the Spotted Owl. This debate raged for nearly two decades until a series of lawsuits eventually culminated in a court injunction in 1992 that halted logging of old forests on federal lands within the range of the Northern Spotted Owl (Marcot and Thomas 1997). To address this impasse, the Clinton administration directed a technical team made up of specialists from a wide variety of scientific fields to develop recommendations for managing forest ecosystems on federal lands within the range of the Northern Spotted Owl (Forest Ecosystem Management Assessment Team 1993). This effort strove to address forest management from a broad ecological perspective and produced 10 potential management scenarios that ultimately informed the development of the Northwest Forest Plan (NWFP) (US Department of Agriculture and US Department of the Interior 1994). This plan was developed to balance the need for forest products while also protecting ecosystem functions and sensitive old-growth-forest-dependent species, including the Northern Spotted Owl (USDA and USD 1994). The NWFP was adopted in 1994 and has remained the management paradigm on US Forest Service lands within the range of the Northern Spotted Owl (USDA and USDI 1994). The BLM was originally a party to this agreement but

The Northern Spotted Owl is listed as Threatened under the US Endangered Species Act and the Oregon Endangered Species Act, and it has been proposed for up-listing to Endangered owing to persistent population declines throughout its range in Oregon. Photo by Alan Dyck

began efforts to abandon the NWFP as early as 2002 and eventually adopted a group of modified plans in 2016 with less restrictive limits on logging in mature and old forests (Bureau of Land Management 2016).

With unprecedented foresight, provisions were included in the NWFP for long-term monitoring of Northern Spotted Owls, Marbled Murrelets, late-successional old-growth forest, and aquatic and riparian areas, as well as economic impacts to communities, including tribal interests (USDA and USDI 1994). The long-term monitoring program for Northern Spotted Owls incorporated eight study areas on primarily federal lands where mark-recapture studies of owl demography were ongoing (Lint et al. 1999). The result is a longitudinal database on multiple study areas across the owl's range with information about known age and known breeding history of individuals spanning more than 30 years on some study areas (e.g., Franklin et al. 2021). Most importantly, data from these long-term demography studies as well as three to six other study areas on private or tribal lands are analyzed in a meta-analysis framework about every five years to estimate trends and status of Northern Spotted Owls across their range (e.g., Burnham et al. 1996, Franklin et al. 1999, Anthony et al. 2006, Forsman et al. 2011, Dugger et al. 2016, Franklin et al. 2021). This collaboration and coordination among principal investigators, crew leaders, and analysts is unique, particularly in its long-term nature, and has resulted in a rigorous scientific process within a uniting framework that has addressed key questions regarding the long-term status of a threatened species.

For example, results from the earliest meta-analyses identified baseline rates of population decline and estimates of annual survival and reproductive success the first five years following implementation of the NWFP (Anderson and Burnham 1992, Burnham et al. 1996, Franklin et al. 1999). Ten to fifteen years later, Spotted Owl populations were still declining, but the rate of decline was slowing (Anthony et al. 2006, Forsman et al. 2011), as predicted when the plan was initially implemented (Lint et al. 1999). Loss of suitable nesting and roosting habitat for Spotted Owls due to harvest had greatly declined on federal lands by this time (Davis et al. 2011, 2016), but it was expected that the recovery of Northern Spotted Owls would lag conservation of suitable habitat as population dynamics of the species rebalanced after the extensive habitat loss in the 1970s and 1980s (Lint et al. 1999). During the first 15 years after implementation of the NWFP, however, a new problem was identified as the Barred Owl, a congeneric generalist species, continued to expand its native range from the eastern United States into the Pacific Northwest (Livezey 2009a, b; see also the Barred Owl History in Oregon section on p. 145). As Barred Owls expanded their range southward from British Columbia and Washington, and densities increased behind the invasion front, evidence of the negative competitive and interference effects of Barred Owls on Northern Spotted Owl demographics began to emerge (see review in Long and Wolfe 2019). The negative effects of competition with Barred Owls on survival, reproduction, recruitment, and occupancy dynamics of Northern Spotted Owls have now been documented range-wide (e.g., Dugger et al. 2016, Franklin et al. 2021). The most current meta-analysis documented Northern Spotted Owl population declines of more than 75% in Washington, more than 60% in Oregon, and between 30% and 60% in California since 1995 (Franklin et al. 2021). By 2018 the probability of historical territory occupancy by Spotted Owls

**TABLE 4. History of Northern Spotted Owl conservation in Oregon**

| Year | Event | Reference |
| --- | --- | --- |
| ca. 1970 to 1980s | Earliest research conducted on basic ecology and habitat associations of Northern Spotted Owls in Oregon. | Forsman et al. (1984), Thomas et al. (1990) |
| 1990 | Northern Spotted Owl is listed as Threatened under the Endangered Species Act, with loss of old-growth forest habitat considered the biggest threat. | USFWS (1990) |
| 1993 | Forest Ecosystem Management Assessment Team report produced by a large technical team from a wide variety of disciplines and focused on developing recommendations for managing federal forest lands using a wholistic ecosystem approach. | Forest Ecosystem Management Assessment Team (1993) |
| 1994 | The Northwest Forest Plan (NWFP) is adopted on federal forest lands across the range of the Northern Spotted Owl to conserve late successional and old-growth forest and the species that rely on them. | US Forest Service and Bureau of Land Management (1994a, b) |
| | First demographic meta-analysis of Northern Spotted Owls estimating survival, reproductive success, and rates of population change on multiple study areas. | Anderson and Burnham (1994) |
| 1996 | Second demographic meta-analysis of Northern Spotted Owls estimating survival, reproductive success, and rates of population change on multiple study areas. | Burnham et al. (1996) |
| 1999 | Third demographic meta-analysis of Northern Spotted Owls estimating survival, reproductive success, and rates of population change on multiple study areas. | Franklin et al. (1999) |
| 2004 | Threatened status of Northern Spotted Owl upheld by US Fish and Wildlife Service. | USFWS (2004) |
| 2006 | Fourth demographic meta-analysis, conducted 10 years after implementation of the NWFP, estimating trends in survival, reproductive success, and rates of population change on 14 study areas; declines in survival on three study areas in Washington linked to Barred Owl presence. | Anthony et al. (2006) |
| 2011 | Fifth demographic meta-analysis, conducted 15 years after implementation of the NWFP, estimating trends in survival, reproductive success, and rates of population change in relation to Barred Owl presence, amount of suitable habitat, and weather and climate covariates on 11 study areas. | Forsman et al. (2011) |
| | Revised recovery plan for Northern Spotted Owls adopted. | USFWS (2011) |
| 2012 | More than 3.8 million ha of habitat designated as critical for Northern Spotted Owls in California, Oregon, and Washington. | USFWS (2012b) |
| 2016 | Sixth demographic meta-analysis, conducted 20 years after implementation of the NWFP, estimating trends in survival, reproductive success, recruitment, rates of population change, and occupancy dynamics in relation to Barred Owl presence, amount of suitable habitat, and weather and climate covariates on 11 study areas. | Dugger et al. (2016) |
| | First published evidence that large-scale Barred Owl removals could be implemented and provide positive impacts on Northern Spotted Owl demographics in northern California. | Diller et al. (2016) |
| 2021 | Seventh demographic meta-analysis conducted 26 years after implementation of the NWFP estimating trends in survival, reproductive success, recruitment, rates of population change, and occupancy dynamics in relation to Barred Owl presence, landscape characteristics, and weather and climate effects. Negative effects of Barred Owl presence identified for all vital rates examined. | Franklin et al. (2021) |
| | Meta-analysis of data from a large-scale Barred Owl removal experiment conducted on four study areas in California, Oregon, and Washington describing the positive effects of Barred Owl removal on Northern Spotted Owl demographics. | Wiens et al. (2021) |

A Northern Spotted Owl stoops on its prey. Long-term research as part of the Northwest Forest Plan has resulted in the ecology of this species being as well understood as that of any North American bird species. Photo by Jeffrey M. Wells

ranged from a low of 6.7% on a study area in Washington to a high of 55% in California, whereas occupancy rates of Barred Owls on Northern Spotted Owl territories ranged from 49% to 97%, depending on the study area (Franklin et al. 2021). Continued loss of suitable nesting and roosting habitat through logging, wildfire, or tree disease was also associated with population declines of Spotted Owls in some parts of the range, but the presence of Barred Owls was the dominant factor associated with Northern Spotted Owl population declines in the most recent analysis (Franklin et al. 2021). Recent experimental research has shown that the removal of Barred Owls from the landscape can have positive effects on Northern Spotted Owl population dynamics (Diller et al. 2016, Wiens et al. 2021, Hofstader et al. 2022). Given the strong negative impact

Barred Owls are having on Northern Spotted Owl populations, the long-term persistence of Spotted Owls will likely require active management intervention to ameliorate the negative effects of Barred Owls (Franklin et al. 2021).

In addition to understanding Northern Spotted Owl population status and trends, the long-term monitoring program has generated an amazing wealth of information on the behavior and ecology of the species (e.g., Forsman et al. 2002, 2004; Sovern et al. 2015, 2019; Jenkins et al. 2019, 2021), genetics (Haig et al. 2004a, b; M. P. Miller et al. 2018a), dispersal (Hollenbeck et al. 2018), interactions with Barred Owls (Dugger et al. 2011; Yackulic et al. 2014, 2019), and new survey approaches (Duchac et al. 2020, Lesmeister et al. 2021), resulting in numerous theses, dissertations, and more than 100 peer-reviewed publications to date. This long-term research has played a key role in informing a variety of management and conservation actions, including (1) the Northern Spotted Owl status review in 2004, in which Threatened status under the ESA was upheld (USFWS 2004); (2) the development and adoption of the revised recovery plan (USFWS 2011); (3) the establishment of a critical habitat ruling (USFWS 2012b); and (4) the most recent status review finding in 2020 in which up-listing to Endangered status was considered "warranted but precluded" (USFWS 2020d). Current Northern Spotted Owl population trends are cause for great concern. But the legacy of long-term research has provided key information that continues to aid land managers and conservation agencies in designing management actions to ensure the persistence of this iconic species in the Pacific Northwest.

## Eric Forsman

Dr. Eric Forsman's career as a biologist began when he stumbled onto a pair of Spotted Owls while working as a fire guard for the US Forest Service in summer 1969. In spring 1970, he documented the first Spotted Owl nest in Oregon and began to think about graduate study on the species. His academic career was interrupted when he was drafted into the Vietnam War as an army medic in 1970.

Leaving the army in 1972, Forsman began masters research on Spotted Owls under Oregon State University professor Howard Wight. This pioneering work led to the realization that their old-growth forest habitats were quickly being diminished because of logging. Forsman completed his M.S. in 1975 and then, with Charles Meslow, launched a Ph.D. study on home range characteristics of Spotted Owls on the H. J. Andrews Experimental Forest in the Oregon Cascades.

Forsman went on to work as a research wildlife biologist at the US Forest Service Pacific Northwest Forest Research Station in 1987 and became a courtesy faculty member at OSU. He spent the next 30 years working on issues related to Spotted Owls and other species as well as mentoring graduate students. During this time, he remained a key contributor to the Northwest Forest Plan and Northern Spotted Owl recovery. Among many awards, Eric was presented with the American Ornithologists' Union Ralph W. Schreiber Conservation Award for his decades of work on Northern Spotted Owls.

*Photo by Susan Haig*

## Robert Anthony

Robert Anthony joined Oregon State University's Department of Fisheries and Wildlife in 1977 as the assistant unit leader and then leader of the US Geological Survey Oregon Cooperative Fish and Wildlife Research Unit. Professor Anthony was especially known for his research on the demography of Bald Eagles and Northern Spotted Owls. He served on the US Fish and Wildlife Service Northern Spotted Owl and Bald Eagle Recovery Teams and was a key contributor to development of the Northwest Forest Plan. He was an expert witness for federal hearings dealing with contaminants, endangered species, and other topics. Bob mentored 48 of his own graduate students and many others.

Every four years, Anthony sequestered the Northern Spotted Owl field crew leaders for a week to carry out intensive demographic analyses that set recovery goals for the species. In addition, Anthony was president of The Wildlife Society, president of the Oregon Chapter of The Wildlife Society, and served on the Northern Spotted Owl Long-Term Demography Research Team. Bob's co-authored book *Population Demography of Northern Spotted Owls* received the Outstanding Publication Award from The Wildlife Society. He was also awarded the Ralph W. Schreiber Conservation Award from the American Ornithologists' Union. He was a fellow of The Wildlife Society and a fellow of the American Ornithologists' Union. OSU established the Robert Anthony Graduate Scholarship in Population Ecology to honor Anthony after he passed away suddenly in 2013.

*Photo by Dan Roby*

## Katie Dugger

Katie Dugger is a quantitative population ecologist who models vital rates of wild birds, including survival, reproductive success, and occupancy rates as key elements in the conservation and management of species and their ecosystems. For more than 20 years, she has carried out demographic analyses on Northern Spotted Owls, the world's largest avian demographic data set, and more recently on the demographics and habitat use of Greater Sage-Grouse in southeastern Oregon. In 2001, she came to Oregon State University as a postdoctoral researcher in Robert Anthony's lab with the Oregon Cooperative Fish and Wildlife Research Unit. In 2002, Katie became an assistant professor in the Department of Fisheries and Wildlife at OSU and has since become a professor of wildlife ecology. In 2011, she became the first woman assistant unit leader in the Oregon Cooperative Fish and Wildlife Research Unit. Professor Dugger is a fellow of the American Ornithological Society and has served on the American Ornithological Society Council. In 2014, she was bestowed with an honorary membership in the Cooper Ornithological Society for her service as a COS board member, treasurer, and chair of the Mewaldt-King Award Committee.

*Photo by Katie Dugger*

## Barred Owl History in Oregon

DAVID WIENS

The Barred Owl is a large and highly adaptable North American owl that currently inhabits forests and woodlands throughout Oregon. A native species of eastern North America, the Barred Owl emerged as a widespread and prominent member of Oregon's avifaunal community within a short time frame of just 40–60 years (Table 5). Prior to arriving in Oregon, Barred Owls had expanded their geographic range westward to include forests of Montana (1909–22), British Columbia (1943), southeastern Alaska (1967), and Washington (1965; Livezey 2009a). The underlying causes of the Barred Owl's continental-scale range expansion remain unclear. One hypothesis suggests that historical changes in climatic conditions in the boreal forests of Canada facilitated the expansion (Monahan and Hijmans 2007), while others propose that ecological changes associated with Euro-American settlement of the Great Plains enabled the Barred Owl's movement west (Livezey 2009b). Regardless, Barred Owls have successfully expanded their geographic range across much of the North American continent in a relatively short period, demonstrating this owl's remarkable capabilities of population growth in an era when many other forest-dependent birds of prey are facing declines.

The first reported observation of a Barred Owl in Oregon was in June 1922, when Albert Prill, a noted physician and avid birder, recorded a pair of Barred Owls at the north end of Crump Lake in the Warner Valley of south-central Oregon (Prill 1924; Table 5). The record presents somewhat of a mystery, however, as it lacks observation details of a sighting that places Barred Owls in Oregon 40–60 years before any other known observation west of Montana (Livezey 2009a, Long and Wolfe 2019). More recently, Prill's account has been disputed (see Contreras et al. 2022).

Prill's outlier record aside, the first reliable observation of a Barred Owl in Oregon occurred in the summer of 1972, when Eric Forsman heard the classic eight-note call of a Barred Owl in the H. J. Andrews Experimental Forest of west-central Oregon. Eric originally identified the call as that of a Barred Owl, but he quickly dismissed it, as he did not hear the call again and thought that he may have been mistaken. Two years later in 1974, however, another observer (Butch Taylor) encountered two pairs of Barred Owls in northeastern Oregon, including one pair in the Blue Mountains and another pair just 56 km to the southwest near Pendleton (Taylor and Forsman 1976). It wasn't until after the sightings by Taylor that Forsman realized that the eight-note call he heard in 1972 had almost certainly been that of a Barred Owl—a precursor to one of the most complex conservation issues in Oregon's ornithological history.

Barred Owl observations in Oregon began to accumulate rapidly over the next two decades, but the novel species remained uncommon and discontinuously distributed. In 1979, there were sightings of Barred Owls in the Oregon Cascades Range near Mt. Hood (Harrington-Tweit et al. 1979), and in 1981, there were observations in Lane County along the west slope of the Cascades as well as in the Mountain Lakes Wilderness of the southern Cascades (Nehls 1998). The distribution of observations during this period of early settlement by Barred Owls suggests that the birds colonized Oregon via the Cascade Range and more mesic forests of the Coast Range, while also moving, perhaps to a lesser extent, southwest from Idaho and northeastern Oregon into forests of the Cascade Range before pressing farther south into the Sierra Nevada of California (Kelly et al. 2003, Long and Wolfe 2019).

Barred Owls are a recent addition to the avifauna of Oregon, having dispersed across the Great Plains from eastern North America and reached Oregon by the 1970s. Barred Owls are now widespread in Oregon and common in many forested habitats. The Barred Owl is closely related to the Northern Spotted Owl, which it has largely displaced in much of Oregon because of its larger size, broader food habits, and more aggressive behavior. Photo by Scott Carpenter

Coincident with increasing, yet still low, numbers of Barred Owls in Oregon was the 1990 listing of the Northern Spotted Owl as a Threatened subspecies under the Endangered Species Act (USFWS 1990) and subsequent implementation of the Northwest Forest Plan (US Department of Agriculture and US Department of the Interior 1994). At the time, Barred Owls—a congeneric species to Spotted Owls and ecologically similar in terms of habitat use and diets—were identified as a potential, yet unknown competitive threat to Spotted Owls. Barred Owl populations continued to expand, however, and between 1989 and 1998, an estimated average of 60 new territorial pairs of Barred Owls were identified each year in Oregon alone (Kelly et al. 2003). As Barred Owls thrived in West Coast forests, they displaced Spotted Owls from their historical breeding territories and subsequently excluded them from protected older forests. Long-term demographic studies of Spotted Owls, used to monitor the effectiveness of the Northwest Forest Plan in arresting population

## TABLE 5. History of the Barred Owl in Oregon

| Year | Event | Reference |
|---|---|---|
| ca. 1900– 1970 | Barred Owls expand their native range from eastern North America westward, reaching northern British Columbia (1943), southeastern Alaska (1967), and southwestern British Columbia (1977), then moving southward into Washington (1965), Oregon (1972), and California (1976). | Livezey (2009a) |
| 1922 | A. G. Prill publishes a list of bird species observed in Lake County, Oregon, that includes a pair of Barred Owls at the north end of Crump Lake. Researchers later question if Prill identified the birds correctly given the well-documented time line of the Barred Owl range expansion into the Pacific Northwest. | Prill (1924) |
| 1972 | E. Forsman documents the first reliable observation of a Barred Owl in Oregon, located in H. J. Andrews Experimental Forest. | Pers. comm.; Livezey (2009a) |
| 1974 | Two territorial pairs of Barred Owls are documented on repeated occasions in northeastern Oregon. | Taylor and Forsman (1976) |
| 1990 | The Northern Spotted Owl is listed as Threatened under the US Endangered Species Act. Barred owls are identified as a potential, yet unknown, competitive threat to ecologically similar Spotted Owls. | USFWS (1990) |
| 2003 | Researchers estimate an average of 60 additional pairs of Barred Owls were detected each year in Oregon during 1989–98. They also publish the first evidence that Barred Owls are displacing Spotted Owls. | Kelly et al. (2003) |
| 2006 | A five-year demographic meta-analysis of Northern Spotted Owls shows that rates of population decline are steepest where Barred Owls are most abundant and have been present the longest. | Anthony et al. (2006) |
| 2006 –12 | A multiagency study is completed, *Competitive Interactions and Resource Partitioning between Northern Spotted Owls and Barred Owls in Western Oregon*, documenting negative impacts of Barred Owls to Spotted Owls and new findings on habitat use, home range, diets, and demography of Barred Owls. | Wiens et al. (2014) |
| 2011 | A five-year demographic meta-analysis of Northern Spotted Owls shows negative effects of Barred Owls on survival across the threatened subspecies' range. | Forsman et al. (2011) |
| | The first systematic survey of Barred Owls in the Pacific Northwest indicates that they outnumber Spotted Owls by 3:1 in the Oregon Coast Range in 2011, and by 7:1 in 2016. | Wiens et al. (2011), Zipkin et al. (2017) |
| 2013 | The USFWS publishes a final environmental impact statement for the experimental removal of Barred Owls to benefit threatened Northern Spotted Owls, outlining experimental research on Barred Owls "necessary for the conservation of the Spotted Owl under the U.S. Endangered Species Act." | USFWS (2013a) |
| 2015 | A Barred Owl large-scale removal experiment is initiated in northern California, western Oregon, and Washington to determine the efficacy of removal as a management tool in benefiting declining populations of Spotted Owls. | Wiens et al. (2019a) |
| 2016 | A five-year demographic meta-analysis indicates that Barred Owls are the leading cause of population declines in Northern Spotted Owls across the threatened subspecies' range, including Oregon. | Dugger et al. (2016) |
| 2019 | Barred Owl specimens collected during removal experiments in older forests of western Oregon show that up to 39% of owls examined tested positive for exposure to anticoagulant rodenticides. | Wiens et al. (2019) |
| 2020 | Federal Barred Owl removal experiments are completed. Results show removals increased survival of territorial Northern Spotted Owls and arrested long-term population declines in California, Oregon, and Washington. | Wiens et al. (2021) |
| 2021 | A five-year demographic meta-analysis indicates that Spotted Owl populations in Oregon have declined by 60% to 75% since population monitoring began in 1993. Evidence is strengthened that Barred Owls are a primary cause of those declines. | Franklin et al. (2021) |
| | The USFWS begins a two-year process to develop a Barred Owl adaptive management strategy that can be implemented to reduce the negative impacts of Barred Owls on Spotted Owls and associated old-forest wildlife. | David Wiens, pers. comm. |

## David Wiens

David Wiens is an avian ecologist and supervisory research wildlife biologist with the US Geological Survey Forest and Rangeland Ecosystem Science Center in Corvallis. Wiens's areas of expertise include conservation and management of threatened and endangered species, wildlife survey and monitoring design, and population dynamics of birds of prey. David received his Ph.D. in wildlife science from Oregon State University in 2012 working with Bob Anthony and Eric Forsman. His doctoral research revealed that invasive Barred Owls were outcompeting native Northern Spotted Owls for critical shared resources, such as old-forest nesting habitat and prey. Wiens was awarded The Wildlife Society's outstanding publication award for this work in 2015. His work further prompted a large-scale, multiagency field experiment to determine whether removal of Barred Owls would improve declining population trends of Spotted Owls—a study that he currently leads with Katie Dugger and others. As the son of noted ornithologist John Wiens, Dave grew up heavily exposed to field research on birds in Oregon and has worked as an active ornithologist in the state for most of his professional career. In addition to owls, David spends much of his time studying Northern Goshawks, Golden Eagles, and other raptors.

declines of Spotted Owls, documented increasing rates of population declines of Spotted Owls despite widespread protection of older forest on federal lands (Anthony et al. 2006, Forsman et al. 2011, Dugger et al. 2016). These studies identified increasing populations of Barred Owls as a primary reason for those declines, especially in recent years.

Seminal research on competitive interactions and resource partitioning between Barred Owls and Spotted Owls in western Oregon confirmed that Barred Owls had broadly overlapping patterns of resource use and habitat requirements with Spotted Owls (Wiens et al. 2014). It also confirmed that Barred Owls had a much broader (generalist) diet that included many terrestrial and aquatic prey species that were rare or absent in the diets of Spotted Owls, including shrews, moles, frogs, salamanders, snails, insects, crayfish, and small fish. The study further revealed that Barred Owls exhibited demographic superiority over Spotted Owls in terms of survival and reproductive rates. The first standardized surveys of Barred Owls were completed in the Oregon Coast Range, which, when later combined with historical detections of Barred Owls, indicated an average increase from 0.13 Barred Owls per Spotted Owl territory in 1995 to 7.5 Barred Owls per Spotted Owl territory in 2016 (Wiens et al. 2011, Zipkin et al. 2017).

Mounting concerns about the dire threat that Barred Owls posed to Spotted Owls (USFWS 2011), and perhaps many other sensitive forest species (Holm et al. 2016), prompted consideration of several potential research and management options. Among these, removal experiments were identified as having the greatest value in determining the role of Barred Owls in population declines of Spotted Owls, plus the experiments would provide a means of directly testing the effectiveness of removals as a possible management tool for Spotted Owl recovery (USFWS 2011, 2013a). In late 2015, after several years of deliberations, public outreach, and ethical considerations, the USFWS, USGS, USFS, and BLM implemented a Barred Owl removal experiment within the range of the Northern Spotted Owl (Wiens et al. 2019a). By the conclusion of the study in late 2020, 3,633 Barred Owls had been lethally removed from more than 3,100 km² of Spotted Owl habitat, including 2,055 Barred Owls removed from the Coast Range and Klamath regions of western Oregon. Spotted Owl survival increased rapidly in areas where Barred Owls were removed, which ultimately arrested long-term population declines of Spotted Owls (Wiens et al. 2021). In areas without Barred Owl removal, Spotted Owl populations continued to decline sharply (Franklin et al. 2021).

Now, with results of the Barred Owl removal experiment in hand, the USFWS has begun rapid development of an adaptive management strategy for Barred Owls in the Pacific Northwest. Even if Barred Owls can be managed to reduce their ecological impacts in some areas, it is likely that this abundant and fiercely territorial predatory bird will continue to exert substantial ecological pressure on Spotted Owls and other native wildlife in Oregon. The Barred Owl will undoubtedly play a dominant role in the continuing conservation saga of the Northern Spotted Owl and management of older forests in the Pacific Northwest.

## Endemic and Iconic Prairie-Oak Birds within a Human-Dominated Landscape

BOB ALTMAN

The landscape of prairie-oak habitats in western Oregon has been the desired place for humans to live, play, and farm since Euro-American settlement. Bird species associated with these habitats have faced numerous challenges, most of which are either directly or indirectly associated with an expanding human footprint. Habitat losses primarily from urban and rural development and conversion to agriculture have been profound from a historical perspective and continue in recent times. In the Willamette Valley, prairie and savannah habitat have been reduced from the most abundant vegetative community (45%) to a few small, scattered parcels of seminatural remnants amid a sea of agricultural lands and development. There has been more than 99% loss of the historical extent of prairie and savannah habitats (Alverson 2005). Further, less than 10% of the historical oak woodlands and savannah remains (Willamette Valley Oak and Prairie Cooperative 2020), including less than 1% of the presettlement oak savannah (Hulse et al. 2000). In the Umpqua Valley, there has been a reduction in grassland, savannah, and oak woodland from 80% land cover historically to 31% cover currently (The Nature Conservancy and Krueger 2013). This includes a 64% loss of prairie/savannah and a 56% loss of oak habitat.

Habitat degradation has largely resulted from enhanced natural succession and encroachment due to extensive alteration of ecological and historical cultural processes that formerly maintained prairie-oak vegetative communities, such as fire and flooding. Fire suppression efforts have contributed to encroachment of conifers, particularly Douglas-fir, and development of dense forests

from previous oak savannahs and open woodlands (Franklin and Dyrness 1973). These habitats typically lack the structurally diverse traits (e.g., extensive branching, large lateral limbs) and microhabitat features (e.g., numerous cavities) required by many oak-associated bird species. There also has been a general loss of large, legacy oak trees owing to increased competition from conifers and a lack of recruitment of young trees in many areas.

The consequences of prairie-oak habitat loss and degradation on many bird species has been extensive population declines, several regional extirpations, and federal or state listing status or consideration for two endemic subspecies (USFWS 2013b, 2018). Among bird species highly associated with prairie-oak habitats, 11 are experiencing significant population declines based on Breeding Bird Survey data, while only one species has a significantly increasing population trend (Rockwell et al. 2021). Extirpations of regularly occurring breeding species at the scale of ecoregion or subregion include (Altman 2011)

- Burrowing Owl from the Willamette Valley and Umpqua Valley by the mid-1950s and Rogue Valley by the early 1980s.
- Lewis's Woodpecker from the Willamette Valley by the early 1970s and the Klamath Mountains by the late 1980s.
- Streaked Horned Lark (*Eremophila alpestris strigata*) from the Klamath Mountains by the mid-1970s.
- Lark Sparrow from the Willamette Valley by the late 1940s and the Umpqua Valley by the mid-1990s.

The earliest anecdotal recognition of population declines of prairie-oak birds was for the Western Bluebird (Gullion 1951, Eltzroth 1983). This sparked the initial nest box trail conservation efforts by citizen scientists in the 1970s in two areas of the Willamette Valley (Eltzroth 1983, Sims 1983; see the biography of Elzy and Elsie Eltzroth on p. 21). These efforts were highly successful in recovering precipitously declining populations mostly attributed to habitat loss and competition from exotic species for nest cavities (Keyser et al. 2004).

Widespread anecdotal recognition of population declines of prairie bird species was reported in the early 1990s by Gilligan et al. (1994) for several species in the Willamette Valley, including Common Nighthawk, Oregon Vesper Sparrow (*Pooecetes gramineus affinis*), Streaked Horned Lark, and Western Bluebird, and Lark Sparrow in the Umpqua Valley.

By the mid-1990s, recognition of conservation concern for many prairie bird species resulted in the initiation of status and distribution monitoring and research by ODFW throughout the Willamette Valley (Altman 1999). This resulted in enhanced interest in the two endemic and most imperiled birds, Streaked Horned Lark and Oregon Vesper Sparrow. They have been the focus of intensive long-term studies and a series of reports on distribution, abundance, and reproductive success, and limiting factors for Streaked Horned Lark in the Willamette Valley (see also section on Streaked Horned Lark on p. 154), Oregon Vesper Sparrow in the Willamette Valley (Altman 2015a, 2021), and Oregon Vesper Sparrow in the Rogue Basin (Stephens and Rockwell 2020). Additional research in the early 2000s included Western Meadowlark throughout the Willamette Valley (Altman and Blakeley-Smith 2011), Western Meadowlark and Grasshopper Sparrow in the

OPPOSITE: Adult Lewis's Woodpecker at nest cavity entrance with food for nestlings. Lewis's Woodpeckers were extirpated as a breeding species from the Willamette Valley in the early 1970s owing to loss and degradation of prairie-oak habitat. Photo by Dan Roby

Western Bluebird male. Western Bluebirds are a species characteristic of prairie-oak habitats in western Oregon that declined dramatically during the mid-twentieth century. But the species has staged a remarkable comeback in Oregon largely because of the widespread provisioning of nest boxes in open habitats, nest boxes that have entrance holes too small for European Starlings, a major competitor for nest sites. Photo by Dan Roby

West Eugene Wetlands (Altman 2015b), and Grasshopper Sparrow in the Rogue Valley (Stephens 2016).

The earliest research efforts on oak bird-habitat relationships and populations in the Willamette Valley were graduate student theses (Evenden 1949, Eddy 1953). Later studies of the oak bird community included population and habitat assessments in the Willamette Valley (Anderson 1970b) and Umpqua Valley (Cross and Simmons 1983), and Acorn Woodpecker ecology in the Willamette Valley (Doerge 1978). Emphasis on monitoring and research in oak habitats in the Willamette Valley has accelerated since the late 1990s, starting with an assessment of

the oak avifauna at multiple sites (Hagar and Stern 2001), Acorn Woodpecker granary site selection (Johnson and Rosenberg 2006), and a series of graduate student efforts on Slender-billed White-breasted Nuthatch ecology (*Sitta carolinensis aculeata*; Viste-Sparkman 2005), isolated legacy oak trees (DeMars et al. 2010), the effects of mistletoe (Pritchard 2015), avian community change in oak habitats over 60 years (Curtis 2014), and modeling distribution and densities now and backdated to pre-Euro-American settlement (Hallman 2018).

A similar breadth of research on oak birds and habitats has been completed by Klamath Bird Observatory in the Rogue Basin. A series of publications provide ecological context for oak habitats within the diverse forests of this region (Alexander 1999, Stephens et al. 2019). A robust avian inventory across the gradient of oak habitats (Halstead et al. 2012) facilitated the development of distribution models and predictions of bird community responses to habitat fragmentation (Halstead et al. 2019). A number of studies have examined bird response to various natural and anthropogenic changes, including effects of cattle grazing (Alexander et al. 2008), wildfire (Stephens et al. 2015a), chaparral removal to reduce fuels (Seavy et al. 2008, Stephens and Gillespie 2016, Gillespie et al. 2017, Gillespie and Stephens 2020), and restoring oak habitats by removing encroaching conifers (Stephens et al. 2021).

Broadscale bird monitoring in oak habitats throughout western Oregon resulted in a multiscale approach focused on GAP analyses (Stockenberg et al. 2008), population estimates (Altman 2008), and habitat relationships (Altman and Lloyd 2012). The summation of all the aforementioned efforts resulted in publication of the *Land Managers Guide to Bird Habitat and Populations in Oak Ecosystems of the Pacific Northwest* (Altman and Stephens 2012). This document emphasized descriptions of optimal habitats and presentation of species abundance by habitat type and ecoregion. Additional online decision support tools based on this document are available as interactive maps and summations (Altman et al. 2017a).

Despite significant challenges for prairie-oak birds in a human-dominated landscape, natural resource agencies and conservation entities have been highly invested in understanding prairie-oak bird species population status as indicated above and implementing numerous conservation actions to support species recovery. Most of the conservation actions have been within the context of broader goals of prairie-oak habitat restoration. But some innovative efforts have included nest exclosure fences (Pearson et al. 2012) and genetic rescue via egg translocations (Wolf 2012) for Streaked Horned Lark, artificial conspecific attraction for Streaked Horned Lark (Anderson et al. 2013) and Oregon Vesper Sparrow (Altman 2020), and most recently the use of Motus Wildlife Tracking System towers for tracking post-fledging survival in Oregon Vesper Sparrow (Stephens and Rockwell 2020).

The recent emphasis on research and monitoring of prairie-oak birds has resulted in extensive knowledge that provides some clear conservation opportunities, but also a sense of urgency to act as development and habitat degradation further limit options for the many declining bird species. The greatest challenge, particularly in the Willamette Valley, for prairie birds is the need for large landscapes of suitable habitat to support resilient populations, such as hundreds of acres for Streaked Horned Lark (USFWS 2019b) and Western Meadowlark (Altman and Blakeley-Smith 2011).

Other noteworthy concerns for Oregon Vesper Sparrow in the Willamette Valley include the near-complete occurrence of populations on private farmland, the stability of relativity large populations but the extirpation of many small populations, and the high site fidelity of adults and to a lesser extent first-year birds; the latter limits establishment of new populations (Altman 2021). The most urgent challenge for many oak bird species is habitat management to release suppressed oak trees, especially older open-grown trees, from conifer encroachment before they are lost (Altman and Stephens 2012). The other primary challenge is restoration and maintenance of ecologically appropriate oak habitat types and conditions, especially to address the dramatic loss of savannah habitats.

The future for many bird species in the prairie-oak ecosystem will be determined by our ability to protect and manage sufficient places

## Bob Altman

Bob Altman has spent the last 30-plus years (10 with Avifauna Northwest and 20 with American Bird Conservancy) working on bird conservation projects throughout Oregon. His efforts have focused on landbirds, and few Oregonians are responsible for protecting more bird habitat in the state than Bob. He prepared the first version of the Partners in Flight landbird conservation plans for five ecoregions in Oregon in the late 1990s. He also developed the initial status survey and demographic research on several declining grassland bird species in the Willamette Valley in the mid-1990s. These results were critical in establishing federal candidate status for Streaked Horned Larks and Oregon Vesper Sparrows. He was a member of the Streaked Horned Lark Recovery Team and prepared the petition to list Oregon Vesper Sparrow under the Endangered Species Act.

Bob has developed and coordinated many avian conservation efforts, including a Black Swift survey at waterfalls from British Columbia to northern California; the international program Quercus and Aves, which worked with partners from British Columbia to Guatemala on habitat for oak-nesting birds; the Flammulated Friends Program, which provided resources for habitat management in ponderosa pine forests for Flammulated Owl, especially on private lands; and he led the effort to establish the Oregon Cascades Birding Trail, the first birding trail in Oregon. He recently initiated development of the first conservation business plan in Oregon, for the prairie-oak ecosystem. Although semiretired since 2019, he currently is leading a long-term metapopulation study on potential demographic factors that limit populations of Oregon Vesper Sparrow.

*Photo by Carol Schuler*

to sustain populations in this human-dominated landscape. The vision to accomplish this is presented in a recent landmark document, *Prairie, Oaks, and People: A Conservation Business Plan to Revitalize the Prairie-Oak Habitats of the Pacific Northwest* (Altman et al. 2017b). This conservation business plan provides an overarching strategic framework for a 10- to 15-year investment strategy for prairie-oak conservation and the recovery of 41 imperiled species, including five bird species.

## History and Ecology of Streaked Horned Larks

### RANDY MOORE

Horned Larks come in a bewildering variety in the New World, where 26 subspecies have been described for North, Central, and South America's only native lark (family Alaudidae), an impressive haul for a species whose populations often are not all that well separated geographically. No fewer than 20 of these subspecies reside, at least in part, in the western third of North America (Beason 2020), and they are all enchanting creatures, with derring-do display flights, costume party masks, and a charming bubbly-tinkly song that utterly belies the devil's horns (actually erectile feathers) worn by males and older females. They are widely distributed in open landscapes, from naturally occurring beaches, prairies, tundra, deserts, and grasslands to anthropogenic habitats in a wide variety of agricultural and industrial settings.

Streaked Horned Larks (*E. a. strigata*) of the lowland Pacific Northwest have developed unique characteristics in biogeographically interesting circumstances. They do not occupy a vast, isolated breeding range but are instead tucked away in the slender lowland cradle formed by the Pacific Ranges of the Coast Mountains (British Columbia, Canada) in the north, the Cascade Range (Washington and

Oregon) to the east, the Siskiyou Mountains (Oregon and California) to the south, and the Pacific Ocean to the west. Anyone who has spent time in this area will know that it is not exactly filled with prime Horned Lark habitat. In fact, the region is currently so climatically moist that it has supplied the bulk of US construction timber for decades, and Douglas-fir forests (indeed, any forests) are 100% incompatible with Horned Larks of any stripe. At some point in geological history, however, perhaps during a period of glaciation when there was more (because of falling sea level) and drier (because of glacially induced climate change) land along the immediate coast of the Pacific Northwest, a population of Horned Larks from California expanded into the coastal beaches / accretion flats of what would become Oregon, Washington, and southwestern British Columbia. As glaciers retreated and the climate became cooler and more humid, the connection between the southern parent population and the new Pacific Northwest coast population was broken as larks were squeezed back toward the south as rising sea levels and inhospitable forests reclaimed much of the coastal northwest (Whitlock 1992, Mason et al. 2014). Crucially, the ancestors of todays' Streaked Horned Lark population retained a foothold in their new Pacific Northwest home by colonizing the wide swathes of outwash left in the southern Puget Trough by enormous streams of meltwater into which Puget Lobe glaciers had transformed as the climate warmed. To this day, what's left of that gravelly glacial outwash soil plays host to the majority of Washington State's remaining Streaked Horned Larks because the drainage characteristics limit vegetation growth to a sparse, low-profile bunchgrass prairie community (Hansen 1947, Stinson 2005). Small subpopulations of these newly isolated Pacific Northwest larks may also have retained a

presence on the beach foredune / accretion flat communities of coastal Cascadia, where they still breed today in small numbers (Pacific and Grays Harbor counties, Washington). This pattern of distribution—pockets of occupancy in a larger unsuitable landscape—was probably the norm when the ancestors of Streaked Horned Larks colonized the region. At intervals since the glacial maximum of the Cordilleran Ice Sheet, ~14,500 years before present (yr BP), the climate of the Pacific Northwest was cooler/drier and warmer/drier until the cooler/wet conditions that persist today began to develop about 7,000 yr BP (Whitlock and Bartlein 1997). Although it is unlikely that these drier periods would have transformed the west Cascadia lowlands into a uniformly treeless lark wonderland, it is likely that patches of tundra/steppe/oak savannah with sparse trees did develop in lowland valleys between the Coast Range and the Cascades (Hansen 1947, Barnosky 1985), and that summer drought and attendant fire activity intensified in an early Holocene dry period that lasted for well more than a millennium, beginning about 10,000 yr BP (Whitlock 1992). Where ample moisture is troublesome for Horned Larks because of its propensity for discouraging sparse vegetation / bare ground, fire can be a best pal because of its propensity for doing exactly the opposite. There is ample evidence that glacial outwash prairies of the south Puget Trough were historically maintained in low-stature vegetation by natural and anthropogenic fire disturbance (Stinson 2005).

The upshot here is that prior to the arrival of Euro-American settlers, the ancestors of modern Streaked Horned Larks were likely able to colonize parts of the west Cascadia lowlands that were historically much more hospitable to larks than they are today. Their numbers likely waxed and waned, and their geographic range expanded and contracted,

Streaked Horned Lark adult male in breeding plumage. This subspecies of Horned Lark is restricted to western Oregon and western Washington. It is listed as Threatened under the US Endangered Species Act, but it is not listed as Threatened by the state of Oregon. Photo by Rod Gilbert

as plant communities shifted with changing climatic conditions and disturbance regimes (like Native American burning practices that removed thatch from outwash prairies and kept them relatively tree- and shrub-free), but they were able to persist in a landscape laden with challenges for a species that prizes open space and sparse vegetation. Protected by the walls of their geographic cradle, they also managed it without significant interbreeding with neighboring Horned Larks. Two analyses of Streaked Horned Lark mitochondrial deoxyribonucleic acid (mtDNA) have shown that the modern population has gone through a severe bottleneck at some point in its evolutionary history, leaving little variation in a mitochondrial genome that is also completely distinct from any surrounding forms of Horned Lark (Drovetski et al. 2006, Mason et al. 2014).

Well-informed guesswork about the historical ecology of Streaked Horned Larks is supplemented by written records after the arrival of Euro-Americans to the Pacific Northwest. The records are typically thin on details but do provide solid information on the distribution of the subspecies from middle to late nineteenth century onward. Up to that point, the cool, wet climate we associate with today's Pacific Northwest had been parked over the region for several millennia (Whitlock 1992), likely providing larks with annually available breeding habitat only in places where substrate conditions limited the development of robust plant communities on a relatively broad scale (like those glacial outwash prairies mentioned above). It is unlikely that the Willamette Valley, the current center of Streaked Horned Lark abundance, hosted significant populations prior to Euro-American colonization simply because things grow too well under the current moisture regime. The rich sediments deposited by the Missoula Floods ~15,000 yr BP, though probably great for Horned Lark habitat the year in which they were laid down, are far too fertile to provide open landscapes with significant bare ground that larks require when you add gobs of water to them. The valley's remaining oak savannahs and mesic prairies of today provide insight; they require precisely timed and relatively severe annual disturbance if they are to serve as Streaked Horned Lark habitat, and there is little indication that Native American landscape management practices would have provided sufficient disturbance prior to 1850. Late-summer and early-fall fires in today's Willamette Valley typically do not set prairie succession back far enough to provide suitable habitat through early June, when Streaked Horned Larks typically fledge their first nests (R. Moore, unpublished data).

This local biogeographic pattern—center of abundance in the Puget Trough with smaller and widely spaced populations north and south—began to change with Euro-American development of the region. There were mixed consequences for Streaked Horned Larks. In a nutshell, the extensive glacial outwash prairies of the southern Puget Trough began to disappear under Euro-American plows and homesteads, while the vegetation structure of the remaining outwash prairies changed with the introduction of non-native invasive species and the suppression of natural and anthropogenic fire regimes (Weiser and Lepofsky 2009). Today, only 2% to 3% of the original outwash prairies are still dominated by native vegetation, and only about 8% remain in prairie form (Stinson 2005). At the same time and for the same reason, the Willamette Valley's more extensive but less suitable prairie

ecosystems started to receive, on a grand scale, the frequent disturbance required to convert these systems into good Horned Lark habitat: European-style agriculture. Intensive row crop, hay, or pasture agriculture is nearly always attractive to Horned Larks as breeding habitat because it often occurs in historically open landscapes, and it encourages consistently available sparse low-stature vegetation and abundant bare ground (Beason 2020). From the 1870s through the 1950s, there are multiple but vague published records of Streaked Horned Lark occurrence in the Willamette Valley, almost all describing larks as "common" to "abundant." Where habitat type is discussed in these reports, it is invariably described as cultivated land (Altman 2011). It is currently unclear why Streaked Horned Larks disappeared from Oregon's Rogue Valley, the subspecies' southernmost outpost, but it seems likely that the culprit was the polar opposite of what occurred in the Willamette Valley. In the Rogue Valley, written records suggest that larks were common breeders in native bunchgrass prairie (Gabrielson 1931), which was suitably sparse because of the valley's relatively low rainfall and short fire return intervals. The advent of heavy livestock grazing regimes may have altered those bunchgrass communities so that they were no longer suitably structured. It is also possible that collecting played a role, as the last breeding records coincide almost exactly with a substantial series of larks collected from the valley in the mid-1970s (pers. obs. of Oregon State University vertebrate collections). There is solid evidence, then, that Euro-American colonization of the Pacific Northwest reversed a millennia-long pattern of Streaked Horned Lark distribution, moving the center of abundance to the Willamette Valley, where it remains today.

Streaked Horned Lark abundance, historical and current, is poorly known. The post

Euro-American colonization demographic flip described in the previous paragraph has helped clarify patterns of abundance north of the Columbia River and in the Rogue Valley because the limited number of breeding sites allows near-complete access to the various agencies that do annual surveys. In the Puget Trough, Lower Columbia River Basin, and Washington's southwest coast, those surveys have consistently turned up somewhere between 200 and 300 singing males at 24 sites in the past decade. Although historical sites in Canada and the Rogue Valley are not surveyed annually, those breeding populations are presumed extinct, making abundance estimates somewhat straightforward. The Willamette Valley is the sticking point in developing a robust population estimate. Although it does have some contained breeding sites with good access, such as the Corvallis Municipal Airport, which routinely hosts the largest known breeding aggregation, which typically vacillates between 40 and more than 100 pairs (Moore and Kotaich 2010), the majority of potential breeding habitat occurs on large tracts of private agricultural land and cannot be accessed except from public rights-of-way. As a result, there has never been an attempt to derive a rigorous global population estimate, though it is in the planning stages as of this writing. Take any numbers you see about Streaked Horned Lark abundance in the Willamette Valley with a healthy grain of salt.

Despite this lack of basic abundance information, it was apparently assumed that populations of this distinctive subspecies had declined in concert with the documented retraction of geographic range and loss of habitat in the Puget Trough. Relying heavily on those facts, in late 2002, a group of conservation organizations filed a petition to list Streaked Horned Larks under the US Endangered Species Act of 1973 (Center for Biological Diversity et al. 2002). Although the petition contained a good deal of inaccurate information, the USFWS responded with a ruling that listing was "warranted but precluded" by higher-priority listing actions. That can was kicked down the road annually until a legal decision forced the USFWS to give a definitive thumbs-up or thumbs-down decision for all "warranted but precluded" species that had

## Randy Moore

 Randy Moore grew up in coastal Virginia and now resides in western Oregon, where he is currently on the teaching faculty in the Department of Fisheries, Wildlife, and Conservation Sciences at Oregon State University. His research efforts in Oregon have focused on the ecology of the Streaked Horned Lark, a subspecies that is listed as Threatened under the US Endangered Species Act and is endemic to western Oregon and Washington. Randy serves on the US Fish and Wildlife Service's Streaked Horned Lark Science Advisory Committee and has spent much of the preceding decade investigating their biogeography, reproductive ecology, and important vital rates in an effort to provide the basic ecological information required to effectively conserve the small populations of this uniquely vulnerable subspecies. He has also worked on the biogeography of lowland rainforest bird communities and their responses to fragmentation in the Republic of Panama as well as applied conservation projects in southern Arizona.

*Photo by Joe Fontaine*

been stacking up on its candidate list for some time. Streaked Horned Larks made the cut; they were proposed for listing on 11 October 2012 and officially listed as "Threatened wherever the species is found" in November 2013. Since that time, in consultation with the Streaked Horned Lark Working Group and the Streaked Horned Lark Science Advisory Team made up of university/agency/NGO ecologists from Oregon and Washington who had made good strides in filling in some of the information missing at the time of listing, the USFWS developed a draft recovery plan for Streaked Horned Larks that will guide efforts to keep this lovely and unique version of a common bird from disappearing before its time.

## Western Purple Martin Distribution and Habitat Use in Western Oregon

### JOAN HAGAR

The Western Purple Martin (*Progne subis arboricola*), an avian insectivore, is a species of conservation concern throughout the Pacific Northwest (Horvath 1999). Compared to the well-studied eastern subspecies (*P. s. subis*), little is known of the life history and biology of the western subspecies. Furthermore, there are great differences in population estimates between the eastern and western subspecies. Rich et al. (2004) estimated more than 10 million birds in eastern North America, whereas western estimates are less than 10,000 birds widely distributed from southwestern British Columbia to southern California (Western Purple Martin Working Group 2018). The eastern subspecies nests in human-provided artificial housing (Tarof and Brown 2013), whereas a large portion of the western birds can still be found nesting in standing live and dead trees in the forests of Oregon and California (Hagar and Sherman 2018).

Availability of nest sites in suitable habitat is a limiting factor for Western Purple Martin

Mated pair of Purple Martins defending their nest cavity in a snag from a prospecting European Starling, an invasive non-native songbird that competes intensely for cavity nest sites with native birds. Photo by David Leonard

populations (Sherman and Hagar 2020). This species prefers to nest in open, post-disturbance (early seral) forests, including clear-cuts with legacy trees and snags. But availability of nesting habitat in post-disturbance forests has been reduced over many decades of snag removal during timber harvest and timber salvage following wildfire (Rose et al. 2001). Furthermore, although the regeneration harvest regime (removing all or nearly all mature trees to establish the next stand) typically used on private lands in western Oregon may provide early seral conditions in the few years immediately following clear-cutting, availability of nest snags is likely to decrease as any

remaining legacy snags decay and are not replaced under short-rotation harvest schedules (trees are typically harvested within 40 years of planting). The opposite set of conditions currently occurs on federal forests, where snags and large trees are more abundant, but historical policies of fire suppression and curtailment of timber harvesting under the Northwest Forest Plan in the early 1990s have reduced the availability of early seral habitat (Phalan et al. 2019).

Sherman and Hagar (2020) found current habitat for nesting Purple Martins was defined by the presence of moderately decayed snags with nest cavities, located well away (more

## Joan Hagar

Joan Hagar has been active in Oregon ornithology since 1988, beginning with her M.S. thesis work examining the effects of forest thinning on songbirds. She received her Ph.D. in forest ecology from Oregon State University in 2004 and is currently an affiliate faculty member in OSU's College of Forestry Department of Forest Ecosystems and Society. Hagar's work as a research wildlife biologist for the US Geological Survey Forest and Rangeland Ecosystem Science Center focuses on forest wildlife-habitat relationships and restoration of native habitats, particularly in Oregon white oak communities. As a researcher for the Forest Biodiversity Research Network, Hagar contributes to their mission of raising public awareness around the importance of science and research to better understand and conserve forest ecosystems. Joan has set up several bird banding stations to study bird population demographics and dynamics, and she conducts banding workshops to transfer knowledge and skills to future or-nithologists. She was awarded an American Ornithological Society Elective Membership in 2020 and has served as an associate editor for the *Northwestern Naturalist* since 2006.

*Photo by Dave Vesely*

than 100 m) from closed-canopy forest in sufficiently large open areas (>15 ha [37 acres]). They also found that a disturbance regime characterized by infrequent but major stand-replacing events, such as fire or tim-ber harvest, is likely the key to maintaining breeding habitat for Purple Martins in western Oregon upland forests. Recent revisions to the BLM Forest and Woodlands Management Plan in western Oregon have reintroduced regeneration harvesting as a management tool (Bureau of Land Management 2016). The revisions offer a unique opportunity to restore potential nesting habitat for the Purple Martin while maintaining the agency's timber harvest goals. But fundamental information on the characteristics of snags, surrounding patches, and landscapes used by breeding Purple Martins is necessary to guide design of harvest units intended to provide nesting habitat.

## History of Avian Conservation Issues in the Klamath-Siskiyou Ecoregion

JOHN ALEXANDER

The Klamath-Siskiyou Ecoregion of southwestern Oregon and northern California encompasses the Klamath and Siskiyou Mountain ranges and includes the Rogue and Umpqua River basins in Oregon, as well as portions of the Klamath River. The landscape is characterized by a complex geology that is diverse in structure. Combined with a mixed-severity fire regime, this geology underlies a heterogeneous mosaic of mixed-conifer, mixed-conifer hardwood, oak, riparian, and chaparral habitats that is unique to Oregon (Whittaker 1960, Agee 1991, Jimerson et al. 1996, Huff et al. 2005, Altman and Alexander 2012, Alexander et al. 2020a). Trail et al. (1997) showed that the ecoregion has a high diversity of birds that reflects the area's status as a global biodiversity hotspot and region of conservation concern (Coleman and Kruckeberg 1999, DellaSala et al. 1999, Ricketts 1999, Dunk et al. 2006).

Historically, birds of the Klamath-Siskiyou Ecoregion have been relatively understudied. This began to change in 1990 when the Partners in Flight (PIF) Landbird Conservation Initiative emerged in the region. The PIF is a comprehensive bird monitoring partnership among many federal and state agencies and private organizations (Alexander et al. 2004a, Alexander 2011). Originally developed in collaboration with the US Forest Service Pacific Southwest Research Station Redwood Sciences Laboratory, ongoing monitoring in the ecoregion is coordinated by the Klamath Bird Observatory (KBO) and focuses on using birds as ecological indicators to inform natural resource management and ecosystem conservation, with data curated at Avian Knowledge Northwest (AKN) a node of the Avian Knowledge Network.

M. Ralph Browning began documenting details about the ecoregion's avian diversity in 1975. Browning (1975) noted 10 species that had expanded their ranges into southern Oregon and 6 species, including 4 oak- and chaparral-associated birds (Acorn Woodpecker, Blue-gray Gnatcatcher, Oak Titmouse, and California Towhee) that reach their northern range limits in southern Oregon's Jackson County. These species have now expanded beyond these northern limits. Trail et al. (1997) further elaborated, listing 37 breeding species that reach their range limits in the Klamath-Siskiyou Ecoregion, where species from throughout the West have overlapping distributions, adding to the avian diversity of the area. The authors noted that several of the species are undergoing range expansions (e.g., Red-shouldered Hawk and Blue-gray Gnatcatcher) or shrinking ranges (e.g., Burrowing Owl and Lewis's Woodpecker). More recently, Veloz et al. (2015) used AKN data to further demonstrate the Klamath-Siskiyou Ecoregion as an important area for conservation based on the predicted abundance and diversity of species that breed in the region. The ecoregion's avian diversity is also illustrated in a species-richness map that was published in the *Partners in Flight Tri-National Vision for Landbird Conservation*; the map shows that a high diversity of western forest migrant birds breed in the area (Berlanga et al. 2010). Because of the Klamath-Siskiyou Ecoregion's complex geology, patchwork of habitats, and avian diversity, the region is a biological crossroads and may serve as a possible refugium in a time of habitat loss and changing climate; it is therefore an important area for studying climate effects, adaptation, speciation, and hybridization (Trail et al. 1997, DellaSala et al. 1999).

Stralberg et al. (2009) used data from the AKN and the Oregon State University PRISM

Great Gray Owl in the Klamath-Siskiyou Ecoregion of Oregon. The species has a stronghold in the mountains of this ecoregion, which is a biodiversity hotspot that received a substantial boost in habitat protection from the designation of the Cascade-Siskiyou National Monument in 2000. Photo by Mel Clements

Group to predict how California species distributions may ebb and flow in the future as a result of climate change. They predicted that parts of the Klamath-Siskiyou Ecoregion may see a reshuffling of species, likely because of the region's geologic, plant community, and avian diversity. This reshuffling could result in novel avian communities as birds respond to climate change, further showing how the area may be an important crossroads for connectivity and a possible climate refugium.

Stewart Janes from Southern Oregon University studies Black-throated Gray Warbler and Hermit Warbler song dialects in the diverse habitats of southern Oregon. Janes and Ryker (2011) showed that these western wood warblers have considerable variation in Type

I songs in the Klamath-Siskiyou Ecoregion. This variation contrasts with the uniformity of first-category singing among North America's eastern wood warblers. Results suggest that topographic diversity and fire play important roles in the production and maintenance of song variants that represent an observable cultural evolution of birdsong. While tall ridges separated distinct song forms in their study, Janes and Ryker (2011) also suggested that stand-replacing fires contribute to periodic isolation of populations and a diversity of variants at finer scales.

Researchers from the University of California, Riverside, are exploring Allen's Hummingbird and Rufous Hummingbird hybridization along the coast of Oregon. The hybrid zone runs in a north–south direction along the coast, and inland in an east–west direction along the Klamath River (Myers et al. 2019). This confluence of coastal and inland gradations leads to some confusing genetic patterns in the Klamath-Siskiyou Ecoregion. For example, birds near O'Brien, Oregon, appeared to be hybrids, but this locality has pure Allen's populations just west and south of it. In contrast, birds in the Cascade-Siskiyou National Monument appeared to be completely Rufous, while the closest nearby populations along the Klamath River were hybrid in origin. The researchers did not find any populations between Horse Creek, California, and the monument, where there may be a 50-mile break in their distribution (Chris Clark, pers. comm.).

AKN data from extensive monitoring in the Klamath-Siskiyou Ecoregion are shedding light on aspects of bird-habitat relationships and demographics (Alexander 1999, 2011; Alexander et al. 2004a), with additional research focused on understanding wildfire and related forest management effects (Alexander et al. 2004b, Seavy 2006), as well as timber management effects (Stephens 2005). In a collaboration with KBO and the National Park Service (NPS) Klamath Inventory and Monitoring Network, monitoring data from within and outside of the national parks that occur in or near the ecoregion were used to show how climate, geography, and vegetation are ecological drivers at ecoregion, vegetation formation, and park unit scales (Stephens et al. 2016). At smaller scales, succession and disturbance are also important drivers. Building on the PIF Focal Species approach using birds as indicators of important ecological conditions (Chase and Geupel 2005, Alexander 2011), the KBO-NPS team developed an empirical approach for testing and refining the use of focal species as ecological indicators at management-relevant scales (Stephens et al. 2019).

KBO continues to work with the Pacific Southwest Research Station and many partners to maintain demographic monitoring during the breeding and fall migration seasons at constant-effort banding stations throughout the Klamath-Siskiyou Ecoregion and beyond (Alexander et al. 2004a). Banding data collected at 206 stations from throughout northern California and southern Oregon operated over 1- to 22-year time periods have been curated into the AKN. Of these, 83 are from within or immediately adjacent to the Klamath-Siskiyou Ecoregion; more than 144,000 captures of more than 150 species have been recorded at these 83 stations. The data from this massive banding effort have been used to examine postbreeding and molt-related elevational movements (Wiegardt et al. 2017a, b), evaluate the effects of breeding and molting activities on songbird site fidelity (Figueira et al. 2020), and demonstrate how, compared to more traditionally used naive measures of diversity, diversity estimates that are informed with demographic information about breeding and molting behaviors may offer an improved approach for identifying ecologically valuable wildlife habitat (Wolfe et al. 2019). In combination, this demographic monitoring effort and initial set of publications provide insight into the importance of sites during breeding and postbreeding seasons and contribute to a growing understanding of spatial and elevational population dynamics in the Klamath-Siskiyou Ecoregion. The data will also serve as a valuable tool for identifying cases of western forest bird population declines.

Historically, wildfires burned throughout the Klamath-Siskiyou Ecoregion within a mixed-severity fire regime. One hundred years of fire suppression and forest management focused on timber extraction have disrupted the ecoregion's natural fire regime, and the effects of this are now being exacerbated by climate change (Huff et al. 2005). KBO's partner-driven monitoring collaboration with the US Forest Service and Bureau of Land Management is focused on better understanding the effects of wildfires in order to improve forest restoration efforts intended to restore the region's historical mixed-severity fire regime.

Seavy and Alexander (2014) used data collected before and after the 2014 Quartz Fire to show how birds responded to changes in vegetation composition, structure, and/or the interaction between these simple metrics. Results from this fire study showed that, in the four years following the fire, eight species appeared to decrease as a result of the fire (Chestnut-backed Chickadee, Red-breasted Nuthatch, Hermit Thrush, Nashville Warbler, Black-throated Gray Warbler, and Hermit Warbler), while only one, the Lazuli Bunting, increased postfire (Seavy and Alexander 2014). When monitoring continued nine years postfire, however, Stephens et al. (2015a) found that bird species that initially decreased were increasing by the end of the study, doing so with greater magnitude in areas that were more severely burned. In this follow-up study, the Olive-sided Flycatcher, a species of conservation concern in the West, initially decreased immediately after the fire. Over time, however, it increased because the high-severity fire resulted in standing dead trees in which this flycatcher nests and shrub understory regrowth that provides the flycatchers with ample insect food. In a wildfire study conducted by OSU, Fontaine at al. (2009) also showed long-term beneficial wildfire effects. Their study looked at the effects of single and repeat high-severity fires in the area of the Biscuit Fire that burned in the Klamath-Siskiyou Ecoregion in 2002. While bird density was down two years postfire, densities were at their highest seventeen and eighteen years after fire. Both fire studies suggest that early seral broadleaf dominance

may be important to the conservation of avian biodiversity in the Klamath-Siskiyou Ecoregion.

A recent paper published in the journal *Science* documented an alarming decline in North America's avifauna, showing that 50% of western forest birds are in decline (Rosenberg et al. 2019). Given the diversity of western forest birds in the Klamath-Siskiyou Ecoregion, this paper adds a sense of urgency to the need for science-driven conservation in the region. Using Breeding Bird Survey trend data (Sauer et al. 2008) and a large avian point-count data set from northwest and southwest Oregon, including AKN data from the Klamath-Siskiyou Ecoregion, researchers from OSU found a strong negative relationship between 42-year bird population trends and species associations with broadleaf-dominated forest conditions (Betts et al. 2010). Such conditions are typically associated with natural patterns of disturbance (i.e., fire) and regrowth. The researchers hypothesized that intensive forest management and fire suppression had resulted in reductions in broadleaf-dominated and early seral forest conditions associated with natural patterns of succession, leading to population declines of constituent species in southwestern Oregon.

Rockwell et al. (2017) used data from 10 bird banding sites operated by KBO between 2002 and 2013, including five located in or just adjacent to the ecoregion, to estimate the abundance of 12 songbird species, all either of regional conservation concern or indicators of coniferous or riparian habitat quality. Population trend results from this regional analysis contrasted with BBS trend results from the broader Northern Pacific Rainforest (Sauer and Link 2011). Two coniferous forest indicator species (Audubon's Warbler [*Setophaga coronata auduboni*] and Purple Finch; Altman and Alexander 2012) that are not showing significant declines in the Northern Pacific Rainforest Bird Conservation Region appear

to be in significant decline in and around the Klamath-Siskiyou Ecoregion. This result further highlights concerns about forest-associated species in the ecoregion where addressing timber harvest, the restoration of mixed-severity fire regimes, and endangered species management continues to present complex conservation challenges. Rockwell et al. (2017) did find that two riparian bird species that are stable or declining insignificantly in the Northern Pacific Rainforest Bird Conservation Region (Yellow-breasted Chat and Black-headed Grosbeak were significantly increasing in the Klamath-Siskiyou Ecoregion. They suggest that this result highlights the importance of preserving riparian habitat conditions in the ecoregion where riparian zones may have remained more intact compared to other areas in Oregon where riparian habitats have been subjected to more substantial anthropogenic impacts.

Increasingly, bird monitoring and research results are being used to inform conservation and restoration efforts that are aimed at addressing the threats that most degrade important forest bird habitats. In the Klamath-Siskiyou Ecoregion, much of this work is focused on the oak ecosystems that many declining western forest birds are associated with (North American Bird Conservation Initiative 2014, Rosenberg et al. 2019, Alexander et al. 2020a, b). Threats to the region's oak ecosystems include agricultural conversion, fire exclusion, conifer encroachment, incompatible cattle grazing, and non-native grasses and forbs (Alexander et al. 2020a, b). Highlighting the need for conservation efforts in these Klamath-Siskiyou Ecoregion habitats, Altman (2011) compared the historical and current distributions of oak- and grassland-associated species. He showed that many were no longer as abundant as when Browning (1975) documented the birds of Jackson County in southern Oregon (e.g., White-breasted Nuthatch), and that some had even been extirpated as breeders (e.g., Lewis's Woodpecker).

Within this context, the Cascade-Siskiyou National Monument represents a science-driven bird conservation success story (Gillespie et al. 2018). In 2000, President Clinton signed a proclamation that established the Cascade-Siskiyou National Monument and directed the BLM to protect specific "objects of biological interest," including the region's high diversity of migratory birds that was originally pointed out by Browning (1975) and then by Trail et al. (1997). When the monument was created, livestock grazing was widespread across its lands, and the proclamation directed the BLM to determine how the grazing activity affected the monument's objects, including birds. In 2003 and 2004, KBO, BLM, World Wildlife Fund, OSU, and many other partners conducted a comprehensive study to evaluate the ecological impacts grazing was having in the monument. Alexander et al. (2008) found significant grazing impacts on migratory birds in the monument's oak woodland habitats. This and the other studies informed BLM and partner efforts to develop a voluntary buyout program, resulting in 93% of the authorized livestock grazing being eliminated from the monument. KBO, BLM, NPS, and Point Blue

OPPOSITE: The Yellow-breasted Chat was formerly classified as a member of the New World Warbler family, but its vocalizations, size, and behavior are so different from other warbler species that it has now been placed in its own family. The habitat of this species is shrubby thickets and dense riparian vegetation in the Klamath-Siskiyou Ecoregion, parts of the Willamette Valley, and parts of eastern Oregon. Photo by David Leonard

Conservation Science then collaborated, using distribution models to map the abundance of bird species on protected public lands within the Klamath-Siskiyou Ecoregion and adjacent areas (Alexander et al. 2017). Results showed that relatively few oak and grassland habitats occur in the region's protected areas and highlighted the Cascade-Siskiyou National Monument and surrounding areas as important because they did include these priority bird conservation habitats. This research informed a monument expansion in 2017 through a new proclamation signed by President Obama (2017). The proclamation expanded the monument by 47,660 acres, thereby protecting more oak and grassland habitats and associated at-risk species.

Building on this oak woodland bird conservation success story, KBO is collaborating with many partners to model PIF's science-based approach to using land-management- and policy-relevant science as a catalyst for improving ecosystem management (Rosenberg et al. 2016). The approach involves using PIF population and habitat objectives and data from bird monitoring to assess management needs, set measurable targets, design management to meet these targets, and measure the effectiveness of conservation actions. The approach lends itself to developing unique relationships between natural resource management agencies and science-based nongovernmental organizations that focus on applying PIF's approach to meeting conservation challenges in the United States (Alexander et al. 2009). Through such partnerships, a strategy for using land-management- and policy-relevant science to integrate PIF's approach into forest management planning and implementation has been developed and tested in the Klamath-Siskiyou Ecoregion (Alexander 2011). This approach is now being used to ensure the effectiveness of accelerated forest restoration efforts and thereby conserve western forest birds of high conservation concern (North American Bird Conservation Initiative 2011, 2016).

The history of ornithology in the Klamath-Siskiyou Ecoregion was originally underrepresented in this region where biodiversity studies do have a rich history. But in recent decades, ornithological work in the region has increased extensively, with a focus on using birds as indicators, and ornithological research and monitoring results as a catalyst for improved ecosystem management and bird conservation throughout the region. This growth of ornithology in the region has also brought a recent growth of public engagement. Public education efforts and community science bird monitoring projects are fostering increased public support for science-driven bird conservation in the Klamath-Siskiyou Ecoregion.

## Dry Forest Ponderosa Pine and Cavity-Nesting Birds: Habitat Restoration and the Need for Snags

BOB ALTMAN

There is a perception among many people that standing dead trees (snags) are unattractive, wasteful, and indicate unhealthy forest conditions. But dead and dying trees are an essential component of natural forest ecosystems that provide irreplaceable wildlife habitat, especially for birds such as woodpeckers that nest in cavities and forage on insects beneath the bark of infested trees. While too many snags may indicate unhealthy conditions, a truly healthy forest contains some amounts of diseased, dying, and dead trees.

In ponderosa pine forests prior to Euro-American settlement, regular understory fires and bark beetles were the primary factors maintaining open forest understories with

White-headed Woodpecker male. This species is characteristic of dry conifer forests, especially ponderosa pine forest with openings, and is an increasingly uncommon species in Oregon. The species and several others are in decline throughout the Pacific Northwest owing to a scarcity of snags as nesting habitat. Photo by Dan Roby

singular or small patches of snags (Franklin et al. 2013). Thus the periodic "disturbance" of fire and beetles ensured a healthy ponderosa pine forest with a continuous supply of snags over time. Further, the oldest trees usually became the snags because they were most susceptible to mortality, thus creating a forest with mostly large snags. Large ponderosa pines generally produce excellent snags for cavity-nesting birds, in part because of their high proportion of sapwood (the outer tree layer). Sapwood decays fast in dead pine trees, and this thick layer provides a deep area of material for cavity excavation.

Fire and bark beetles are still influencing the ponderosa pine landscape, but the forest has changed, and with it the patterns of snag creation, persistence, and value. Fire suppression initiated in the early twentieth century was supposed to protect forests from a perceived enemy but has succeeded in creating the perfect conditions for severe wildfires and beetle infestations. Fire suppression has resulted in crowded forests, more flammable material, and greater competition. This weakens tree growth and vigor and produces forests where younger trees dominate. Fires now reach into the canopy and may kill large areas of forest. Likewise, beetles now cause mortality in larger patches of trees because the forest is so dense and overcrowded with stressed trees. But snags created by these circumstances are often small, uncharacteristic of historical conditions, and of limited value to wildlife. Additionally, snags now are often removed, especially on private lands, to reduce the fire hazard and for economic value.

The degradation of ponderosa pine forests and especially the reduction in mature trees and snags has been consequential for many cavity-nesting bird species. Several of the ponderosa pine cavity-nesting bird species are considered high priorities for conservation because they are experiencing local and/or regional population declines. In the Pacific Northwest, these species include the Flammulated Owl, Lewis's Woodpecker, White-headed Woodpecker, Williamson's Sapsucker, Pygmy Nuthatch, White-breasted Nuthatch, Mountain Bluebird, and Western Bluebird.

There has been limited ornithological research on cavity-nesting birds in ponderosa pine forests in Oregon, with the exception of White-headed Woodpecker. One of the early studies to document populations of birds in ponderosa pine forests was in 1971 in central Oregon (Gashwiler 1977). There have been some general studies of cavity-nesting woodpeckers in forests in northeastern Oregon dominated by ponderosa pine, including resource partitioning (Bull 1980, Bull et al. 1986) and nest ecology and density (Nielsen-Pincus 2005). Galen (1989) conducted an assessment of habitat and populations of Lewis's Woodpecker in ponderosa pine and other forest habitats in north-central Oregon.

The cavity-nesting bird with the most conservation interest and research and most emblematic of mature forests of this forest type is the White-headed Woodpecker. In the Oregon Cascade Mountains, it has been the focus of studies on abundance and habitat (Bate 1995), ecology (Dixon 1995a, b), densities (Frenzel and Popper 1998), occurrence (Lindstrand and Humes 2009), and a long-term examination of habitat suitability and nesting status and

OPPOSITE: Mountain Bluebird male at the entrance to its nest cavity in a snag at the edge of a clear-cut in the Blue Mountains of eastern Oregon. This species has experienced local population declines owing to a scarcity of suitable nesting cavities in snags. Photo by Dan Roby

ecology (Frenzel 1998, 2000, 2004; Hollenbeck et al. 2011). Prior to most of this research, Marshall (1997) prepared a status assessment for Oregon, and Dixon (1998) prepared an assessment of distribution, density, and habitat characteristics in three states, including Oregon.

One of the most unique of the cavity-nesting birds, the Lewis's Woodpecker, was the focus of an innovative citizen science conservation effort by volunteers with the East Cascades Bird Conservancy (Shunk 2011a). They were able to enhance a declining population of birds in postfire habitat where the snags were old and falling by attracting them to nest in boxes, a previously undocumented occurrence. The American Bird Conservancy and several other partners extended the technique with some success to private lands in multiple states in the Pacific Northwest where habitat restoration and snag creation were occurring (Shunk 2011b).

Dry forest ponderosa pine restoration has been the focus of an extensive effort throughout the West, including Oregon (Franklin et al. 2013). Most silvicultural restoration prescriptions don't include the maintenance or creation of appropriate size or quantity of snags, however, and there have been no studies in Oregon to understand the population response of cavity-nesting birds to the typical thinning, understory removal, and prescribed fire of recent ponderosa pine restoration activities. Thus, there has been an absence of the documentation of response to restoration for key indicator species for the health of this ecosystem. This is despite significant studies elsewhere on this topic, especially in the northern Rocky Mountains and the Southwest.

The ecological health and functionality of dry ponderosa pine forest, which includes components of standing tree decadence, will largely dictate the fortunes of many cavity-nesting bird species. The assumption that ongoing

## Evelyn Bull

Evelyn Bull is one of the few avian researchers working on forests in northeastern Oregon. Starting with her M.S. program working with Chuck Meslow, she studied Pileated Woodpeckers and forest management, finishing in 1975. She went on attain her Ph.D. in 1980 at the University of Idaho. She spent her career working as a research biologist for the US Forest Service Pacific Northwest Research Station in LaGrande. She was primarily an avian researcher but had remarkably diverse taxonomic interests that ranged from tailed frogs to American martens. She continued to work on Pileated Woodpeckers throughout her career and furthered her interest in conservation and management of old-growth forests by working on Great Gray Owls, Flammulated Owls, Long-eared Owls, and Vaux's Swifts. She was active in the Oregon Chapter of the Wildlife Society and served as president. She received numerous awards during her career, and in 2002 she was added to the Registry of Distinguished Graduates by the Department of Fisheries, Wildlife, and Conservation Sciences at Oregon State University.

*Photo by Michael Snider*

restoration in Oregon will provide suitable habitat for cavity-nesting birds needs to be evaluated for the short and long term, especially where restoration prescriptions do not specify the conditions for standing tree decadence.

## Northern Saw-whet Owls

JEFFREY MARKS

Northern Saw-whet Owls are common and fairly widespread in North America, but their breeding biology has received little attention. We studied Northern Saw-whet Owls that nested in boxes on the Boardman Tree Farm, Morrow County, in north-central Oregon from 2012 to 2017. At the time our study began, no published information existed on age of first breeding, age-specific productivity, breeding dispersal, and natal dispersal. Moreover, the species was thought to be nomadic based on low return rates of adults banded at nests in southwestern Idaho (Marks and Doremus 2000). We monitored 46 nesting attempts, including 38 attempts where we determined the age of both parents. We also assessed diet by identifying prey items found in boxes or recovered in mist nests while trapping adults.

The Boardman Tree Farm was a 10,480 ha hybrid poplar plantation owned by GreenWood Resources. Trees were fast-growing and harvested 10–12 years after planting. We put up wooden nest boxes on live trees each winter before the nesting season: 126 boxes in 2012 and 2013, 112 in 2014, 81 in 2015, 46 in 2016, and 30 in 2017. Natural cavities were virtually absent on the study area, and we assumed that all nesting attempts by Northern Saw-whet Owls occurred in our boxes.

The Boardman Tree Farm was sold in 2016, at which time harvested trees were not replanted. Consequently, the amount of forested habitat declined substantially relative to previous years, and fewer boxes were erected during the last two years of study.

The number of nesting attempts per year ranged from 25 in 2012 to only 1 in 2017. At the same time, the proportion of voles (*Microtus montanus*) found among all prey items (n = 422) dropped steadily from 54% in 2012 to

23% in 2013, 19% in 2014, 2% in 2015, and 0% in 2016 and 2017. Owls began nesting in their second calendar year when about 11 months old. Nesting success was high, with 82.6% of all attempts resulting in at least one young surviving to fledging age. Over all years, owls produced an average of 3.5 fledglings per nesting attempt and 4.3 fledglings per successful nesting attempt.

At 38 nests in which we captured both adults, both pair members were yearlings at 12 nests (mean clutch size 5.6, mean number of young fledged 4.2), the female was at least two years old and the male a yearling at 9 nests (mean clutch size 5.8, mean number of young fledged 4.1), the female was a yearling and the male was at least two years old at 11 nests (mean clutch size 5.6, mean number of young fledged 4.5), and both pair members were at least two years old at 6 nests (mean clutch size 5.2, mean number of young fledged 3.3). Neither clutch size nor number of young fledged differed significantly among the four combinations of parental age and sex.

Prior to our work, few banded adult Northern Saw-whet Owls had been captured at nests, and no adult banded as a nestling (thus with known natal origin) had been captured at a nest (Marks and Doremus 2000, Marks et al. 2015). From 2013 to 2015, we recaptured only three breeding adults that had nested at the Boardman Tree Farm in a previous year (one male and two females) and recaptured one breeding female that we had banded as a nestling.

In 2016, four of the eight breeding males had been banded in a previous year, and by the end of the nesting season, we had captured two breeding adults (one of each sex) that had been banded as a nestling and five breeding adults (three males and two females) that had nested at the Boardman Tree Farm in two different years. On balance, however, most of the

adults that we banded nested at Boardman Tree Farm in only one year, supporting the notion that the species tends to be nomadic and apparently flexible enough in habitat selection to use an artificial, irrigated softwood plantation in the middle of what once was a grassland in the high desert.

# Urban Avian Ecology: The New Frontier

MICHAEL T. MURPHY

The growth of megacities worldwide, and shift in population from rural to urban environments, dominated the second half of the twentieth century. The consequences for wildlife, and birds in particular, are uncertain but likely negative, and thus when I began work at Portland State University in 2000, I took the university's motto, "Let knowledge serve the city," to heart and began to see Portland's expansive park system as an important research system. Naysayers existed, and my first proposal to the National Science Foundation to study urban birds had mixed reviews, including comments that the study of urban ecology would lead nowhere. Two decades later, and we find urban ecology as mainstream ecology!

Portland was a bit ahead of the curve as Houle's *One City's Wilderness* (2010), Houck and Cody's *Wild in the City* (2009), and The Friends of Tryon Creek's *A Forest in the City* (1994) described plants and wildlife generally, and birds specifically, in Portland's parks. Nancy Broshot's (1999) dissertation research in Forest Park, among other things, quantitatively examined variation in avian community composition in relation to distance from city center. The first peer-reviewed publication on urban birds from Portland emerged from Lori Hennings's M.S. thesis from Oregon State University. Her research indicated that greater canopy cover along urban riparian zones favored native bird species over non-native species. Wood and Yezerinac (2006) studied song behavior of Song Sparrows and showed

OPPOSITE: Northern Saw-whet Owl at nest cavity entrance in an old quaking aspen. This small, nocturnal species of owl is frequently overlooked, and its nesting biology is poorly understood. Northern Saw-whet Owls prey primarily on deer mice, voles, and other small mammals. Photo by Jared Hobbs

Adult male Black-throated Gray Warbler in an alder thicket within Portland city limits. This Neotropical migrant songbird has persisted in some of Portland's larger parks and greenspaces. Photo by David Leonard

that they shifted the dominant frequency of song to avoid overlap with anthropogenic noise. The latter study, to my knowledge, was the first North American study to document bird responses to anthropogenic noise and has been cited 320 times in the primary literature.

My contributions began in 2001 when I initiated study of the determinants of variation in vertebrate species richness in Portland's parks and greenspaces (P&GSs), research funded by the USFWS. Given Hennings and Edge's (2003) focus on riparian habitats, my students and I directed our efforts toward undeveloped

(beyond trails) upland forests, and we used 48 P&GSs within Portland, ranging in size from small 1 ha sites to >2,000 ha (4,900 acres) Forest Park. Three students, including David Bailey, a gifted local birder, conducted the research. David surveyed birds in all P&GSs in 2003, and along with the others measured area, shape, and connectivity of the P&GSs, quantified plant community structure within each, and determined landscape composition that surrounded all sites. Separate analyses of residents and Nearctic-Neotropical migrants showed that resident and migrant species richness declined and increased, respectively, with increases in park area. A minimum area of 10 ha (25 acres) appeared to be the critical break point for "forest-dependent species." Landscape composition surrounding P&GSs was also important; trees in the landscape benefited forest-dependent species (residents and migrants). Later work by Adam Baz (2018) showed that some woodpecker species (Hairy Woodpecker and Pileated Woodpecker), a group poorly sampled by our original study, required 40 to 50 ha (100- to 125-acre) forest reserves if they are to persist in the urban landscape.

Spotted Towhees along with Song Sparrows were the most widely distributed and abundant species based on our community study, and thus Sarah Bartos Smith began a population study of towhees to determine whether they have self-sustaining populations (Bartos Smith et al. 2016). Jenny McKay began as her field assistant but then pursued a graduate degree, and together they conducted a detailed population study based on individually marked birds to quantify annual survival and seasonal reproductive success. Towhee populations in the southwest hills of Portland did indeed appear to self-sustaining (i.e., no need for immigrants from outside; Bartos Smith et al. 2016). Unlike many species, towhees do well along forest edges, and pseudo-edges like trails through parks had no adverse effects on them. Amy Shipley later used radiotelemetry to show that fledgling towhees survived poorly along edges because of predation by native species such as Western Screech-Owls and Cooper's Hawks. Additive losses to domestic cats, however, turned parks into population sinks (Shipley et al. 2013).

The existence of individually marked populations made towhees a species for numerous additional studies. Bartos Smith, for instance, found extra-pair fertilizations to be rampant in this socially monogamous songbird (60% of social pairs). Frequent interactions among unpaired birds at bird feeders at residences that abutted P&GSs possibly explained the unusual behavior (Bartos Smith et al. 2016). Jennifer Richardson showed that towhee males had song repertoires of 6 to 10 songs and that males with the most extensive repertoires had smaller tail spots. Sampling of towhee cloacal bacteria communities by Jennifer Klomp also demonstrated, surprisingly, that towhees from our four main parks had bacterial assemblages distinctive enough to allow individuals to be classified to park of origin (Klomp et al. 2008). Why birds in different parks carry distinctive cloacal bacterial communities is unknown.

Research in other areas continued, and against my advice (fortunately!), John Deshler pursued his personal interest in Northern Pygmy-Owls and conducted a multiyear population study of the diminutive owls in Forest Park. His thorough and extensive investigations revealed an apparent thriving population, but one with considerable annual turnover owing possibly to either low annual survival or frequent long-distance dispersal (Deshler and Murphy 2012). John has continued his work with Northern Pygmy-Owls in Oregon and has become one of the foremost experts on the species in North America. He is currently working on a book on pygmy-owls and is rewriting the North American species account for Northern Pygmy-Owls for the *Birds of North America* series.

Urban ornithology also occurs outside of P&GSs, and an important, albeit not yet published, work on Portland's urban bird community was carried out by Andrew Gibbs. Gibbs conducted vegetation analyses and bird surveys in backyards to evaluate how well Portland Audubon Society's Backyard Birds Program accurately assessed vegetation composition, and whether backyard

vegetation had measurable effects on birds. Portland Audubon's classification scheme for highest (platinum) and lower (gold and silver) levels of native vegetation were reasonably accurate, but the larger landscape surrounding backyards was a better predictor of bird species composition in backyards than the onsite vegetation. If native birds are to profit, Gibbs's work suggests that a cultural change needs to take place wherein neighborhoods become dominated by native vegetation.

The surface of urban avian ecology in Portland has just been scratched. Much remains to be learned, and there is urgency to our need to fill the gaps in our knowledge because the projection for Portland's growth is unremittingly upward. More research is needed to help inform management agencies of present and future needs.

OPPOSITE: Northern Pygmy-Owls are the smallest owl in Oregon and fierce predators for their size that sometimes prey on small birds and mammals that weigh more than they do. The species frequently hunts during the day and has adapted to foraging and nesting in city parks such as Portland's Forest Park. Photo by Jared Hobbs

# Wetlands and Waterbirds

SUSAN HAIG

At first glance, Oregon does not strike a person as a place with many wetlands. Of course, there are coastal estuaries, but the state is mostly desert. However, Oregon currently has approximately 570,000 ha (1.4 million acres) of wetlands, 1,400 named lakes, and an additional 3,800 ponds and reservoirs. Oregon's tidal and nontidal wetlands once covered as much as 930,000 ha (2.3 million acres) in the late 1700s (Dahl 1990). Even so, extensive agricultural and urban development have greatly affected the extent and quality of Oregon's waters and wetlands (Table 6; Taft and Haig 2003, Haig et al. 2019). Currently, agricultural, residential, and commercial development have replaced more than 35 million ha (86 million acres) of former wetlands in Oregon. Oregon lost over half of the wetlands present when the first Euro-American settlers arrived in the 1800s. The roughly 570,000 ha (1.4 million acres) of wetlands that remain represent just 2% of Oregon's total land surface (Varva and Wood 2000), and their existence is greatly threatened by the climate crisis.

Although considerably reduced in extent, the wetlands we do have are essential for success of a great diversity of flora and fauna throughout the annual cycle and are largely managed by the USFWS, ODFW, BLM, and nongovernment groups such as The Nature Conservancy at Sycan Marsh or The Wetlands Conservancy. A tremendous amount of funding has been put into Oregon wetland management by Ducks Unlimited and waterfowl hunters via the annual purchase of Duck Stamps.

Oregon's wetlands are protected by federal laws such as the Clean Water Act (1972) and the Swampbuster Act (1985), plus state laws that prohibit dumping, draining, and dredging (Table 6). The birds inhabiting these areas in Oregon, like almost all native birds, are protected by the Migratory Bird Treaty Act (1918), which states that native species (not invasives such as European Starlings, House Sparrows, and Rock Pigeons) cannot be harmed, harassed, or killed unless they are being hunted as part of a federally or state-sanctioned hunt. For waterfowl hunting, there is also a law that prohibits the use of lead ammunition for a number of reasons (Haig et al. 2014). First, lead pellets or bullets tend to shatter and scatter when they hit the bird. Thus, it is impossible to remove all the lead from the carcass of a bird shot with lead ammunition and thereby make it safe for human consumption. Next, if a bird is shot and wounded, chances are that a predator or scavenger, such as a raptor, coyote, or wolf, will find it, eat it, and die from lead poisoning, or feed it to their young with the same outcome. Finally, if a hunter shoots at and misses the target bird altogether, the lead pellets go into the wetland where waterfowl (ducks, geese, and swans) and diving birds like loons and grebes have a good chance of ingesting lead in the food and grit they consume. A similar issue arises when anglers use lead sinkers that can accumulate in lake and river sediments and can have a devastating impact on waterfowl and other diving birds (Rattner et al. 2008). There have been a number of local and federal efforts to ban lead fishing tackle, but to date none have been successful in Oregon even though lead tackle has been banned in a number of eastern states.

In 1947, after many years of research, Canada, Mexico, and the United States identified the primary routes, or "flyways," that birds use during migration in North America (see Figure 5 on p. 89). Oregon falls midway down the Pacific Flyway, with some birds following coastal pathways, some following the Willamette Valley, and some flying through the Great Basin in eastern Oregon. Regardless of the path, there are a great number of waterbirds living or moving through Oregon during all periods of the year. In winter, we have large numbers of Canada Geese, Cackling Geese, loons, Dunlin, swans, and other Arctic breeders, while in summer, waterbirds spread out to breed and are not as obvious.

Surprisingly, the diversity and number of birds present in winter wetlands might be greater than in the summer. There are several reasons for this. For one, the high winter rainfall, especially in western Oregon, makes for great habitat for overwintering waterbirds. Next, Oregon's position on the Pacific Flyway is such that it is far enough south to preclude the harsh winter conditions that Arctic breeders are escaping, and far enough north that it is a relatively short trip back to their nesting grounds in spring. And for those migrants that go much further south, Oregon is a convenient stopover location. One migratory mystery that has yet to be understood is why, given the vast mudflats along the coast at Tillamook Bay, Yaquina Bay, and farther south, does coastal Oregon not support large overwintering flocks of shorebirds, such as those just to the north in

Washington's Willapa Bay and Grays Harbor, or just to the south in California's Humboldt Bay? Ornithologists have hypothesized that it has to do with migratory traditions determined before the mudflats in Oregon were established, or perhaps the Washington and California coastal estuaries are just more productive, or perhaps shorebirds encounter the sites farther north first and then do not need to stop again until they reach Humboldt Bay. In any event, it is a puzzlement.

While they are diverse and abundant with birds, the remoteness of many wetlands in eastern Oregon has made it easy for them to be forgotten in some ways; hence there has been relatively little funding for wetlands and waterbird research over the years. This is especially true for research on non-waterfowl species. For example, over the past 50–70 years, there have only been five (Oregon) faculty members who have studied Oregon wetlands and waterbirds: Robert Jarvis (OSU) in the 1970s, Susan Haig (OSU/USGS) since 1994, Katie O'Reilly (University of Portland) since 1997, Dan Roby (OSU/USGS) since 2000, and Bruce Dugger (OSU) since 2002. Among these, only Dan Roby and Susan Haig studied non-waterfowl waterbirds east of the Cascades (and Mark Stern as a nonacademic). In eastern Oregon, Marty St. Louis, the ODFW manager at Summer Lake Wildlife Area, was the only wetlands and waterbird expert in the region for decades until he retired in 2018.

Our changing climate and impinging human developments are the silent crises currently altering wetland viability and water availability for waterbirds and humans. Lack of water or poor water quality is already an ever-growing crisis du jour or de annum. With waterbirds as indicators of the status of wetlands and water bodies, it is even more crucial to monitor and manage for their conservation.

Belted Kingfisher adult female hunting for fish. Formerly considered pests and persecuted for their fish-eating habits, kingfishers were not protected in Oregon until 1958. Oregon's only kingfisher, Belted Kingfishers occur in a wide variety of wetland habitats in Oregon that support fish. Photo by Timothy Lawes

## Table 6.   Key conservation acts and events affecting Oregon waterbirds and wetlands

| Date | Conservation Action |
|---|---|
| 1849 | Swamp Lands Act turned federal land over to states that agreed to drain it. |
| 1850 | |
| 1860 | |
| 1900 | Lacey Act banned interstate trafficking of wildlife taken illegally; the first game protection act. |
| 1903 | National wildlife refuges established. |
| 1907 | Three Arch Rocks National Wildlife Refuge established. |
| 1908 | President Theodore Roosevelt establishes Lower Klamath National Wildlife Refuge as the first waterfowl refuge. Malheur National Wildlife Refuge established. |
| 1909 | Deer Flat National Wildlife Refuge established. |
| 1918 | Migratory Bird Treaty Act protects birds migrating through Canada, Mexico, and the United States. |
| 1928 | Upper Klamath National Wildlife Refuge established. |
| 1929 | Migratory Bird Conservation Act of 1929 formed a commission to acquire and conserve land for national wildlife refuges. |
| 1934 | Migratory Bird Hunting and Conservation Stamp Act (Duck Stamp Act) is passed, enabling the federal government to purchase lands that will be managed as migratory bird sanctuaries by the Bureau of Biological Survey. The Duck Stamps program leads to development of the National Wildlife Refuge system and Waterfowl Production Areas Program (WPA). |
| 1935 | Oregon Islands National Wildlife Refuge established. |
| 1938 | Cape Meares National Wildlife Refuge established. |
| 1944 | Summer Lake Wildlife Area established. |
| 1947 | Flyway concept is defined; Sauvie Island Wildlife Area established. |
| 1949 | Ladd Marsh Wildlife Area established. |
| 1950 | E. E. Wilson Wildlife Area established. |
| 1953 | Wenaha Wildlife Area established. |
| 1954 | Denman Wildlife Area established. |
| 1957 | Fern Ridge Wildlife Area established. |
| 1958 | Klamath Marsh National Wildlife Refuge and Klamath Wildlife Area established. |
| 1964 | William L. Finley National Wildlife Refuge established. |
| 1965 | Ankeny and Baskett Slough National Wildlife Refuges established. |
| 1969 | Jewell Meadows Wildlife Area established. |

1971    National Environmental Policy Act (NEPA) is passed; Columbia Basin Wildlife Area established.

1972    Clean Water Act passed with wetland protection provisions. Lewis and Clark National Wildlife Refuge as well as Phillip W. Schneider Wildlife Area established.

1976    Riverside Wildlife Area established.

1977    Swamp Lands Act rescinded (Executive Orders 11988, 11990); wetlands preservation measures instituted.

1978    Bear Valley National Wildlife Refuge established.

1980    Yaquina Head Outstanding Natural Area established by the BLM.

1983    Bandon Marsh National Wildlife Refuge and the Lower Deschutes Wildlife Area established.

1985    Swampbuster Act (1985 Food Safety Act) defined wetlands and established policy of no net loss of wetlands. This has been variably interpreted by different administrations.

1991    Lead ammunition banned for use in waterfowl hunting. Nestucca, Siletz Bay, and McKay Creek National Wildlife Refuges established.

1992    Tualatin River National Wildlife Refuge established.

1995    Lake Abert declared an Area of Critical Environmental Concern by the BLM.

1997    Sycan Marsh acquired by The Nature Conservancy.

2013    Wapato Lake National Wildlife Refuge and Coquille Valley Wildlife Area established.

2016    A group of armed right-wing militants occupies the headquarters buildings at Malheur National Wildlife Refuge for 41 days.

2017    US Fish and Wildlife Service Director's Order 219 states regional USFWS supervisors must work with state hunting officials to replace all lead-based gear used on agency property with nontoxic alternatives by 2022.

        Secretarial Order 3346 removes USFWS Director's Order 219 and states that lead ammunition can still be used to hunt mammals and fishers can use lead tackle on federal land; lead ban for waterfowl hunting remains.

2021    Saline Lake Ecosystems in the Great Basin States Program Act of 2021 introduced to protect Lake Abert and Goose Lake.

*Sources*: "A History of Wetlands Protection in the United States," Berkshire Environmental Action Team, accessed 25 March 2022, https://www.thebeatnews.org/BeatTeam/history–federal–wetland–protection/; Wikipedia, s.v. "List of National Wildlife Refuges of the United States: Oregon," last modified 11 March 2022, https://en.wikipedia.org/wiki/List_of_National_Wildlife_Refuges_of_the_United_States#Oregon; "Visit ODFW Wildlife Areas," Oregon Department of Fish and Wildlife, accessed 25 March 2022, https://myodfw.com/visit–odfw–wildlife–areas.

American White Pelicans are surface-feeding, piscivorous colonial waterbirds that can form large breeding colonies on islands when foraging and nesting conditions are favorable. They are capable of preying on fish the size of adult common carp and will commute hundreds of miles from their nesting colony to forage in wetland areas with abundant fish resources. Until California Condors were reintroduced to Oregon, American White Pelicans had the greatest wingspan of any Oregon bird. Photo by Jared Hobbs

## Willamette Valley Wetlands and Agricultural Changes

ORIANE TAFT,
PETER SANZENBACHER,
AND SUSAN HAIG

### History

Wetlands in the Willamette Valley loosely follow the north-flowing Willamette River and stretch from Eugene to Portland, the most densely populated region of Oregon. Whether in urban or rural areas, these water resources have always been integral to the inhabitants of the valley. Prior to Euro-American invasion, the flora, fauna, and water in Willamette Valley wetlands were key to survival of the native Kalapuya (Taft and Haig 2003). This was particularly true during winter, when seasonal rains filled the wetlands and birds migrating south from the Arctic spent the majority of the annual cycle in the Willamette Valley.

Historical accounts and more recent analyses indicate that open prairie was the dominant feature of the Willamette Valley in the early 1800s, with half the prairie areas containing wetlands (Hulse et al. 1998, Taft and Haig 2003). In summer and autumn the Kalapuya set fire to prairie vegetation to promote maintenance of food plants and hunting for their estimated population of 13,000 (Boyd 1999). The combination of aboriginal fire and ponding from winter rainfall likely shaped this wetland prairie into valuable habitat for many waterbirds. And while mosquitos and the malaria they transmitted were deadly to local human residents, birds flourished on abundant invertebrate food resources in the shallow wetlands.

The best reports of which avian species were present historically came from fur trappers in the 1820s and 1830s. The Kalapuya recognized and hunted one swan species, three goose species, and three duck species (Jacobs 1945, Zenk 1976). Among Euro-Americans, most

swans were thought to be Tundra Swans, although it is likely that some were Trumpeter Swans. Snow Geese, White-fronted Geese, and Sandhill Cranes were widespread and abundant during migration and winter (Hartlaub 1852, Douglas 1959). Among the dabbling ducks, Northern Shoveler and Gadwall seemed to be less abundant than other species. Shorebirds were not well described, but accounts included Long-billed Curlew, Killdeer, and Wilson's Snipe.

For many Euro-American trappers and explorers in the early 1800s, the Willamette Valley was viewed as a paradise, given the access to huntable species and predictable water (Taft and Haig 2003). But the Euro-American invasion was a catastrophe for the native Kalapuya people, who were decimated by smallpox and malaria introduced by the early invaders (the population declined to about 600 people by 1841). After the few survivors were herded onto reservations in 1856, seasonal burning of prairies in the valley ceased, and Euro-American farmers began draining the wetlands and converting them to agriculture. While estimates vary, the resulting loss of habitat and species has been significant. In 2003, Taft and Haig estimated 67% loss of wetland habitat over more than 100 years. The species that have decreased the most dramatically in valley wetlands are Long-billed Curlews, Trumpeter Swans, Snow Geese, Sandhill Cranes, and Killdeer (Sanzenbacher and Haig 2001, Taft and Haig 2003).

Over the past 50 years, much of the agricultural land in the valley has been used for grass seed and sheep production. The flooded grass fields are still useable by birds during part of the year. In 1964–65, the USFWS recognized the value of the wetlands, particularly for Dusky Canada Geese (*B. c. occidentalis*), a subspecies of Canada Goose that nests on the Copper River Delta of Alaska and winters

Great Blue Herons and many other wetland birds overwinter in fields in the Willamette Valley that are planted with turfgrass for seed production. Many of these grass seed fields are being converted to hazelnut groves, threatening the availability of wetland habitat for overwintering birds. Photo by Dan Roby

primarily in the Willamette Valley. Under the leadership of Dave Marshall, three national wildlife refuges were developed (i.e., Finley, Baskett Slough, and Ankeny National Wildlife Refuges) that came to be known as the Willamette Valley National Wildlife Refuge Complex. Managing for geese had the side benefit of providing winter habitat for other waterbirds such as other waterfowl, shorebirds, herons, egrets, and more.

The Willamette Valley NWR Complex is a significant source of habitat for waterbirds, but the importance of agricultural wetlands in the valley as habitat for waterbirds cannot be overlooked. Looking at food abundance for wintering shorebirds and other waterbirds in Willamette Valley agricultural wetlands, Taft and Haig (2005) found aquatic invertebrate density and diversity to be equivalent to low compared to other similar and important winter sites in the United States. Wyss et al. (2013) more directly compared invertebrate communities in Willamette Valley agricultural areas (grass seed production) with more natural valley wetlands and found invertebrate biomass in perennial-grass wetlands was greater than in either annual-grass wetlands or native-prairie wetlands. In a third study, Evans-Peters et al. (2012) compared aquatic plant production (important for waterfowl) on private lands enrolled in the federal Wetland Reserve Program with similar wetlands on federal lands, such as wildlife refuges. Not surprisingly, they found better plant production in the more managed wetlands than those without active management. This study was undertaken to demonstrate the importance of active management on the federal Wetland Reserve Program lands. In the final analysis, while agricultural wetlands may have lower waterbird food production compared to natural areas, they are still critical producers of waterbird food (Taft and Haig 2006a).

In a comparison of wetlands gained or lost between 1972 and 2012, Fickas et al. (2016) measured the net change in the area of wetlands in the Willamette Valley before and after the no-net-loss policy established under Section 404 of the Clean Water Act in 1990 and found vegetated wetlands experienced a net loss of 314 ha (776 acres), and non-vegetated wetlands experienced a net gain of 393 ha (971 acres). These results indicated that higher-functioning wetlands were being replaced by non-vegetated wetland habitats, such as agricultural and quarry ponds. Wetland gains and losses were attributed to gains and losses of agricultural land. After the 1990 policy implementations, the rate of loss of wetland area slowed for some wetland categories and reversed into trends of gain in wetland area for others, perhaps representative of the success of increased regulation.

Today, the remaining wetlands are primarily dispersed among small urban remnant wetlands, hundreds of scattered agricultural wetlands, duck hunting clubs, holdings of The Nature Conservancy and the McKenzie River Trust, as well as the Willamette Valley National Wildlife Refuge Complex and E. E. Wilson State Wildlife Area. Unlike refuge wetlands, however, agricultural wetlands receive no formal protection. It is estimated that more

than 200 ha (500 acres) of valley wetlands are lost each year, according to the Oregon State of the Environment Report (Varva and Wood 2000). Recent and ongoing conversion of many Willamette Valley grass seed fields into hazelnut groves takes even more of these areas out of use for waterbirds and other wetland species. Prior to this conversion, Valley wetlands annually supported at least 200,000 waterfowl and 40,000 or more shorebirds in winter and an unknown number of migrants in spring and fall (Sanzenbacher and Haig 2002a, b; Taft and Haig 2003). The rapid increase in hazelnut production in the Willamette Valley makes it impossible to accurately determine the additional loss of wetlands at this time. But there is no recovery of the lost wetlands as a result of this change.

## Contaminants

The Willamette Valley National Wildlife Refuge Complex is managed in part for grass seed production, which includes use of the usual agricultural applications of pesticides, fertilizers, and dairy manure. These products can enter aquatic habitats and pose a risk to aquatic life. Materna and Buck (2007) and Buck (2017) investigated chemical use on the refuges and the potential effect they had on wildlife, particularly waterfowl. They found that while pesticides were rarely detected in refuge waterways, traces of the illegal pesticides atrazine and diuron exceeded levels considered safe for aquatic life and likely came from sources outside the refuge. Among nutrients, nitrate and phosphorus exceeded recommended levels in most places sampled, and the bacteria *Escherichia coli* was found in most samples. All the chemicals mentioned above are commonly used in the valley but are dangerous in water sources. Buck (2017) recommended enhancing vegetative buffers to preclude leaching of contaminants into waterways.

Looking more closely at the Willamette River, Chuck Henny (see Henny's biography on p. 54) spent much of his career examining contaminants along the Willamette River as they related to recovery of Osprey and Bald Eagles. More recently, Jackson et al. (2020) examined the presence of mercury in songbirds at various sites along and near the river; this included a mercury-contaminated Superfund site (Black Butte Mercury Mine), a reservoir known to contain methyl mercury (Cottage Grove Reservoir), and all downstream reaches of the Coast Fork and Willamette River and the Willamette Valley National Wildlife Refuge Complex. Not surprisingly, the highest mercury concentrations occurred near the mine, but mercury did not decline linearly with distance from the source of contamination. Jackson et al. (2020) also found that birds had elevated mercury levels in habitats such as wetlands commonly associated with enhanced methyl mercury production. Thus, the mercury risk to birds must be sampled far from the source to truly determine the extent of contamination.

## Migratory Connectivity

An obvious feature of the Willamette Valley from fall through spring is the daily movement of the seven subspecies of Canada Geese and Cackling Geese that fly from nighttime wetland roosts to foraging fields at daybreak and then back to the roosts before dark. Yet it might surprise people to know that the valley is equally important to shorebirds that also winter in the valley (Taft and Haig 2006b). Because it is critical to understand where a species spends each phase of the annual cycle and because birds spend most of their time away from breeding areas, we spent a number of years researching how three shorebird species, representing varying habitat and space needs, spent their time in the Willamette

Valley. In the final analysis, growth of crop vegetation, declining monthly rainfall, and evaporative water loss toward spring each contributed to a dramatic decrease in the availability of shorebird habitat between early and late winter (Taft et al. 2004). Whereas shorebird habitat could be found on 16% of the landscape in early winter, only 4% of the valley provided this habitat by late winter, representing a 75% reduction in shorebird habitat between winter periods (Taft and Haig 2006a, b).

Dunlin may be the easiest migratory shorebird to observe in winter, as they can form huge flocks that perform murmurations like European Starlings. Dunlin have a circumpolar distribution, and our genetics work indicated that Dunlin in the valley are likely from western Alaska (e.g., the Nome area) and belong to the subspecies *C. a. pacifica* (M. P. Miller et al. 2015). Depending on seasonal rains, they generally arrive in the valley in November and remain as late as April, thus spending most of the year in Oregon. Our radiotelemetry data suggested that they begin winter by using one to a few large wetland foraging areas, but by spring, when most wetlands (75%) have dried, they use a number of small wetlands for foraging and roosting (Sanzenbacher and Haig 2002a, Taft et al. 2008). These late-winter sites are farther apart than earlier sites and point to the need for wetlands to remain viable (i.e., not drained) throughout the spring.

In contrast, Killdeer, while present throughout the year, include a migrant population (89%) and a year-round resident population (11%; Sanzenbacher and Haig 2002b). Year-round residents were more sedentary than migrants, but neither had fidelity to particular sites in the valley. Radio-tagged individuals were found at an average of 12 sites with a home range size of about 13 km$^2$ (5 miles$^2$; Sanzenbacher and Haig 2002a). Like Dunlin,

A large flock of Cackling Geese, a species that has recently overwintered by the hundreds of thousands in the Willamette Valley, where grass seed fields have provided forage. Loss of grass seed fields and other wetland habitats in western Oregon poses an increasing threat to large numbers of overwintering shorebirds, waterfowl, and raptors. Photo by Scott Carpenter

home ranges of Killdeer increased as spring approached owing to having to fly farther to find appropriate wetland habitat. Killdeer are one of the species most people might be able to readily identify and consider common. Their population size has been declining for several decades, however, as Killdeer habitat succumbs to wetland development, conversion to agriculture, and pesticide use (Sanzenbacher and Haig 2001).

The mystery shorebird in the valley is Wilson's Snipe. This species is not well studied, yet it remains one of two shorebirds still legally hunted in North America. There is mounting evidence for significant decline throughout their range, particularly in the Pacific Northwest (Sauer et al. 2017 and others). As early as 1951, OSU alum Gordon Gullion suggested snipe were common year-round residents in the Willamette Valley, while more recently, David Marshall et al. (2003) suggested otherwise. We did not find that snipe were as common as previously described, but we did find both authors were correct about residency (Cline and Haig 2011).

Our snipe study spanned every month of the annual cycle to determine whether they did in fact remain throughout the year. Using radio-tracked birds (n = 37), we found 74% were winter residents, 14% were transients during winter, 9% were summer residents, and one remained throughout the entire annual cycle. Like Killdeer, Wilson's Snipe moved around the valley a bit but basically maintained a home range of less than 7.8 km² (3 miles²), expanding in early spring as surface water areas subsided and crops began growing.

While we have gained some insight into the annual life cycle of Wilson's Snipe, they remain a mystery. Answering the question about their annual residency would help prepare more effective conservation plans for a species that is likely in decline.

## Robert Jarvis

 Professor Jarvis came to Oregon State University in 1971 as the university started to expand its wildlife program. A waterfowl biologist, he developed graduate research programs primarily in the Willamette Valley and Klamath Basin. While known principally for his work on waterfowl, he also directed students in studies of Yellow Rails, Snowy Plovers, and the movement of White-faced Ibis into Oregon. His students Mark Stern, Maura Naughton, Ruth Wilson Jacobs, and many others went on to make important contributions to Oregon avian conservation efforts. For his many contributions, Bob was awarded OSU's Outstanding Service Award in 2000 and the David B. Marshall Lifetime Achievement Award in 2001 from the Oregon Chapter of The Wildlife Society. He retired from OSU in 2001.

*Photo by Trina McGaughy*

Taken together, results of our Willamette Valley shorebird connectivity studies highlight the need to consider differences among species, even closely related species, when developing regional wetland conservation plans. Each study above illustrated the dependence on agricultural fields for success of shorebirds throughout the winter and migratory periods. If these wetlands are converted to other purposes, the species will be entirely dependent on federal and state lands, which given the distance between them, may put undue stress on the birds as they try to move among them as conditions change.

## Bruce Dugger

Bruce Dugger joined the faculty of Oregon State University's Department of Fisheries, Wildlife, and Conservation Sciences in 2002 and is now professor and associate department head. Research by Dugger and his students focuses on the ecology and conservation of waterbirds, particularly waterfowl, and the wetland habitats they depend on. In Oregon, his work in the Klamath Basin has provided an understanding of how waterfowl are affected by water shortages and water allocation decisions by regulatory agencies and is demonstrating the importance of basin wetlands to continental populations of migratory waterfowl. Work in the Willamette Valley has focused on understanding the value of restored wetlands for migratory ducks as well as how Canada and Cackling Geese use agricultural habitats to meet their needs in winter. Formerly the Mace Watchable Wildlife Chair at OSU, Dugger also served on the board of directors for the Oregon Chapter of The Wildlife Society, was an associate editor of the *Journal of Wildlife Management*, is on the editorial board for the journal *Wildfowl*, and is an elective member of the American Ornithological Society.

*Photo by Bruce Dugger*

## Klamath Wetlands

MARK STERN

The vast Klamath Wetlands, nestled in the north-south-trending fault block mountains of south-central Oregon, are remnant waterbodies from the Pleistocene-era Lake Modoc, a lake that once spanned more than 2,500 km$^2$ (1,000 miles$^2$) south from Fort Klamath in the Wood River Valley well beyond Tule Lake in northeastern California (Dickens 1985). Unlike other Pleistocene lakes to the east that evolved in the arid Great Basin environment, Lake Modoc and its successor water bodies and wetlands benefited from snowmelt off the Cascades, Yamsi Mountain, Winter Rim, and a basaltic volcanic substrate with a rich groundwater aquifer, generating an abundance of springs and seeps and spring-fed rivers among a forested and sagebrush landscape; hence the genesis of Klamath wetlands (Benke and Crushing 2005).

Over time, Klamath wetlands in the northern half of the upper basin evolved into lush marshes, wet meadows, fens, springs, riparian corridors, and shallow lakes amid forests of stately ponderosa pine and other conifers. Only on the north side of Klamath Marsh can one still sit among the huge legacy 300-year-old ponderosa pines and quaking aspen and see and hear a White-headed Woodpecker and other woodland species while simultaneously looking into a vibrant tule marsh less than 100 feet away to see a Yellow-headed Blackbird chortling away. Klamath is, and always has been, rich in birds, and by all accounts, it is a cornerstone of the Pacific Flyway. In the southern half of the Upper Klamath Basin, water from the Lost River meandered to a terminal basin that filled Tule Lake and adjoining lands with vast marshes and lakes amid an arid sagebrush setting.

Nearly 11,000 years ago, Indigenous people of Klamath and Modoc ancestry lived near and along the springs, marshes, rivers, and lakes in the upper basin, hunting and gathering from the abundance of native foods, fish, wildlife, and birds. One can only imagine the wealth of natural diversity that thrived then, including Chinook salmon (*Oncorhynchus tshawytscha*) that migrated up the Klamath River from the Pacific Ocean, and resident endemic native fishes, especially Lost River suckers (*Deltistes luxatus*), shortnose suckers (*Chasmistes brevirostris*), and Great Basin redband trout (*O. mykiss newberrii*). The numerous waterbirds included migrants passing through to the Subarctic and breeding birds that found homes and nested in these Klamath wetlands, contributed to this diversity of wildlife.

Fast-forwarding to the late nineteenth and early twentieth centuries, with the arrival and settlement of Euro-Americans, ranching became a dominant activity in the meadows, wetlands, and riverways of the Upper Klamath Lake watershed. Below Upper Klamath Lake, irrigated agriculture was launched with the Bureau of Reclamation's Klamath Project in 1905. Engineers, farmers, and ranchers busily diked and drained much of Lower Klamath and Tule Lakes, building an expansive system of weirs, water controls structures, canals, dikes, and ditches. Farmers, assisted and encouraged by local, state, and federal governments, developed more than 210,000 acres of irrigated agriculture, growing alfalfa, grains, potatoes, onions, mint, and other row crops. The resultant combination of regulated water and a hot growing season created a bountiful agricultural economy at the expense of an enormous loss and alteration of historical wetlands.

In the 1900s, renowned conservationist William Finley and others recognized the abundance of colonial waterbird populations in the Klamath Basin, including nesting American White Pelicans (4,000–9,000

pelicans in Lower Klamath alone), egrets, herons, grebes, and terns, and began to decry the impacts to these species from market hunters who were busily shipping huge numbers of skins and dead birds to markets in San Francisco and elsewhere. In 1903, more than 30,000 grebe skins were taken and shipped from one location in Klamath (Foster 2002). In addition to plume hunters, ducks were commercially harvested. In 1907, over 120 tons of ducks were shipped to meat markets in San Francisco (Foster 2002). In response to the outcry from Finley and others, in 1908, President Theodore Roosevelt designated Lower Klamath NWR, the first refuge set aside for migratory waterfowl and other marsh birds. But Lower Klamath NWR was superimposed on the footprint of the Klamath Project, as was Tule Lake NWR in 1928. For many decades there has been ongoing tension and disagreement among stakeholders about how to prioritize agriculture versus wildlife on these two refuges. The Kuchel Act of 1964 attempted to clarify this issue, codifying commercial farming on 22,000 acres of refuge lands, requiring the secretary of the interior to administer these refuge lands for "the major purpose of waterfowl management, but with full consideration to optimum agricultural use that is consistent herewith." The wording of this legislation is open to interpretation, however, and conflicting viewpoints remain unresolved today.

While ongoing discord occurred in the lower basin around the management of Tule Lake and Lower Klamath refuges, there was also continued recognition of the value and opportunity to protect waterfowl and waterbirds in the basin, and four additional national wildlife refuges were designated between 1911 and 1978. These were Clear Lake NWR (1911), Upper Klamath NWR (1928), Klamath Marsh NWR (1958), and Bear Valley

A breeding pair of Bald Eagles, larger female on the right, in the Upper Klamath Basin, which has the highest nesting densities of Bald Eagles in Oregon and where large numbers of Bald Eagles congregate in winter owing to the concentration of overwintering waterfowl. Photo by Mel Clements

NWR (1978; see Table 1 on p. 6), leading to a combined 190,000 acres of what is now called the Klamath Basin National Wildlife Refuge Complex. The refuge habitats in the basin are diverse, from deep water marshes to alkaline flats and, along with adjacent private meadows and farmlands and wetlands restored by The Nature Conservancy and others, support large numbers of birds. In the 1950s, nearly 6 million waterfowl migrated annually through the Klamath Basin, the Klamath refuges, and especially Lower Klamath and Tule Lake refuges, which are renowned and favored by duck and goose hunters.

Today, over 75% of all waterfowl in the Pacific Flyway stops in the Klamath Basin, and while peak numbers have declined since the 1950s, peak fall numbers for migrating waterfowl average 1 million birds annually (Gilmer et al. 2004), though recently the numbers of waterfowl may be lower owing to low-water conditions on Lower Klamath NWR. In addition to migratory waterfowl, more than 280 avian species have been recorded on or near the refuges (USFWS 2016b), and 345 species of birds have been reported throughout Klamath County, Oregon (eBird 2021). Klamath remains rich in birds.

The abundance of fall and winter waterfowl also attracts wintering Bald Eagles that migrate from more northerly breeding areas, congregating into what has long been recognized as the largest gathering of wintering Bald Eagles in the United States outside of Alaska. Wintering eagles often roost in conifers in the Bear Valley NWR, foraging on waterfowl during the day on Tule Lake and Lower Klamath NWRs, and around the basin generally. The Klamath Basin Audubon Society hosts an annual Winter Wings Festival that originated in 1980 in honor of and as an opportunity to view the more than 500 Bald Eagles that winter in the basin. Hundreds of people attend this festival annually, generating a local economic benefit of an estimated $500,000 or more annually to the Klamath Falls community. The Winter Wings Festival has been recognized by birding magazines as the oldest birding festival in the United States. The Klamath Basin is also home to the largest number of breeding Bald Eagles in Oregon (see the Bald Eagle section on p. 56).

Research on Klamath birds is surprisingly limited in contrast to the diversity and abundance of birds in the Klamath Basin. In addition to the early work of Finley (1907a, b), seminal efforts were led by the USFWS, which undertook its first aerial survey in 1943, subsequently conducting annual basin-wide waterbird surveys from 1953 to 2001, with a focus on waterfowl. Jim Hainline, USFWS wildlife biologist for the Klamath Refuge Complex (1978–2001), and his predecessors and successors were instrumental in this work, thoroughly documenting annual waterfowl numbers during all seasons (Gilmer et al. 2004).

In the early 2000s, David Shuford, Point Blue Conservation Science (formerly Point Reyes Bird Observatory), led a basin-wide survey for all nongame waterbirds, affirming the impressive diversity and abundance of this bird resource (Shuford et al. 2004). Songbirds and Neotropical migrants also abound in the basin. Since 1997, the Klamath Bird Observatory has orchestrated a long-term monitoring and research effort in the Klamath-Siskiyou Ecoregion in partnership with the USFS Redwood Sciences Laboratory and others, capturing and banding more than 100,000 birds of 150 species from a network of 39 banding stations throughout riparian habitats in the Upper Klamath Basin (see Figure 5 on p. 63, Alexander 2011, Rockwell et al. 2017).

Localized inventory and research within the basin have highlighted locations of Bald Eagles and characterized their winter roost sites and foraging patterns (Keister and Anthony 1983, Keister et al. 1987), and diet and environmental contaminants (Frenzel and Anthony 1989). Other research efforts include the study of Mallard broods on the refuges (Mauser et al. 1994) and have focused on various prominent birds in the Klamath, including Sandhill Cranes (Stern et al. 1987a, Ivey and Herziger 2000), Yellow Rails (Stern et al. 1993, Popper and Stern 2000, M. P. Miller et al. 2012b), Black Terns (Stern and Jarvis 1991, Stephens et al. 2015b), Western and Clark's Grebes (Ratti 1985, Nuechterlein and Buitron 1989), Red-necked Grebes (Watkins 1988), and the movement of fall migrant songbirds (Wiegardt et al. 2017), among other species (Farner 1952, Stern et al. 1987b, Summers 1993). And downriver in the Lower Klamath Basin, Tiana Williams, the director of the Yurok Wildlife Department, is now helping to lead efforts to reintroduce the California Condor to its historical Oregon range.

Of note, the basin hosts at least three breeding colonies of American White Pelicans and an array of colonial nesting waterbirds, including Great Egrets, Great Blue Herons, Black-crowned Night-Herons, Double-crested Cormorants, Snowy Egrets, Black Terns, Forster's Terns, and White-faced Ibis, the latter species having expanded its breeding range into the Klamath in the 1980s when the Great Salt Lake and other Intermountain Lake areas incurred high waters and flooded out many marshes and wetlands in Utah and elsewhere (Ivey et al. 1988). Western Grebes, Clark's Grebes, Eared Grebes, Horned Grebes, and Red-necked Grebes have all nested in the Klamath wetlands, at least in small numbers. Sandhill Cranes are found throughout the basin, feeding in meadows and nesting in adjacent marshlands. Sycan Marsh and Klamath Marsh NWR populations are two of

the larger concentrations of breeding cranes, numbering over 150 breeding pairs (Ivey and Herziger 2000). Yellow Rails, generally found in wetland systems northeast of the Rockies, occur as a disjunct population of more than 200 pairs, with the largest concentrations found at Klamath Marsh NWR, the Wood River Valley, and Sycan Marsh (Lundsten and Popper 2002; see the Yellow Rails section on p. 201). Landbirds also abound in Klamath, too many to enumerate.

The critical ingredient to Klamath wetlands and bird habitat has been and always will be water, of which there is a growing scarcity owing to a combination of competing needs among fisheries, waterfowl, irrigated agriculture, and ranching, compounded by downward trends in annual snowpack attributable to changes in climate and weather. Tule Lake and Lower Klamath NWRs are part of the Klamath Reclamation Project and are at the terminal end of the extensive irrigation system, with water rights that are "junior" to the Klamath Project, essentially receiving the "runoff water" from the massive Reclamation Project. Even in an average water year, late-season deliveries of water to Tule Lake and Lower Klamath refuges can be tenuous. In the past several years, little to no water has reached Lower Klamath NWR. That means in years of low water the refuges get less water which, compounded by hot temperatures, in some years can result in the drying out and exposing of large areas of mudflats, concentrating molting ducks in the remaining stagnant pools and leading to outbreaks of avian botulism and generally less habitat for fall migrants. In a recent year, the California Waterfowl Association estimated the loss of 60,000 birds to avian botulism in the Klamath Basin.

On the bright side, in the 1990s, Tule Lake and Lower Klamath NWRs initiated a program called Walking Wetlands, where crop fields

Black Tern in breeding plumage. Black Terns are a small tern that nest in wetlands of the Upper Klamath Basin, where they lay their eggs on floating mats of vegetation, especially tules, and prey mostly on large insects. Photo by Mick Thompson

and wetlands management units are interchangeably managed, rotating from wetland to cropland and back on a frequency of five years or so. The rotating interchange of crops and wetlands benefits waterbirds and farmers. Waterbirds get young, developing marshes that have a more heterogeneous mix of marsh vegetation and open water and sloughs, unlike the more established adjoining wetlands that tend to grow into dense, continuous stands of hard-stemmed bulrush and cattails. These younger seral marshes are highly productive for waterbirds—a win. Likewise, recently drained wetlands are typically free of nematodes and other soil pathogens, reducing the need for the use of pesticides and herbicides, and regenerating soils that are highly productive, all leading to increased yields, less cost,

and the ability to grow organic crops more expediently. Agriculture on the refuge and adjacent private lands also provides an attractive abundance of "food" to fall and wintering migratory waterfowl, particularly recently harvested and partially flooded fields of barley, wheat, and other grains; these "ag" fields are a key factor in drawing peak numbers of waterfowl in the Pacific Flyway to Klamath.

Fast-forward again to the 1980s, when native freshwater fish populations in Upper Klamath Lake, especially the Lost River sucker and the shortnose sucker, were on a steep decline as a result of habitat loss and poor water quality in Upper Klamath Lake. In 1988, both fish species were federally listed as Endangered by the USFWS. Subsequently, federally mandated biological opinions set annual minimum water level requirements in Upper Klamath Lake, which effectively governed how much water could be allocated to the Klamath Project, and the refuges, for irrigation and waterbirds. In 1997, coho salmon (*O. kisutch*), which occur in the Klamath River and the Lower Klamath Basin, were listed as Threatened by the National Marine Fisheries Service. Again, a federally mandated biological opinion set minimum flow requirements for the Klamath River, flows that originate primarily from Upper Klamath Lake. This set the stage for an ongoing tug-of-war between water needs for endangered suckers in the upper basin versus threatened coho salmon downstream in the Klamath River, with the Klamath Project and two refuges caught in the middle—third in line for water, with refuges last in line.

Long story short, water requirements for threatened and endangered fishes in Upper Klamath Lake and the Klamath River take legal precedence over the federally operated Klamath Project and its 210,000 acres of irrigated cropland and the Tule Lake and Lower Klamath refuges that are last in line to receive water after the Klamath Project (USFWS 2019c). During the severe drought of 2001, water flows into Upper Klamath Lake were so low that the irrigation season for the entire Klamath Project was canceled, and neither the irrigated crops nor the refuges received any water, except for a nominal last minute late-season delivery in July.

In 2016, after a lengthy 40-year process, the water rights in the Klamath Basin, Oregon, were adjudicated when the Oregon Department of Water Resources issued their Final Order of Determination, determining the seniority date and quantity of water that was the legal water right for all landowners. With this determination, the Klamath tribes received the most senior water right for

"instream flows." These tribal water rights prioritize keeping water in the tributaries to and in Upper Klamath Lake for the benefit of fisheries, which are part of their treaty rights, and to maintain ecologically appropriate water levels for Upper Klamath Lake and Klamath Marsh NWR.

In spring 2021, below-average snowfall and projection of low runoff to Upper Klamath Lake and the Klamath River again led to the nearly complete curtailment of delivery of irrigation water to the Klamath Project and Lower Klamath and Tule Lake NWRs. In short, it is not possible to honor or fulfill all the commitments to water, particularly in drought years, that have been made over time—to the Klamath tribes for their instream water rights, the Lost River and shortnose suckers and coho salmon, the farmers and ranchers of the Klamath Project, and lastly, the Tule Lake and Lower Klamath refuges.

A changing climate and predicted higher temperatures, reduced snowpack, earlier snowmelt, and longer and hotter summers do not bode well for a Klamath wetland system that has already reached a point of stress and impacts to a wide range of biological, cultural, economic, and social values, including Klamath birds (Barr et al. 2010). The Klamath Basin Restoration Agreement signed by Governors Arnold Schwarzenegger (California) and Ted Kulongoski (Oregon) and Secretary of the Interior Ken Salazar in 2010 called for removal of four large dams on the Klamath River, plus massive investments in restoration and other changes in the upper basin that would have helped build ecological and social resilience into the entire 10-million-acre Klamath River basin. The Klamath Basin Restoration Agreement, however, needed companion federal legislation for enactment and lapsed when Congress failed to act within a five-year timeline, in part because of the inaction and opposition of Republican congressmen Greg Walden (Oregon) and Doug LaMalfa (California). Even so, PacifiCorp, owner of the four large dams on the mainstem Klamath River, has proceeded on a path to decommission and remove these dams, which is a first step toward addressing the downstream fishery needs and further restoration in the Upper Klamath Basin in Oregon and northern California.

Meanwhile, the Klamath Basin remains an essential part of the Pacific Flyway, welcoming migrants from the north during their fall and spring journeys. Long skeins of white geese, Snow Geese, and their cousin Ross's Geese take flight from their feasting on recently harvested farm fields with a thunderous roar of wingbeat and high-pitched voices. White-fronted Geese, with their colorful speckled bellies and a similarly distinct high-pitched squeak, group together among the other geese. In winter, flocks of a 1,000 or more Tundra Swans pass gracefully, cooing as they fill the skyline. And majestic Bald Eagles gather and soar above, their presence often causing an instantaneous ruckus and flight of large flocks of ducks and geese below. In breeding and nesting seasons, Klamath is home to a bountiful array of waterbirds and landbirds—American White Pelicans foraging on lakes and soaring gracefully while catching thermals; Western and Clark's Grebes with their elegant courtship dances; Black Terns in delicate flight, gleaning insects atop the water surface; Sandhill Cranes standing tall and announcing their presence in raucous duets; and the colorful array of Cinnamon Teals, Northern Shovelers, Northern Pintails, Gadwalls, Canvasbacks, and other marsh birds, all in brilliant breeding plumage adorning the wetlands and open waters. Secretive marsh birds—Virginia Rail, Sora, Yellow Rail, and

Least Bittern—call uniquely, often at night but rarely observed. Distinguished-looking Great Egrets and herons, with breeding plumage once drooled over by plume hunters, traverse about the landscape with smooth, confident, long wingbeats. White-faced Ibis fly low above the water, with individuals jostling and maneuvering in tight squadrons. And in the nearby coniferous trees, brightly colored yellow-and-red Western Tanagers announce their presence in a distinctive, melodic robin-like call—one of the innumerable landbirds that call Klamath home. Klamath wetlands and Klamath birds are one of the many great marvels of Oregon's rich and diverse ornithological history.

### Yellow Rails

MARK STERN AND SUSAN HAIG

The Yellow Rail has had a checkered history in Oregon ornithology and will likely continue in that vein, given climate-related changes to its southeastern Oregon marshy habitat. In general, Yellow Rails are found across shallow marshy areas on the US-Canada border and along Atlantic and Gulf of Mexico coastal areas, almost entirely east of the Rocky Mountains except for the disjunct breeding populations in south-central Oregon and a few sightings in eastern California (Leston and Bookhout 2020). Although they are widely distributed, Yellow Rails are secretive, rarely observed in daylight, and mostly detected at night by their distinctive, metallic-sounding "click-click, click-click-click" call. While first reported in Oregon in the early 1900s (e.g., Prill 1924, Griffee 1944), they were never considered to be common, and none were seen between 1950 and 1980 (Stern et al. 1993). In 1983 the American Ornithologists' Union determined that the species had been extirpated from the western United States. Soon after, however,

A small, local, disjunct population of Yellow Rails nesting in south-central Oregon was redis-
covered in the early 1990s at a number of sites with shallow, flooded sedge meadows. This
population is at risk of extirpation because of the increasing frequency and severity of drought
in the region. Photo by Ken Popper / The Nature Conservancy

scattered reports emerged of Yellow Rails
near Fort Klamath, and eventually there were
enough reports for The Nature Conservancy
(TNC) biologist Mark Stern and colleagues to
launch an organized effort over the next few
years to search for the rails. And they found
them, locating Yellow Rails at 14 sites and documenting the first Yellow Rail nest in Oregon
in 65 years (Stern et al. 1993).

Subsequently, Ken Popper and Mark Stern
studied the breeding ecology of Yellow Rails in
south-central Oregon during 1995–98, finding
34 nests in shallow, flooded sedge meadows
(Popper and Stern 2000). Most nests were ex-
tremely well hidden, swirled among bundled-up

sedges, with a slight canopy and an opening
along the side, in shallow water of 5–8 cm
depth. Popper had read about an egg collector,
Reverend P. B. Peabody, who scoured the sedge
meadow swales of the Dakotas in the early
1900s and had a knack for finding Yellow Rail
nests. Ken, who is naturally inquisitive, deliberate in manner, and generally not in a rush,
seemed to absorb and transcend Peabody's
demeanor, honed his search image, and clearly
mastered the art of finding Yellow Rail nests.
His discoveries were remarkable. Using radio
transmitters, Ken tracked nine male Yellow Rails
for three to seven weeks each, documenting
home ranges that averaged 19.3 ha (48 acres),

with birds moving their territories to stay in ideal water depths of around 7 cm (Popper and Stern 1996). From 1995 to 2006, Popper, Stern, and colleagues also banded 473 adult male Yellow Rails at various sites in the Klamath Basin and Big Marsh in northern Klamath County; 31 individuals were recaptured at the same site in the following year, and 2 were captured again in the third year (Lundsten and Popper 2002). These three-year-old adult males hold the longevity record for this species. Locations used by Yellow Rails were generally consistent among years but varied periodically owing to annual differences in water levels. Most movements of Yellow Rails during the breeding season were localized within breeding sites; however, there were a few longer movements, the longest occurring in 2000 when a male Yellow Rail captured and banded at Sycan Marsh was recaptured at Klamath Marsh NWR within the same breeding season. The largest concentrations of Yellow Rails occurred at Klamath Marsh NWR, and in total there were an estimated 200–230 pairs of Yellow Rails in Oregon in 2002. In 2021, a coordinated interagency survey found only 125 Yellow Rails in Oregon; 75% were in Klamath Marsh NWR. The drought conditions of 2019–21 may explain the diminished numbers. So, while Popper and Stern spent many a night traipsing through Klamath Marsh clicking stones together to call out the rails, other questions arose, like, where else in Oregon can Yellow Rails be found, where did these rails come from, and where do they go (if anywhere)?

From 2012 to 2014, USGS biologists Susan Haig and Sean Murphy, USFWS biologist Mike Green, and TNC's Ken Popper attached VHF radios to Yellow Rails at Klamath Marsh to address the movement questions. The radiotelemetry information revealed that the birds left the refuge during winter and moved around the refuge a bit prior to leaving in September and October. Where the rails went

to during winter was a mystery, although the researchers did hear radio signals from two birds in the Wood River Valley in the Upper Klamath Basin in October 2013. As recently as October 2021, a road-killed Yellow Rail was found in Benton County, Oregon, near Corvallis, highlighting once again that the distribution and migration behavior of Yellow Rails is poorly understood.

In 2008, Susan Haig's lab at the Forest and Rangeland Ecosystem Science Center (FRESC) in Corvallis conducted a molecular study in which they examined the population status and subspecific relationships of Yellow Rails (M. P. Miller et al. 2012b). Ken Popper and Mike Green collected blood samples from birds in Manitoba, Michigan, Minnesota, Quebec, and Wisconsin to complement Oregon samples already collected by Haig's field crew. The molecular data yielded surprising results, as they showed that even though there was a huge gap in the species distribution between Oregon and birds east of the Mississippi and there were significant population genetic differences across their range, the results could not be explained based on geographic separation (M. P. Miller et al. 2012b). That is, there were no subspecific differences. Rather, there were a number of instances (50% of sites sampled) of population bottlenecks indicating significant loss of genetic diversity in various regions. Across all comparisons, Oregon's Yellow Rails had the lowest genetic diversity and the strongest evidence of a major population bottleneck, likely occurring when the Oregon population was founded and/or became quite small and was sustained by only a few individuals.

The levels of differentiation among Yellow Rail populations across North America were similar to or less than the differentiation seen among geographically closer populations of the Black Rail (*Laterallus jamaicensis*) and Clapper Rail (*Rallus crepitans*) analyzed from

California alone (Fleischer et al. 1995, Girard et al. 2010). Thus, Yellow Rails may disperse more readily than either Black Rails or Clapper Rails. Additional data on movements among all of these species are needed to verify such behavioral differences and would help in planning habitat conservation measures. Either way, the genetic results did not suggest that Oregon Yellow Rails constituted a genetically healthy local population. Increased population sizes and more genetic exchange among populations would help rectify this situation.

Determining the precise causes and timing of past bottlenecks is difficult. As for Yellow Rails, wetland loss is generally considered to be the greatest threat to most rail populations and can create barriers to dispersal, hence genetic and demographic bottlenecks (Eddleman et al. 1988). In Oregon, the Yellow Rail is listed as a Conservation Strategy Species of Concern because of its scarcity and the long period over which it was not observed in the state (Oregon Department of Fish and Wildlife 2016b). It has been estimated that most wetlands in the Klamath Basin of Oregon and northern California, as well as the marshes of San Francisco Bay, have been lost or otherwise altered since 1900 because of development or agricultural conversion (Bottorff 1989, Dedrick 1989, Fleskes 2012). Considering the Yellow Rail's strong reliance on these wetlands for breeding and wintering habitat (Robert et al. 1997, Goldade et al. 2002), these activities may have decimated their local populations and in part account for the 30-year absence of the species in Oregon. If so, this process also likely resulted in the reduction of the Oregon population detected by our genetic bottleneck analyses. Looking forward, and as with many other waterbird species, the current and ongoing drought in the Upper Klamath Basin does not offer a great deal of promise for their recovery without additional conservation action.

## Wetlands and Waterbirds in the Great Basin Region of Oregon

SUSAN HAIG AND LEWIS ORING

Great Basin wetlands in Oregon provide diverse, abundant, and some of the most important regional waterbird sites in the Pacific Flyway, as they support over 70% of the dabbling ducks that pass through during migration (see Figures 5 and 6 on p. 63; Ducks Unlimited, unpublished data, 2021). These wetlands stretch from Malheur Lake in central Oregon south to Goose Lake, which straddles the Oregon-California border (Figure 8). This wetland chain continues south into northeastern California and Nevada, and taken together, it has come to be known as the SONEC (southern Oregon and northeastern California) region of the Great Basin. The Great Basin wetland chain ultimately ends at the Salton Sea, Imperial Valley, or in the Gulf of California, where many Great Basin species spend the winter before heading back north to their breeding range.

Despite the critical importance of Great Basin wetlands to birds in the Pacific Flyway, there has been surprisingly little attention paid to them until recently. In the 1980s, Joe Jehl (Hubbs Sea World, California) studied the movements of Wilson's Phalaropes and Eared Grebes from Lake Abert south to Mono Lake and beyond (Jehl 1988). BLM biologist Walt Devaurs carried out weekly waterbird counts at Lake Abert in the 1980s (Warnock et al. 1998); ODFW Wildlife Manager Marty St. Louis managed Summer Lake Wildlife Area from the 1980s until 2018; as the wildlife biologist at Malheur National Wildlife Refuge, Gary Ivey carried out shorebird surveys and studied Sandhill Cranes; and Ron Larson worked for USFWS out of Klamath Falls (Larson et al. 2016). USGS biologists from California tracked Northern Pintails through the Great Basin region and north to Alaska (M. R. Miller et al.

2005). We worked on shorebirds across the region from 1995 to 2018. All of that expertise has since retired and/or left the area. Recognizing the importance of these eastern Oregon wetlands, the Intermountain West Joint Venture and the National Audubon Society (Stan Senner) have recently become more active in promoting better management of wetlands and waterbirds in the area, although primarily managing for waterfowl (Donnelly et al. 2019). Furthermore, Portland Audubon has now stationed a full-time biologist at the Malheur Field Station to carry out surveys and monitor populations.

In summer 2021, Oregon Senator Jeff Merkley and Utah Senator Mitt Romney as well as congressional representatives from Oregon and Utah sponsored a bill that passed the US Congress, The Saline Lake Ecosystems in the Great Basin States Assessment and Monitoring Program, which calls for USGS to carry out an assessment of the water status and needs for Great Basin wetlands and waterbirds. This is the first major funding program for Great Basin wetlands and promises to be an important step in mitigating the effects of climate change and human alteration of the landscape.

For a number of waterbird species, the SONEC / Great Basin wetlands represent their primary habitat during various phases of the annual cycle. For example, the entire North American population of Eared Grebes, up to 90% of all Wilson's Phalaropes, and more than 50% of American Avocets stop and/or breed at these sites (Oring et al. 2013). A large percentage of American White Pelicans (up to 50% of the global population), Black-necked Stilts, Long-billed Curlews, and Snowy Plovers also breed in the region. Yet extensive evidence already indicates that Great Basin waterbird populations are in decline (Warnock et al. 1998, Haig et al. 2019). Great Basin shorebird

populations have experienced a 70% decline since 1973, and more than half of the nine most important western saline lakes for birds have diminished by 50% to 95% over the past 150 years (Jehl and Johnson 1994, Shuford et al. 2018). Among nine Great Basin breeding waterbirds considered, Langham et al. (2015) found seven that were in danger of losing half or more of their current range by 2050 because of climate change (Eared Grebe, American White Pelican, White-faced Ibis, American Avocet, Snowy Plover, Marbled Godwit, and Wilson's Phalarope).

In our more than 25 years of examining Great Basin waterbird use and movements among various wetlands throughout the annual cycle, we have come to recognize that a key factor in the use of these wetlands is the salinity of the water (Haig et al. 1998, 2019; Figure 8). While all waterbirds need fresh water to drink, young birds (chicks, ducklings, nestlings, etc.) do not have well-developed salt glands at hatch and consequently cannot drink or forage in the saltier water found at many Great Basin wetlands. Great Basin wetlands range from fresh water at Summer Lake Wildlife Area to saline water in Summer Lake, Malheur Lake, Harney Lake, and others to hypersaline water (i.e., water saltier than the ocean) at Lake Abert. Thus, fledglings need fresh water as found at locations such as Summer Lake Wildlife Area and in the freshwater in-stream flows to other lakes. This patchwork of saline and nonsaline wetlands means that adults birds can use them all for foraging, but they need to nest near a freshwater source to raise their young. And if the freshwater source dries up or becomes more saline over the course of

the summer or as temperatures rise because of changing climates, adults must march their nonflying offspring across the desert to a freshwater source, risking dehydration and predation along the way.

Lake Abert is one of only three hypersaline lakes in North America, with Mono Lake, California, and the Great Salt Lake, Utah, being the other two. While Lake Abert is viewed as the waterbird "bread basket" in the region, Senner et al. (2018) found there was a fairly narrow margin of salinity that produced the massive quantities of brine flies that waterbirds feed on at the lake. Thus, very wet years or very dry years can eliminate food production altogether.

During spring migration, some waterfowl and shorebirds have similar issues with needing snowmelt-flooded fields or pastures to obtain food and water. The Intermountain West Joint Venture has launched a large campaign to work with landowners (55% of Great Basin land in Oregon is in private ownership) to leave this water on their fields for as long as possible to ensure safe passage for migrants and residents alike. This program particularly aims to benefit Northern Pintail, a species that is suffering from significant declines. To that end, Donnelly et al. (2019) recently examined the relationship between dabbling duck arrival in the Great Basin region and water availability via spring snowmelt on private land and managed refuges. Not surprisingly, water availability coincided with bird arrival in managed areas. Donnelly et al. (2019) made suggestions for how water permits and other activities could better align water availability on private and public land. While this program

OPPOSITE: A pair of American Avocets feeding on a swarm of brine flies on the shores of hypersaline Lake Abert in the Great Basin. Hypersaline lakes are critical migratory stopovers for many wetland birds within Oregon's portion of the Great Basin. Drought and over allocation of water resources to agriculture are threatening these productive ecosystems. Photo by Susan Haig

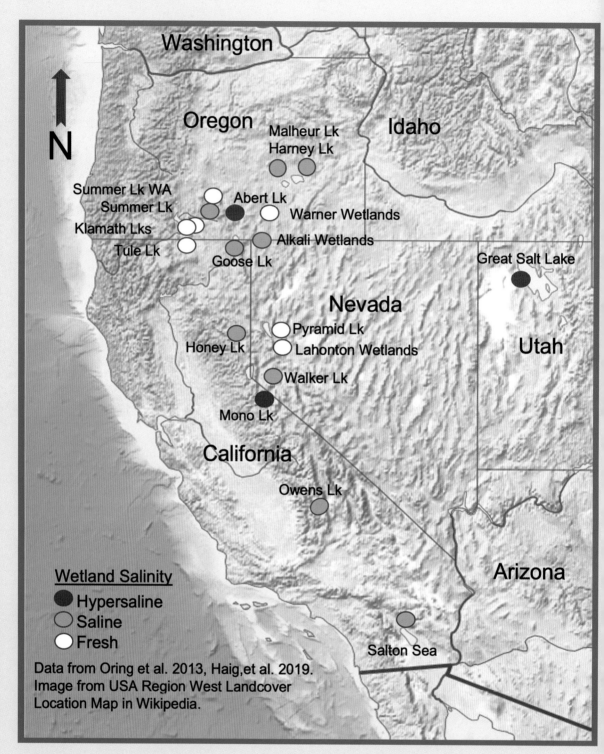

FIGURE 8. Distribution of Great Basin wetlands and other inland Pacific Flyway waterbodies and their relative salinities. Susan Haig, Oregon State University

Wilson's Phalarope female. Phalaropes have a polyandrous breeding system, where one female will mate with multiple males in a breeding season. That is, she lays her eggs in several nests, and leaves a male to incubate the eggs and tend the young once they've hatched. Consequently, female phalaropes are more brightly colored than males. Wilson's Phalaropes nest in wetlands throughout eastern Oregon. Photo by Jared Hobbs

provides early spring habitat for a number of waterfowl species, several conservation groups are concerned with its effect on withholding the spring water in the Chewaucan Basin from running into Lake Abert. They worry that not filling Lake Abert in the spring will lead to excessive salinity or dying up later in the summer, taking away brine flies and brine fly larvae for bird food and for commercial harvest.

Understanding the relationship of wetland salinity or water quality to waterbird use is key in evaluating the potential impact of changing climates (Haig et al. 2019). Precipitation in the Great Basin region comes mostly in winter, in the form of snow. If there is little snow or it melts early, there can be negative impacts on

the subsequent spring migration and summer breeding season for waterbirds and more. Recently, we examined patterns of temperature and precipitation over a 100-year period for the Great Basin and found that over the past 20–30 years, the temperature has increased on average 0.23°C per decade. This warming climate has significantly reduced the amount of water and shifted the seasonality of water flowing into wetlands (Haig et al. 2019). We found that the annual measure of how many days water was present in the region was declining by about one day per decade. Taken over 100 years, this would mean 10 fewer days per year that water would be available for ranchers, birds, and other constituents. There are positive

A pair of Clark's Grebes engaged in the courtship display called "rushing," where two or more birds seemingly walk on water side by side by vigorously paddling their feet while their bodies are lifted vertically above the water surface. Clark's Grebes share this remarkable behavior with the closely related and more numerous Western Grebe. Photo by Mel Clements

and negative aspects to this pattern. On the one hand, as mentioned above, earlier spring snowmelt may be beneficial for early spring migrants such as Northern Pintail and early breeding species such as Wilson's Phalarope, as there are puddles (i.e., hatching invertebrates) for them to forage in. On the other hand, warmer environments in spring and summer can limit foraging options for chicks and adults as the number, quality, and extent of water bodies decline over the summer. A shortened hydroperiod also negatively affected species such as American Avocets, which tend to remain in the Great Basin region for months during the postbreeding period (Plissner et

al. 2000a). When we considered the status of various Great Basin waterbirds, we found that the decline of 11 out of 15 waterbird species was associated with a climate variable (temperature, precipitation, or both), and five of eight species that had already undergone significant population declines were negatively associated with increasing temperatures: Killdeer, Snowy Plovers, Black Terns, Western Grebes, and Clark's Grebes (Haig et al. 2019).

Looking at how these changes affect a particular site, the combination of drought and water diversion for agriculture at Lake Abert has resulted in recent freshwater inflows of less than half what they were in the past (Moore

2016, Senner et al. 2018). In 2014 and 2015, the surface area of the lake declined from historical levels of more than 15,000 ha (37,000 acres) to 236 ha (583 acres) and 666 ha (1,646 acres), respectively. Between 2011 and 2015, counts of phalaropes went from more than 60,000 at Lake Abert to less than 13,000 birds, and counts of Eared Grebes from more than 11,000 to fewer than 100 birds. Senner and colleagues found that as the area of Lake Abert decreased and salinity increased, invertebrate and waterbird numbers declined.

Overall, climate change is predicted to increase in severity, leading to more frequent and more severe droughts in the Great Basin region and shifts in the timing, type, and amount of precipitation (Intergovernmental Panel on Climate Change 2022). This will further threaten Great Basin wetlands and the birds that depend on them during every phase of the annual cycle (Haig et al. 2019).

A final critical aspect for understanding the results of changing climate in the Great Basin region is that it is just one step in the pathway of annual movements that many bird species undertake from Alaska to Mexico and beyond. Thus, there is a compounding effect of changing climates in each area that a species visits such that taking all factors together starts to point to an unraveling of the Pacific Flyway. Mitigation of this potential disaster will not be simple but will be key to survival of humans and the other creatures living in the Great Basin region and beyond.

## Migratory Connectivity among Great Basin Shorebirds

SUSAN HAIG AND LEWIS ORING

Understanding species' challenges and needs within and among the various phases of the annual cycle is the only way to definitively design comprehensive management and/or recovery plans (Haig et al. 1998). This is challenging for one species, let alone a suite of species. Yet even closely related species can have distinct differences in their migration routes, habitat needs, and timing of critical events such as the phenology of breeding. To address this for shorebirds breeding in the Great Basin region, we studied three shorebird species with different life history patterns to understand how they and similar shorebirds might best be accommodated in Oregon and beyond. We spent six years tracking the movement behaviors of American Avocets, Western Willets (*Tringa semipalmata inornata*), and Killdeer throughout their annual cycle, using radiotelemetry and color bands to document movements of individuals within and among dispersed study areas in California, Nevada, and Oregon. We found that movements of all three species were tied to salinities of the mosaic of wetlands that make up the Great Basin region (Figure 8).

American Avocets were the most vagile in their use of Great Basin areas and beyond (Plissner et al. 1999, 2000a). They arrived in early April and often stayed in Oregon until October or November. Birds were often seen at several potential breeding wetlands before they laid eggs, and even birds that were nesting were sometimes found away from their breeding wetlands. Adults laid eggs in nests close to fresh water but foraged in saline or hypersaline wetlands. On average during the breeding season, they were detected at more than two sites (e.g., Summer Lake Wildlife

Area and then Lake Abert) and ranged up to six sites. Almost one-quarter of the 185 adult avocets we tracked moved more than 200 km among Great Basin wetlands during the breeding season. After the chicks fledged, the adults and then the chicks moved to Summer Lake or Lake Abert, where they remained foraging on the ultra-dense food source until they migrated south in the fall. Interestingly, a number of birds that we tracked from Honey Lake in northeastern California showed up at Lake Abert to spend the pre-migration period. Eventually, the birds migrated south to the Salton Sea or wetlands in northern Mexico.

Willets were less vagile than avocets and were fairly rigid in their behaviors (Haig et al. 2002). They arrived at Great Basin breeding sites in Oregon and California, almost always on 15 April. Lakeview ranchers called them the "Tax Day" birds. Willets set up territories and laid eggs almost immediately (weather depending) on sites close to fresh water. While on incubation recess (once or twice a day), the non-incubating adult would fly up to 13 km to forage in an adjacent saline or hypersaline wetland such as Summer Lake or Goose Lake. A day or so after chicks hatched, females flew to their postbreeding sites, often in the San Francisco Bay area, where they would remain in a small home range until the following breeding season. Males reared the chicks until fledging, but then they too flew seemingly directly to California coastal winter territories. Most willets were gone from Oregon by 4 July. We had several marked birds that were seen year after year in the same small coastal winter territory. Overall, their time in the Great Basin seemed to be as short as possible, leading us to conjecture that their reason for bothering to migrate might have something to do with access to fresh water for chicks that they might have difficulty accessing in coastal areas.

Killdeer were the least flexible in their movements and habitat use (Plissner et al. 2000b). They stayed at the same freshwater site where adults used an average of 6 ha (15 acres) over a 6- to 10-month period. They did not migrate, and 73% of the adults moved less than 1 km from their breeding site. Occasionally, when the weather got cold, they would fly to another site for a few days, but then they were back as soon as the weather warmed up.

These varying life history patterns for closely related taxa point to the challenge for managers seeking to provide appropriate resources for all wetland taxa. Yet not knowing the requirements of these species and others for types of water bodies could lead to misguided management actions.

OPPOSITE: Black-necked Stilt on the shores of Summer Lake Wildlife Area in eastern Oregon. Many of these Great Basin lakes are drying out because of drought, climate change, and overallocation of water resources to agriculture, leaving wetland bird species in the Great Basin without adequate foraging habitat. Photo by Susan Haig

## Sandhill Cranes

GARY IVEY AND M. CATHY NOWAK

In the 1800s, Oregon's breeding Sandhill Crane subspecies that share the Pacific Flyway and Oregon (Greater, *A. c. tabida*; Lesser, *A. c. canadensis*; Canadian, *A. c. rowani*) were extirpated from much of their breeding range and reduced to critically low levels because of wetland habitat loss and uncontrolled exploitation by settlers, miners, and market hunting. Passage of the Migratory Bird Treaty Act in 1918 helped a bit, but in the early 1940s, Gabrielson and Jewett (1940) reported only about 100 pairs remained in Harney, Lake, and Klamath Counties. Happily, unregulated hunting ended, and populations have gradually increased to over 50,000 in the flyway.

Carroll (C. D.) Littlefield pioneered Sandhill Crane research in Oregon starting in 1966. He studied Greater Sandhill Cranes at Malheur National Wildlife Refuge through 1989, which resulted in his authorship of many scientific papers that contributed to our knowledge and management of the species. Among his accomplishments, Littlefield conducted the first comprehensive survey of Oregon's breeding population of Greater Sandhill Cranes in the early 1970s and found 604 territorial pairs (Littlefield and Thompson 1979). He found an additional 103 pairs in the early 1980s and, with the assistance of Mark Stern, documented 947 pairs in 1986 (Littlefield et al. 1994). Gary Ivey and Caroline Herziger repeated the Oregon pair survey in 1999 and 2000 and found an increase to 1,151 pairs (Ivey and Herziger 2000).

Mark Stern studied nesting ecology of Sandhill Cranes at Sycan Marsh in the 1980s, detailing hatching success and fledgling production at the second-largest breeding site in Oregon. He also documented use of Langell Valley in the Klamath Basin by Sandhill Cranes as a migration staging area (Stern et al. 1986).

In the early 2000s, Gary Ivey and Caroline Herziger trapped and marked eight Sandhill Cranes along the Lower Columbia River on Sauvie Island Wildlife Area and Ridgefield NWR for a satellite telemetry study of their movements (Ivey et al. 2005). They tracked the marked cranes to breeding areas along the coasts of British Columbia and southeast Alaska and documented their migration through the Willamette Valley to stage at Lower Klamath NWR and Butte Valley Wildlife Area, before continuing to winter areas in the Sacramento Valley. All birds captured were the Canadian subspecies. In 2011, Gary Ivey and Robert Dillinger captured three summering Sandhill Cranes at Sauvie Island that were also measured and genotyped as the Canadian subspecies (G. Ivey, unpublished data). No other subspecies have been documented, even after repeated visits to the Lower Columbia River. In 2020 and 2021, a successful nesting pair was documented on the Washington side of the Columbia River at Ridgefield NWR, confirming a local breeding population of the Canadian subspecies at that location.

Gary Ivey studied factors affecting nest success and mortality of juvenile cranes (colts) at Malheur NWR in the 1990s. He also marked Greater and Lesser Sandhill Cranes with VHF radio and satellite telemetry equipment to study their movements in California in 2007 and 2008. The Greater Sandhill Cranes marked on wintering grounds in the Sacramento Delta region were found summering at Chewaucan Marsh, near Lake Abert, and on the Silvies River floodplain. Lesser Sandhill Cranes migrated though eastern Oregon, stopping at Chewaucan Marsh, Warner Basin, Harney Basin, Jordan Valley, and Ladd Marsh Wildlife Management Area during their movements

A pair of Sandhill Cranes, a species that is highly migratory and recovering from overhunting and draining of many wetlands where it formerly nested in Oregon. Photo by Jared Hobbs

back and forth to breeding grounds in Alaska (G. Ivey, unpublished data).

ODFW conducted research on Greater Sandhill Cranes at Ladd Marsh Wildlife Management Area in northeastern Oregon. An estimated 25 pairs nest at Ladd Marsh, with additional nesting pairs scattered across public and private land throughout that northeast corner of the state. As of the early 2000s, no information was available regarding where those birds wintered or which population they were affiliated with. Speculation among biologists was that Ladd Marsh cranes belonged to the Central Valley of California

population. The winter of 2015–16 was the first with satellite transmitters on Ladd Marsh cranes; all three of those cranes migrated to and wintered in the Central Valley, seemingly validating early speculation about their population affiliation.

As migration commenced in the fall of 2016, four of five cranes satellite-tagged at Ladd Marsh traveled but one flew southeast, first to the Payette River Valley in Idaho and then through Nevada to Arizona, to join the Lower Colorado Valley population. The satellite transmitter failed shortly after its arrival in Arizona (Nowak et al. 2018). This was just the second documented instance of population mixing by Sandhill Cranes in the western United States. The "Arizona Bird" was recaptured in 2018 and received a new transmitter, subsequently documenting its continued residence in the Lower Colorado Valley. In 2019, a newly captured and satellite-tagged crane joined the ranks of Lower Colorado Valley cranes using Ladd Marsh. During winter 2019–20, five Ladd Marsh cranes wintered with the Central Valley population, and two wintered with the Lower Colorado Valley birds.

It remains to be discovered what proportion of Ladd Marsh and/or Sandhill Cranes nesting in northeast Oregon belong to each of the represented populations and whether any are affiliated with other more distant populations. Meanwhile, researchers marking cranes in areas previously "assigned" to the Lower Colorado and the Rocky Mountain populations have documented additional cases of mixing among those populations. These discoveries may lead to revision of population delineations and management for Sandhill Cranes in the western United States.

## Carroll Dwayne Littlefield

Carroll Dwayne "C. D." Littlefield was a passionate wildlife biologist at Malheur National Wildlife Refuge from the late 1960s to early 1989. Littlefield was internationally recognized for his outspoken efforts to protect Sandhill Cranes. In the 1970s, Littlefield began working on his doctorate in ecology at the University of Arizona, which led to his studies of Sandhill Cranes at Malheur. C. D. then worked for the refuge, conducting research on a wide range of species, mentoring a succession of Fish and Wildlife Service biologists and teaching field ornithology classes at Malheur Field Station. C. D., often seen in his long-sleeve woolen shirt, routinely walked in thigh-deep water through the tules, nonchalantly enduring hordes of mosquitos while locating crane nests and documenting their success. He was sought out by many at the Malheur Field Station and was highly respected for his knowledge and love of natural history. Littlefield also strongly advocated for cessation of cattle grazing on the refuge.

In 2002, Littlefield left Malheur to return to Texas and then the Bioresearch Ranch in the Peloncillo Mountains of New Mexico. He passed away unexpectedly in 2019 and left behind his legendary passion for Sandhill Cranes and wildlife conservation as well as his book *Birds of Malheur National Wildlife Refuge* (1990), a treasured resource for birders visiting the refuge.

*Photo courtesy of Malheur Field Station*

## Gary Ivey

Gary Ivey has spent more than 40 years working on bird conservation in Oregon. He began work in Oregon as a biological technician for Malheur National Wildlife Refuge in 1979. Then, after hopscotching between Oregon and California, he served as the wildlife biologist at Malheur National Wildlife Refuge from 1983 to 1998.

In 1999, Ivey started work independently as a wildlife and wetlands consultant, specializing in surveying Sandhill Cranes and authoring wetland habitat recovery and conservation plans. He went on to attain his M.S. (2008) and Ph.D. (2015) under Bruce Dugger at Oregon State University. There he studied Sandhill Cranes in California's Sacramento–San Joaquin Delta and Sauvie Island on the Columbia River in Oregon. Since 2007, Gary has been a research associate of the International Crane Foundation.

Ivey has served in many advisory roles for bird conservation in Oregon. He further served on the board of the Oregon Chapter of The Wildlife Society and was president of the Trumpeter Swan Society and Friends of Malheur National Wildlife Refuge. He leads the Oregon Trumpeter Swan Restoration Project as a representative of the Trumpeter Swan Society.

*Photo by Cathy Nowak*

## Trumpeter Swans

GARY IVEY AND M. CATHY NOVAK

The Trumpeter Swan is the largest North American waterfowl species and was one of the first species hunted to the brink of extinction in the nineteenth century. It had disappeared from Oregon by the late 1800s. For almost 100 years, multiple groups have worked to restore Trumpeter Swans in Oregon. The Trumpeter Swan Society, ODFW, and USFWS have been working to expand the numbers and distribution of breeding Trumpeter Swans in south-central and southeast Oregon, primarily at Malheur National Wildlife Refuge and Summer Lake Wildlife Area.

Trumpeter Swans were released at Malheur National Wildlife Refuge from the late 1930s through the mid-1950s (Cornely et al. 1985). The first nesting occurred in 1958, and the Malheur Trumpeter Swan flock slowly grew until its numbers peaked at 19 breeding pairs and 77 individuals in 1986 (Ivey 1990). Flood conditions in the mid-1980s allowed high numbers of common carp to invade the swan wintering sites at Malheur. This reduced aquatic food supplies, resulting in degraded wetland conditions and a decline in swan numbers. Shortage of winter food caused by degraded wetland conditions and low recruitment were the primary factors limiting the population, which were compounded by the sedentary behavior of the flock. Collisions with powerlines are a continuing source of mortality, as is ingesting lead from sinkers leftover from fishermen. The swans swallow the sinkers when they ingest tubers and other plant material on the bottom of lakes and ponds.

Malheur NWR still provides excellent sites for breeding Trumpeter Swans; however, their wintering area in the south Blitzen Valley is marginal and likely limits the carrying capacity of the refuge for the existing flock. In recent years, brood survival has been minimal, likely because of reduced water availability in the Blitzen Valley. Even so, the Malheur flock persists, although it is tiny, with only three adults remaining in summer 2020 (all banded and sexed as female; G. Ivey, unpublished data). This flock is currently in danger of local extinction.

Efforts to expand the breeding range of Trumpeter Swans at Summer Lake Wildlife Area originally began in 1991, and some progress was made in expanding their flock; however, the project was halted in 1998 owing to political concerns. Unfortunately, most of the breeding pairs from the 1990s releases were gone by 2009 when restoration efforts resumed. The current Pacific Flyway objective for the state is at least 15 breeding pairs and 75 adults. Since 2009, 9 new pairs have nested in this region, which fledged 17 wild hatched cygnets. From 2009 to 2020, 141 swans, mostly cygnets, have been released at Summer Lake Wildlife Area. Early results show about a 50% mortality rate the first year of release owing to predation, powerline collisions, and illegal shootings. A 50% mortality rate is not unusual for Trumpeter Swans during the first year post-release (Mitchell and Eichholz 2020).

New telemetry methods are shedding light on annual movements of Trumpeter Swans in the Summer Lake Wildlife Area flock. In winter 2019, ODFW staff captured and GPS-collared a male swan at Summer Lake Wildlife Area. On 11 March, it left for summer breeding grounds, spending more than five months near a pond in the Peace River Region of British Columbia. This area is one of the core breeding areas for the Rocky Mountain population of Trumpeter Swans. On 21 September, it moved east to the vicinity of Grande Prairie, Alberta, staging for a month before migrating back to Summer Lake. Remarkably, this bird flew from southern Alberta to Baker County, Oregon, crossing

Adult Trumpeter Swan foraging for submerged aquatic vegetation. Trumpeter Swans, which formerly migrated and overwintered in large numbers in Oregon, were nearly driven to extinction early in the twentieth century. Thanks to protection from overhunting, habitat protection, and widespread restoration efforts, the species was removed from the US Endangered Species List in 1968. Efforts to establish Trumpeter Swans as a breeding species in Oregon have met with mixed success, and currently only two small breeding populations persist. Photo by Dan Roby

the Rocky Mountains above the Salmon River region in just 12 hours. Based on its daily movements during the summer, biologists believe the swan and his mate hatched a brood. ODFW continues to monitor this swan's movements.

In February 2020, eight migrant Trumpeter Swans were captured and transmittered at Summer Lake. Subsequently, four were documented in northern Alberta. ODFW continues to work with the Trumpeter Swan Society and others to restore Trumpeter Swans to their native range in Oregon.

## Martin St. Louis

Martin "Marty" St. Louis recently (2020) retired after a nearly 30-year career with the Oregon Department of Fish and Wildlife, most of which included his work as manager of the Summer Lake Wildlife Area in eastern Oregon. Marty was one of few wetland experts east of the Cascades in Oregon and as such spent a career not only developing and conserving waterbird habitat in his area but also taking a regional perspective on the water needs for birds in eastern Oregon. Working with multiple partners, he developed Summer Lake Wildlife Area to include more than 7,500 acres of wetlands. For many years, Marty carried out aerial waterfowl surveys across the region for the US Fish and Wildlife Service and participated in US Geological Survey migratory Snow Goose and shorebird research. Marty also collaborated with the Trumpeter Swan Society to introduce swans to the Summer Lake Wildlife Area. One of his later projects was to work with Oregon State University Professor Dan Roby to create nesting islands and other habitat to entice Caspian Terns to nest at Summer Lake. For his efforts, Marty was recognized with the 2020 Ducks Unlimited Conservation Achievement Award. While retired, Marty St. Louis continues to help out at Summer Lake Wildlife Area.

*Photo by Susan Haig*

# Arid Lands

SUSAN HAIG

Oregon's diverse sagebrush (*Artemisia* spp.) habitats make up a surprisingly high proportion of the state yet receive a surprisingly small proportion of conservation attention relative to the economically valuable forests to the west. Lack of attention can be a blessing and a curse, as we will see in this section.

Sagebrush habitats have been classified by the Oregon Conservation Strategy into four ecoregions: (1) Big Sagebrush steppe communities of the Blue Mountains, (2) shrub habitats in the Columbia Plateau region, (3) shrub habitats of the East Cascades, and (4) Big Sagebrush habitats of the Northern Basin and Range region. Since 1850, the distribution of sagebrush communities has been reduced by 30% across the state, most of which has occurred in the Columbia Plateau region. Limiting factors vary by region but include human expansion and urbanization that takes away or fragments native sagebrush communities; fire suppression that provides for expansion of juniper (*Juniperus occidentalis*) trees and loss of sagebrush habitat; and extensive competition from introduced cheatgrass (*Bromus tectorum*) and other annual invasive grasses.

Major efforts have been launched to address each of these issues, with varying success. Conservation recommendations (Braun et al. 1976, Knick et al. 2003, Davies et al. 2011, Remington et al. 2021) have changed over time. In 1976, sagebrush was (finally) deemed worthwhile, following years of dumping herbicides to reduce the percentage of sagebrush on the landscape and to enhance forage for livestock. Fire suppression and sagebrush control programs were encouraged (Braun et al. 1976). By 2003 (Knick et al. 2003), there was widespread recognition of the value of sagebrush habitat to birds and other natural resources on the landscape. Sadly, so much had been altered that bringing it back to a vibrant ecosystem has not been simple. By 2011, however, Davies et al. (2011) concluded the following about efforts to restore sagebrush ecosystems: (1) efforts to restore higher-elevation sagebrush communities by removing juniper trees were frequently successful and (2) efforts to remove exotic grasses often failed. Davies et al. (2011) recommended measured fire regimes be undertaken to help address both of the issues above and to work with private landowners and developers to mitigate the anthropogenic impacts on the landscape. The latest strategy released in 2021 by the Western Association of Wildlife Agencies (Remington et al. 2021) indicated that success of juniper removal is a solid first step in quickly reopening landscapes to their shrub-steppe state, and additional targeted measures are needed to address the suite of threats facing the ecosystem. Nevertheless, most if not all states in the sagebrush biome are working in coordinated targeted frameworks to address the threats facing the ecosystem and have been doing so for more or less the past 10 years.

Understanding of the importance of sagebrush habitats in Oregon and beyond was significantly enhanced, starting in the 1960s, by OSU Professor John Wiens and colleagues. Over several decades, Wiens and (often) colleague John Rotenberry described the diversity and importance of birds to the shrub-steppe community (summarized in Wiens and Rotenberry 1981). Another early investigator was David Dobkin from Bend, Oregon, whose work focused on assessing the changing status of birds and mammals in the sagebrush steppe community. Using 40-plus years of Breeding Bird Survey data (1968–2001), Dobkin and Sauder (2004) examined trends for 25 upland bird species and 12 riparian species. They found significant declines in 16 of 25 upland species, significant increases in 3 of 25 upland species, and significant declines in 5 of 12 riparian species. They concluded that recovery of the ecosystem to its previous structure and diversity would be difficult. Although conservation of sagebrush habitats is challenging, because over 70% of sagebrush habitats are on federal lands means there are many opportunities for management (Knick et al. 2003). Ellis et al. (2019) examined 20-year changes in riparian bird communities and their habitats in east-central Oregon and found declines in three common shrub-dependent riparian species: Yellow Warbler, Willow Flycatcher, and Song Sparrow. These trends reflected regional Breeding Bird Survey trends rather than changes in local site conditions. Their findings suggested managing lands to increase wetness and extent of riparian zones can be beneficial for grassland and wetland bird species. But managing for riparian shrub cover or volume, important metrics of grazing intensity and riparian system health, may be insufficient to conserve riparian shrub-dependent birds because other unidentified local and/or regional factors are likely contributing to habitat suitability.

Long-billed Curlew female in arid nesting habitat consisting of mixed shortgrass prairie / shrub-steppe vegetation types. This species is in decline largely because of habitat loss on the breeding grounds. Photo by Mick Thompson

## Grazing Impacts on Birds in Oregon's Remnant Prairies

PATRICIA KENNEDY

Livestock graze most rangelands globally, and livestock production is an important contributor to many rural economies. In Oregon, Euro-Americans have been raising cattle since John Quincy Adams was elected president in 1824. Cattle and calves ranked as the state's second-leading agricultural commodity in 2016, with a value estimated at $701 million. Two of the thirteen Oregon State University agricultural experiment stations, Union Station founded in 1901 and Burns Station founded in 2011, are dedicated to providing research to support the Oregon livestock industry.

Research needs of the livestock industry have changed over time (Pumphrey et al. 2001). At the end of the twentieth century, it was apparent that research into the sustainability of livestock practices needed more scientific scrutiny. Although an important agricultural commodity, inappropriate livestock grazing had been identified as a threat to ecosystem health and biodiversity in many western landscapes. Conversely, others argued that grazing was necessary to maintain the health of rangeland systems and that grazed rangelands preserved biodiversity. The "grazing debate" intensified in the 1990s as conservation groups began to focus on livestock grazing as a national issue, and federal agencies were increasingly challenged with respect to their management of livestock on public lands and in their charge to lead the recovery of threatened and endangered species. As sociopolitical debates regarding the effects

and appropriateness of livestock grazing in the United States intensified, the scarcity of available ecological science on which to inform the issue became increasingly apparent (Kennedy et al. 2012, 2017).

In 1999 the Oregon legislature approved additional funding for Union Station to develop multidisciplinary research on the effects of livestock grazing on natural resources (Pumphrey et al. 2001). With these funds, OSU expanded the disciplines traditionally affiliated with agricultural experiment stations to include me, the first wildlife biologist at the Union Station in northeast Oregon. I had expertise on how land management decisions influenced the sustainability of avian populations, and I was tasked with helping to develop and direct a collaborative research program on sustainable livestock grazing in northeastern Oregon. The unique emphasis of this program was its focus on wildlife. At the time I was hired at OSU in 2002, the vast majority of studies conducted on the ecological effects of livestock herbivory involved only plants, the abiotic environment, and the focal livestock. There was little experimental research on the impact of livestock herbivory on the resident fauna in grazed ecosystems, and wildlife were the focus of the contentious debates on the sustainability of livestock grazing. At that time, studies that investigated faunal response to domestic livestock herbivory were typically correlative or "quasi-experiments" where comparisons were made between "grazed" and "ungrazed" sites. An untested assumption of many studies was that vegetative changes resulting from grazing could be used to explain

OPPOSITE: Sage Thrashers are common breeders and summer residents in Oregon east of the Cascade Range wherever sagebrush steppe habitats occur but are rare in Oregon west of the Cascades. The decline of sagebrush habitats in eastern Oregon owing to juniper incursion, cheatgrass invasion, and agricultural conversion is resulting in declines in this iconic species for the sagebrush sea. Photo by Jared Hobbs

The Western Meadowlark is the official State Bird of Oregon and is found year-round through-out the state in suitable open habitats. Although five other states have designated this species as their state bird, the beautiful song of this songbird is characteristic of grasslands throughout Oregon. Photo by Jared Hobbs

varying patterns of biodiversity in grazed and ungrazed areas.

In November 2005, a multidisciplinary team of OSU and Nature Conservancy scientists obtained USDA–National Resources Inventory funding for a unique, large-scale field experiment. This experiment was conducted from 2006 to 2009 in Zumwalt Prairie, located just northeast of the town of Joseph in northeastern Oregon. Zumwalt Prairie has been in private ownership since the area was homesteaded (Kennedy et al. 2009) and is the last large (about 65,000 ha) remnant of the vast Pacific Northwest bunchgrass prairie that once covered ~800,000 ha in the northwestern United States and Canada. Compared to other prairies in North America, relatively little was known about this semiarid temperate grassland, as the majority disappeared quickly after Euro-American settlement.

Zumwalt Prairie is slightly higher, drier, colder, and more geographically isolated than most other bunchgrass prairies in western North America and has remained relatively intact. Little of the prairie was farmed, and spring/summer cattle grazing remains the primary land use (Bartusevige et al. 2012). Our field experiment was located on the portion of this prairie remnant owned by The Nature Conservancy (Zumwalt Prairie Preserve). Using OSU livestock, the study investigators manipulated livestock stocking rates and evaluated the effects of stocking rates on the grassland food web and on livestock performance. The objectives of this project were to determine the ecological and economic sustainability of current stocking rate practices in a system dominated by seed-producing grasses. This was the first study to experimentally examine questions about agricultural sustainability at a spatial scale that was appropriate for wildlife and applicable to livestock management. The experimental unit was a 40 ha pasture with three replicates of four stocking rate treatments (Johnson et al. 2011).

Our study organisms for this experiment were selected members of the grassland food web, which included vascular plants, terrestrial invertebrates, and breeding songbirds and their predators. Invertebrates are involved with a wide variety of supporting services, including providing pollination for native and agriculturally important plants, pest control through the actions of natural predators, and food resources for other organisms, including birds. Breeding grassland birds also provided a number of ecosystem services, including cultural services, because society values their existence for aesthetic reasons, and regulating services, because of their important roles as predators, insectivores, herbivores, granivores, and prey. Birds are sensitive to changes in soil and vegetation characteristics affected by grazing. North American grassland bird populations have shown dramatic declines in recent years and appeared to be declining more markedly than all other avian guilds on this continent (Sauer and Link 2011).

Our experimental results demonstrated that low to moderate stocking rates were ecologically and economically sustainable, but ecological functioning and livestock performance began to deteriorate at high stocking rates. There were dramatic effects on invertebrates and birds. Preferred bird food such as spiders and butterflies showed decreased abundance, diversity, and/or changes in community composition with increased stocking rate (Kimoto et al. 2012). The high stocking rate had a negative effect on bird and nest abundance, as several species (e.g., Grasshopper Sparrows) disappeared and avian community composition changed in high-stocking-rate pastures. Although stocking rate influenced vegetation structure, the only nest failures related to stocking rate were from trampling, which was a common occurrence. Thus the study results did not support the common contention that rangeland health could be evaluated solely on vegetative and abiotic characteristics of rangelands; animals clearly responded to vegetative structural changes caused by grazing even when there was no change in plant composition.

While the experiment was being designed and implemented, we resurveyed the entire Zumwalt Prairie to determine whether the area was continuing to support high nesting densities of the three species of Buteo hawks (Ferruginous Hawk, Swainson's Hawk, Red-tailed Hawk) that were documented in the area in 1979–80 (Houle 2007). Unlike other prairie remnants, land management on the Zumwalt Prairie and surrounding agricultural areas was consistent over the past several decades, so not surprisingly we found territory occupancy of all three Buteo species had not changed over the study period of about 25 years. This suggested that local range management practices—that is, summer-only grazing of livestock at moderate stocking rates (Johnson et al. 2011)—were not negatively affecting these raptors (Kennedy et al. 2014). These results combined with the food web experimental data indicated that cows, bugs, and grassland birds could coexist, but that not all livestock management strategies were equally sustainable or ecologically beneficial for wildlife.

## Patricia Kennedy

Professor Kennedy is among the first small but significant tranche of women to break the gender barrier in academic fish and wildlife departments. She was the first woman hired to be a wildlife faculty member by Oregon State University. OSU hired Pat to develop a research program in sustainable livestock grazing centered at the Eastern Oregon Agricultural Research Station, Union Experiment Station, in northeastern Oregon. She was OSU's first ecologist to be located at an OSU agricultural experiment station. There she taught natural resource management classes in OSU's Eastern Oregon Agriculture and Natural Resource Program and became director of this program in 2016. Pat retired in 2018 and is now professor emerita at Oregon State University. Kennedy has studied a wide range of avian taxa in a variety of arid ecosystems: mixed conifer forests, remnant grasslands, sage-steppe, and riparian habitats.

*Photo by Tom Kennedy*

## David Dobkin

David Dobkin is an international authority on the birds and ecosystems of the western United States. He came to Oregon on sabbatical in 1990 to continue his earlier research on Great Basin riparian bird and vegetation communities. He ended up settling permanently in Bend and established the High Desert Ecological Research Institute, for which he served as executive director from its inception in 1993 through 2017. HDERI has been a regional center for ecological research and policy analysis, focusing on natural resource issues related to the Intermountain West and Pacific Northwest. Dobkin also was an adjunct professor at Lewis and Clark College of Portland, teaching one-week postgraduate field courses based in Bend and the Malheur Field Station. He is the author or coauthor of five major books on the ecology and conservation of birds, most notably *The Birder's Handbook: A Field Guide to the Natural History of North American Birds*, coauthored with Paul Ehrlich and Darryl Wheye. For many years, Dobkin was editor of the ornithological journal *The Condor*, now *Ornithological Applications*. He is a fellow of the American Ornithological Society.

*Photo by Donna Dobkin*

## Greater Sage-Grouse: Icons of the High Desert and Ghosts in the Last Dark Place in Oregon

### CHRISTIAN HAGEN

The Greater Sage-Grouse (hereafter sage-grouse) is interwoven within the fabric of eastern Oregon's sagebrush sea. A sagebrush obligate, sage-grouse are particularly dependent on sagebrush as their primary food item in winter. Sage-grouse are generally well known for their elaborate spring displays, where large numbers of males congregate on traditional breeding arenas known as "leks" to attract females and fight among one another for the right to breed. These elaborate dances are paired with a variety of popping and swooshing sounds, emanating from air being released from the air sacs on their breast, and the swish of the leading edge of their wing on the rigid white feathers surrounding their neck and air sacs. All the while, the male lightly stamps his feet and rotates about, giving nearby females a 360-degree view of his fitness. These dances have captured the imagination of Native peoples for millennia, and more recently the attention of behavioral biologists, population ecologists, and conservationists. For just about everyone involved, including hard-core scientists, birders, ranchers, hunters, poets, writers, soccer moms, and politicians, these odd birds with their weird ritualistic dances have come to be viewed as a bellwether for the overall health of the sagebrush ecosystem and a test of our ability to deal with environmental change (Crawford 1982).

Known to Native Americans of the Great Basin as "seeskadee," this majestic (and goofy at times) bird provided food, feathers, and a cultural anchor. The spring displays of the male sage-grouse and its cousins the Sharp-tailed Grouse and prairie-chickens (*Tympanuchus cupido* and *T. pallidicinctus*) were mimicked in ceremonial dances, and even today many tribes continue to emulate the male display in their contemporary pow-wow competitions. Male dancers often adorn the tail of their dancing costumes with the tail feathers of sage-grouse.

Although Euro-American trappers, traders, and settlers were already invading portions of the sage-grouse range in the late 1700s, it was the Lewis and Clark expedition in 1803–4 that provided the first scientific description of the "Cock of the Plains" (Zwickel and Schroeder 2003). Their observations placed sage-grouse near the Columbia River between the Deschutes and Snake River inlets (Zwickel and Schroeder 2003). It was during Townsend's expedition in 1839 that sage-grouse were recorded near the Powder River. As eastern Oregon became more settled in the 1800s, sage-grouse distribution became better known (Crawford 1982).

By the late 1800s, a series of land-use changes (unfettered livestock grazing), climatic conditions (the mini ice age), and introduction of invasive annual grasses (such as cheatgrass, *Bromus tectorum*) set the stage for many of the conservation challenges of the next two centuries. Sheep herders and cattlemen found their way to Oregon in the late 1800s, by most accounts in numbers hard to envision today (Miller and Rose 1995). It is thought that the sheer volume of animals combined with season-long grazing led to dramatic reductions

OPPOSITE: Male Greater Sage-Grouse displaying at a lek on a cold early spring morning in the Oregon High Desert. Greater Sage-Grouse are a "bellwether for the health of the sagebrush ecosystem," and the poor condition of most of the sagebrush habitats in Oregon are reflected in the population declines of this iconic species of the sagebrush sea. Photos by David Leonard

in herbaceous plant composition and increased the shrub component. Overgrazing also reduced fine fuels to an extent that altered fire regimes at higher elevations, making way for western juniper (*Juniperus occidentalis*) to establish a foothold in vast areas that were once sagebrush grasslands. Today, that equates to 1 million ha of sagebrush habitat replaced by juniper. At this same time, cheatgrass arrived from Eurasia and was quickly spread by humans and their livestock, vehicles, and contaminated seed. In one case, cheatgrass was even deliberately introduced in a laboratory experiment in Washington State. As the climate began to warm, cheatgrass has proliferated largely unabated at lower elevations (Smith et al. 2021). This novel fine fuel has a synergistic relationship with wildfire and appears to have accelerated the frequency and extent of wildfire over the past 30 years (R. F. Miller et al. 2011). The ongoing threat of habitat loss from conifer encroachment at higher elevations and the fire-cheatgrass cycle at lower elevations is referred to as the "big squeeze," leaving sage-grouse with fewer options to eke out a living in the Great Basin.

By the turn of the twentieth century, the alarm bell had been rung for sage-grouse nationally and locally. Hornaday (1916) offered an impassioned plea to the western states to cease hunting of sage-grouse in order to save the species from extinction. He cited the incursion of roads, automobiles, and pump shotguns as the primary cause of the species' demise. As was common at the time, however, he also placed a fair bit of blame on wolves and coyotes, and he pointed out that sage-grouse were often absent from areas where humans had settled. In the state of Oregon, general concern for the species continued into the 1930s (Crawford 1982). In 1941, one of the first scientific field studies of sage-grouse was initiated in Oregon by Batterson and Morse

(1948). Often overshadowed by their contemporary Patterson (1952) and his seminal volume on sage-grouse in Wyoming, the Batterson and Morse study in Baker County nevertheless shed light on the effectiveness of captive rearing, trapping and translocation, and avian predator control (i.e., lethal removal of Common Ravens). The work of Batterson and Morse (1948) and Patterson (1952) established spring lek counts of males as the best survey methodology to monitor population status and trend, and those counts continue today. In fact, some of the leks included in their 1940s Oregon research are still counted by state biologists.

Although sage-grouse populations were being monitored by the Oregon Game Commission (now the Oregon Department of Fish and Wildlife), there was a nearly 40-year hiatus in sage-grouse research in Oregon after the Batterson and Morse study. This changed in the 1980s when John Crawford and his students at Oregon State University began detailed field studies of sage-grouse at Hart Mountain Antelope Refuge that led to seminal works on nest ecology, dietary needs of pre-laying females, diets of chicks, and the first look at individual chick survival using miniaturized telemetry devices. Crawford's work spanned over 20 years and made significant contributions to our understanding of sage-grouse ecology and management.

More than 100 years after Hornaday's warning, in the early 2000s, the alarm bell for sage-grouse was rung once again, as numerous petitions to protect the species under the federal Endangered Species Act were submitted to the USFWS. These petitions, citing habitat loss and lack of regulatory authority, resulted in strong pushback from politicians and the powerful ranching lobby, and motivated state wildlife agencies across the range of the sage-grouse to develop statewide conservation strategies to maintain or enhance populations and avoid federal listing of the species. These strategies were developed in large collaborations that represented stakeholders from across the sagebrush sea. In eastern Oregon the strategy paved the way for local groups to implement conservation actions to reduce threats facing the species and the landscapes in their backyard (Hagen 2011).

Using the ODFW Conservation Strategy (Hagen 2011) as a stepping-off point, the USDA Natural Resources Conservation Service (NRCS) and ODFW devised a sub-strategy to address the threat of invasive juniper into sagebrush ecosystems on private lands. This highly targeted strategy set a goal to reduce the threat by 53,000 acres over a five-year period. Fast-forward to 2022, and 137 km² (53 mi²) of early successional stage juniper had been removed from sage-grouse habitat on private lands across Oregon. The Bureau of Land Management has been tackling this threat as well, but at a pace often dictated by National Environmental Policy Act processes. The

ongoing monitoring and research are providing compelling evidence that the removal of early successional juniper stands has benefits for the ecosystem (Severson et al. 2017a), the broader avian community (Holmes et al. 2017), and vital rates and population growth of sage-grouse, not to mention the ranching community (Severson et al. 2017b, Olsen et al. 2021).

More recently, NRCS and partners have begun to focus their attention on the threat of the wildfire–invasive grass cycle. Recent research has demonstrated that greater than 400,000 ha (1 million acres) of Oregon's sagebrush biome have been altered by fire since 2005, and 13% (26,000 km² or 10,038 mi²) of the Northern Great Basin is infested with cheatgrass (Smith et al. 2021). Sage-grouse and sagebrush do not respond well to wildfire in the near term, and it will require decades to recover the habitat and birds to prefire conditions (R. F. Miller et al. 2011, Coates et al. 2016). Ongoing monitoring of sage-grouse in the Trout Creek Mountains since the Holloway Fire of 2012 (more than 180,000 ha or 444,789

## John Crawford

John Crawford joined the wildlife ecology faculty at Oregon State University in 1974 and subsequently spent 27 years developing a game bird research program for the state. Professor Crawford was especially recognized for his work on recovery of Mountain Quail, sage-grouse, and other upland game birds found in eastern Oregon. In addition to his academic work, John was active in professional societies. He served as the president of the Oregon Chapter of The Wildlife Society and was active in the Society for Range Management, the Wildfowl Trust, the Pacific Northwest Bird and Mammal Society, the World Pheasant Association, and the Game Conservancy. Additionally, Crawford was associate editor for the *Journal of Wildlife Management*, the *Journal of Range Management*, and the *World Pheasant Association Journal*. He was recognized for his dedication to his work with sage-grouse in the Great Basin and on Hart Mountain National Antelope Refuge with awards from The Order of the Antelope. He further received the Arthur M. Einarson Award from the Northwest Section of The Wildlife Society, the Wildlife Society Award from the Oregon Chapter of The Wildlife Society, a Meritorious Service Award from the US Fish and Wildlife Service, and a Special Recognition Award from the Oregon Department of Fish and Wildlife. John Crawford passed away in 2010.

*Photo by Eric Pelren*

acres burned) indicated that sage-grouse suffer from dramatic reductions in annual survival (−75%) and nest success (−50%) in the first two years postfire (Foster et al. 2018). These reductions had direct effects on population growth, and there appears to be lingering effects of fire on population growth seven years postfire (Anthony 2020).

Sage-grouse are a bellwether to the health of the sagebrush ecosystem. While threats persist, the unprecedented collaboration of conservationists is rising to these challenges, as evidenced by more than half a million acres of sagebrush maintained through removal of early successional juniper. Additionally, the advancement of technologies to detect and map invasive grasses and use of selective herbicides hold promise to change the trajectory of invasive grasses and interrupt synergistic relationships with wildfire. Restoration moves slowly in sagebrush systems. It took 150 years to create the conservation challenges that we face in this region; it will require partnership, collaboration, time, and patience to restore some of what has been lost.

## Christian Hagen

Since 2011, Christian Hagen has been a faculty member and senior researcher for Oregon State University's Department of Fisheries, Wildlife, and Conservation Sciences. From 2006 to 2010, Hagen served as an instructor at the Cascades Campus of Oregon State University in Bend. With broad focuses on avian ecology, arid systems, habitat selection, and demography, Hagen's current work specializes in Greater Sage-Grouse population-level response to habitat disruption. The Hagen Lab further works to assess and provide for the needs of Lesser Prairie-Chicken and Greater Sage-Grouse by measuring changes in space use, resource selection, and demographic rates. Additionally, Hagen and his lab strive to translate scientific findings to laypeople to further educate the public and bring awareness of the importance of these birds. Hagen serves as the national science advisor for the Natural Resources Conservation Service's Lesser Prairie-Chicken Initiative, which is a companion effort to the Sage-Grouse Initiative, whose mission is to maintain healthy sagebrush ecosystems in order to recover sage-grouse populations.

*Photo by Elizabeth Schuyler*

## Steven Herman

Steve Herman (1936–2020) taught biology at Evergreen State College outside Tacoma, Washington, for 35 years but spent much of his life conducting research and teaching about birds in eastern Oregon. He was a researcher and early activist for the banning of DDT to protect Peregrine Falcons and California Condors, having also served on early condor recovery programs. He was instrumental in the establishment of Bowerman Basin National Wildlife Refuge in Washington and the removal of cattle from the Hart Mountain National Antelope Refuge in eastern Oregon. He published numerous scientific papers and a book titled *The Naturalist's Field Journal: A Manual of Instruction Based on a System Established by Joseph Grinnell*, which continues to be used by many amateur and professional naturalists. Herman was beloved by his students, and principal among his many achievements was his extraordinary mentoring of hundreds who went on to distinguish themselves as biologists and other professionals.

*Photo courtesy of Evergreen State College*

## Lead Ammunition and Arid Lands Raptors

### SUSAN HAIG

Over the past several decades, since the United States banned the sale or use of DDT in 1972, birders and conservationists have breathed a collective sigh of relief as more and more hawks, owls, vultures, and other raptors show signs of population recovery. If you visit the Portland Audubon Wildlife Care Center, Cascades Raptor Center in Eugene, Badger Run Wildlife Rehab in Klamath Falls, or Chintimini Wildlife Center in Corvallis, however, they will tell you about the numerous raptors, crows, and other predators and scavengers that are brought in because they are blind and/or dying as a result of lead poisoning. Lead poisoning in raptors typically happens when a bird either scavenges or kills a bird or mammal that has been shot with lead ammunition. Ingested lead is deadly in vertebrates, as it is transported throughout the body via the circulatory system and mimics critical metabolic body functions (Herring et al. 2017). This results in a wide range of physiological and neurological responses, including mortality (Haig et al. 2014).

Lead poisoning in scavengers (e.g., ravens, crows, California Condors, Bald Eagles, Golden Eagles, Red-tailed Hawks, coyotes) results mainly from ingesting lead in the carcasses or gut piles of animals killed by hunters, including game animals such as deer, pronghorn, and elk, and nongame animals such as Belding's ground squirrels (*Urocitellus beldingi*), which are shot and poisoned by the thousands to reduce damage to agricultural crops in eastern Oregon (Redig et al. 1991). These scavengers may also be exposed to other contaminants such as anticoagulant rodenticides because they commonly consume animals that may have been poisoned by rodenticides as well (Herring et al. 2017). In Oregon, there are at least 13 bird species that commonly scavenge dead Richardson's (*U. richardsonii*) or Belding's ground squirrels, including Bald Eagles, Common Ravens, Ferruginous Hawks, American Kestrels, Golden Eagles, Red-tailed Hawks, and Swainson's Hawks (Herring et al. 2016). As Oregonians anxiously hope for success of the newly released California Condors back into the state (see below), it is important to understand the impact of using lead ammunition on scavengers, especially condors (Walters et al. 2010, Haig et al. 2014).

Lead bullets have been the mainstay of hunters for centuries. But recent ballistic analyses show that when the bullet hits an animal's body, it shatters into many tiny fragments that can never be completely removed (Haig et al. 2014, Pain et al. 2019). Thus whatever or whomever eats the dead animal ingests lead. The detrimental effect of lead in paint, gasoline, and other products has long been recognized and banned to protect humans. And in 1991, recognizing the negative effects of lead ammunition, the US Fish and Wildlife Service banned the use of lead shot for hunting waterfowl, resulting in an overnight switch to using steel ammunition for all waterfowl hunters in Canada, Mexico, and the United States. California and a few other states other states have also now banned use of lead ammunition (and lead fishing tackle, in some cases) to protect scavengers and humans. In Oregon, however, it remains legal to hunt with lead rifle ammunition and with lead shotgun ammunition for most bird species other than waterfowl. Numerous studies have been carried out to investigate the effects of ingesting lead ammunition on various scavengers in Oregon, and the results are described below. Across agricultural lands in Oregon, there is a long tradition of shooting coyotes or other wildlife that are considered pests, including recreational hunts in eastern

An adult Golden Eagle suffering from poisoning, likely from ingesting carcasses containing lead bullet fragments or shot. This eagle was found in Hart Mountain National Antelope Refuge and was barely able to fly but could not be captured for treatment. Photo by Dan Roby

Oregon, where shooters line up on the back of truck beds to shoot hundreds of ground squirrels per shooter per day (Herring et al. 2016). These hunts are conducted using lead ammunition, and the carcasses are typically left on the ground. Golden Eagles, in particular, are often found eating the carcasses and/or feeding them to nestlings (Stauber et al. 2010). In a recent study, Herring et al. (2020) found that 45% (n = 258) of Golden Eagle nestlings contained lead from ammunition in their systems, and 89% of nestlings with elevated lead levels contained lead that could be sourced to ammunition. Lead concentrations were highest in nestlings near agricultural fields; however, nestling growth rate was highest near the agricultural fields, presumably because of the high densities of prey in agricultural fields

and the ease with which parents could bring scavenged prey to the nest.

Potential solutions to the lead problem would be to stop the use of lead ammunition for hunting (copper ammunition is an alternative) and to utilize different methods to remove unwanted small mammals from agricultural fields. While this sounds simple, the situation is complex and involves many factors, including the relative expense of developing different types of ammunition, availability of non-lead ammunition, ballistics issues, and lack of alternative pest removal methods that are efficient and do not involve rodenticides. These issues continue to be worked on, yet fuel debate among stakeholders regarding the best way forward. Big-game hunters using non-lead ammunition will provide a healthier food source for scavengers and their own families. And success of the California Condor reintroduction in Oregon depends on the availability of lead-free carcasses for birds to feed on (Church et al. 2006).

## Return of the California Condor

JESSE D'ELIA

The California Condor is an indelible part of Oregon's historical avifauna. This species' relationship with humans dates back millennia, as evidenced by unearthed bones from at least 22 California Condors, or possibly a slightly larger ancestral relative, at an archeological site east of The Dalles (Table 7; Miller 1957, Hansel-Kuehn 2003, Syverson and Prothero 2010). Some of these bones, which are estimated to be ~7,000–12,000 years old, have distinct cut marks that indicate condors were likely hunted for their feathers (Hansel-Kuehn 2003). In southwest Oregon, a broken California Condor radius excavated from a Native American shell mound near Brookings was dated to ~1,000 years BP (Miller 1942, Simons 1983). In addition to archeological evidence, ethnographic studies evaluating artwork, language, and mythology—while undoubtedly incomplete—also suggest a deep historical familiarity and reverence for condors by many Native American tribes throughout the Pacific Northwest (D'Elia and Haig 2013).

The first observation of California Condors by Euro-Americans in Oregon occurred in November 1805, when William Clark recorded them in his journal near Astoria (Lewis et al. 2002; Table 7). The Corps of Discovery had observed them in the preceding weeks along the north shore of the Columbia River and would document their occurrence on five more occasions while on the Oregon coast and along the Lower Columbia River during their journey back upstream. John Kirk Townsend, David Douglas, other explorer-naturalists, fur-traders, and settlers also verified the presence of condors in the early 1800s from the Columbia Gorge to the Pacific Ocean and south through the Willamette Valley and Coast Range to northern California (D'Elia and Haig 2013). Evidence of condors in eastern Oregon is more limited, but at least one credible and specific observation of two condors feeding on a sheep carcass in southwestern Idaho in 1879 (Lyon 1918) suggests their historical range likely included portions of eastern Oregon.

Condor observations in Oregon declined precipitously after the mid-nineteenth century. The last reliable observation of condors in Oregon occurred in March 1904 around the town of Drain when Henry Peck—a skilled ornithologist familiar with condors—observed four condors in flight at a close distance (Finley 1908). Several other observers accurately described California Condors in southern Oregon around this time, although details on the exact timing and location of these observations are wanting (Gabrielson and Jewett 1970). A review of the hypotheses on causes for their disappearance from the Pacific Northwest suggested that contaminated food resources from predator poisoning were primarily responsible, although direct persecution also contributed (D'Elia and Haig 2013). Lead poisoning from spent ammunition would eventually cause the condor population in California to further collapse in the twentieth century—after the advent of high-powered rifles that caused significant fragmentation of bullets upon impact—but it is unclear how much of a role lead poisoning played in their disappearance from Oregon in the nineteenth century (D'Elia and Haig 2013). Regardless of its historical role, lead poisoning is currently the primary constraint to condor recovery (Finkelstein et al. 2012).

Seminal research into the status and natural history of California Condors in the mid-1900s

California Condor sunning its enormous, outstretched wings in preparation for taking flight in the morning. California Condors have the greatest wingspan of any bird in North America. This adult female (678) died in the 2020 Dolan Fire in Big Sur, California. Major wildfires are an increasing threat to California Condors. Photo by Tim Huntington, webnectar.com

# Table 7. History of California Condors as they return to the Pacific Northwest

| Year | Event | Number of Wild California Condors |
|------|-------|-----------------------------------|
| 1805–6 | The Corps of Discovery observes condors on numerous occasions along the Lower Columbia River and along the Oregon and Washington coast. | unknown |
| 1850 | California Condors no longer regularly reported from the Columbia River. Still sporadically collected and reported from elsewhere in Oregon and the Pacific Northwest. | unknown |
| 1904 | Last reliable observation of wild condors in Oregon, near Drain. | unknown |
| 1953 | Carl Koford publishes the first detailed natural history study of the California Condor. He estimates only 60 condors remaining, but other researchers later revised this estimate up to 150 based on additional information. | 150 |
| 1967 | California Condors are listed as Endangered under the Endangered Species Preservation Act. | 60 |
| 1975 | The US Fish and Wildlife Service establishes a California Condor Recovery Team and adopts the Condor Recovery Plan. It is the first recovery plan for any species under the US Endangered Species Act. | 25–35 |
| 1980 | The USFWS adopts the first revision to its recovery plan for the California Condor. Recommends captive breeding and identification of release sites by surveying area of former occupation, including areas in Oregon. | 25–35 |
| 1982 | Nadir of the California Condor population, with only 22 condors in existence and only 20 remaining in the wild. | 20 |
| 1987 | The last wild California Condor captured in southern California for a captive breeding program. | 0 |
| 1989 | First successful reproduction of a California Condor in captivity at the San Diego Zoo. | 0 |
| 1992 | Condor reintroductions begin in the southern portion of their historical range. | 7 |
| 2000 | In preparation for the Lewis and Clark bicentennial celebration, the Oregon Zoo initiates discussions of reintroducing condors to Oregon and joining the recovery program. | |
| 2001 | The Oregon Zoo is accepted into the Condor Recovery Program to breed condors. | 58 |
| 2003 | Six condor breeding pairs brought to the Jonsson Center for Wildlife Conservation in Clackamas County. | 83 |
| 2004 | First condor chick, named Kun-Wak-Shun (Thunder and Lightning), hatched in captivity in Oregon. | 96 |
| 2007 | The Yurok Tribe begins efforts to evaluate reintroducing condors to Yurok Ancestral Territory in northwestern California, near the Oregon border. | 144 |

| 2008 | American Ornithologists' Union, Audubon California Blue Ribbon Panel on the status of recovery efforts for the California Condor recommend increased efforts to release condors back to the Pacific Northwest. | 150 |
| 2013 | Oregon State University Press publishes *The California Condor in the Pacific Northwest*, a detailed history of condors in the region (D'Elia and Haig 2013). | 230 |
| 2014 | The Oregon Zoo unveils its California Condor exhibit, *Condors of the Columbia*, to the public. | 228 |
| 2016 | Memorandum of Understanding signed by federal, state, tribal, and nongovernment stakeholders agreeing to evaluate the feasibility of reintroducing condors to the Pacific Northwest. | 276 |
| | The Nez Perce Tribe begins to evaluate condor reintroduction potential in the vicinity of Hells Canyon. | 276 |
| 2017 | The National Park Service, USFWS, and Yurok Tribe formally initiate planning for reintroducing condors to northwestern California. | 290 |
| 2022 | An initial cohort of condors released in the Bald Hills of Redwood National Park within Yurok Ancestral Territory (~8 miles south of the Oregon border) with the expectation of their flying into Oregon. Other releases planned for the Snake River and other sites to restore condors to Oregon. | 329 |

revealed a diminished population confined to a fraction of their historical range in the mountains of central and southern California (Koford 1953). It also confirmed the condor's low reproductive rate, which inhibited rapid recovery from excess mortality. In 1967 the California Condor was listed as endangered under a precursor to the Endangered Species Act (USFWS 1967), and recovery planning soon followed (USFWS 1975). Continued population declines in the 1970s, along with the discovery that condors would lay a replacement egg if their first egg was lost or removed (Snyder and Hamber 1985), led the USFWS to revise its recovery strategy and begin taking eggs from nests to form a captive flock for future reintroductions (Snyder and Snyder 2000). Catastrophic population losses in the early 1980s forced the USFWS to bring all remaining wild condors into captivity despite bitter controversy over whether this was an

ethical or viable solution (Pitelka 1981, Alagona 2004). Ultimately, it proved to be their salvation (Snyder and Snyder 2000).

As captive breeding efforts got under way, there was a need to identify areas where condors could be reintroduced to the wild. Initial habitat assessments included Oregon (USFWS 1980), but recovery efforts were ultimately concentrated in the southern portion of the condor's historical range (USFWS 1984, 1996; Walters et al. 2010). It wasn't until 2000 that discussions of bringing condors back to Oregon were reinvigorated during planning sessions for the Lewis and Clark bicentennial celebration. In 2001 the Oregon Zoo was invited to participate in the recovery program and constructed an additional breeding center, with condors arriving in 2003 and the first chick hatching in 2004.

A confluence of events starting in 2007 further energized deliberations of reintroducing

California Condors to the Pacific Northwest. First, the Yurok Tribe—with grant funding from the USFWS and Bureau of Indian Affairs—began building stakeholder support for reintroducing condors to Yurok Ancestral Territory in northwestern California near Oregon's southern border. Second, a review of the condor recovery program by a Blue Ribbon Panel commissioned by the American Ornithologists' Union and Audubon California recommended assessing other parts of the condor's historical range for possible reintroductions (Walters et al. 2010). Finally, scientific research to assess the biological feasibility of a condor reintroduction to the Pacific Northwest began in earnest. Research included assessing the condor's historical distribution and causes of decline (D'Elia and Haig 2013), modeling habitat suitability and connectivity (D'Elia et al. 2015, 2019), evaluating historical genetic diversity and population structure using ancient DNA from museum specimens (D'Elia et al. 2016), and measuring contaminants in food resources and other scavengers (Gunderson et al. 2012; West et al. 2017; Herring et al. 2016, 2018). These research efforts revealed that Oregon and northern California contained vast areas of suitable habitat (D'Elia et al. 2015); the historical population in the Pacific Northwest had a shared ancestry with condors from the southern United States and Baja California, meaning that the captive population was a genetically suitable source of condors for repopulating the Pacific Northwest (D'Elia et al. 2016); and contaminants were present in food resources and surrogates, likely requiring monitoring and management similar to other parts of the condor's range.

Following publication of key feasibility studies and mustering stakeholder support, official planning for condor reintroductions to northwest California commenced in 2017.

## Jesse D'Elia

Jesse D'Elia works for the Ecological Services branch of the US Fish and Wildlife Service in the Portland Regional Office. Over his two decades with the USFWS, he has helped coordinate conservation efforts in the Pacific Northwest for some of its most iconic species, including the gray wolf, grizzly bear, sage-grouse, and California Condor. Jesse earned his Ph.D. with Susan Haig at Oregon State University, where he worked on designing reintroduction plans for California Condors in the Pacific Northwest, a topic he continues to pursue. His research on endangered species spans landscape ecology, reintroduction science, ecological modeling, and genetics. He is the lead author of the book *California Condors in the Pacific Northwest*.

*Photo by Courtney D'Elia*

Subsequently, the National Park Service, USFWS, and the Yurok Tribe published a draft environmental assessment describing the proposed reintroduction effort and evaluating its impacts (National Park Service et al. 2019). Recognizing the need to balance socioeconomic costs of reintroducing an endangered species with its biological benefits, the USFWS concurrently proposed a special regulation under the Endangered Species Act governing management of reintroduced condors (USFWS 2019d). Publication of a final environmental assessment (National Park Service et al. 2020), final regulations governing the management of reintroduced condors (USFWS 2021), and a monitoring and management plan (Yurok Condor Restoration Program) soon followed. Now, after more than a century of absence and more than a decade of scientific investigations and planning, the return of the California Condor to Oregon has begun. Shifting our gaze to the east, the Nez Perce Tribe has begun to evaluate the feasibility of restoring condors to their ancestral lands around Hells Canyon—an effort that is ongoing and may eventually prove to be another avenue for the California Condor's return to Oregon.

# Long-billed Curlews

ELISE ELLIOTT-SMITH AND
SUSAN HAIG

The Long-billed Curlew is an iconic arid land species whose range and numbers have been shrinking since the early 1900s (Allen 1980, Pampush and Anthony 1993, Dugger and Dugger 2002). Its breeding distribution ranges only through the arid West in short-grass prairie and mixed-grass habitats of the United States and Canada. Long-billed Curlews winter along the Oregon and California coasts, Central Valley of California, and as far south as Mexico and Central America (Dugger and Dugger 2002, Marshall et al. 2003). There are no current or older reliable population estimates for Long-billed Curlews in Oregon, but they are currently a species of great concern throughout their range because of urbanization, contaminants, and agricultural development (Dugger and Dugger 2002). They are considered highly imperiled by the US Shorebird Conservation Plan, in part because of historical and current population declines (Brown et al. 2001). They are a USFWS Bird of Conservation Concern, were identified by the North American Bird Conservation Initiative as a species of concern in the Great Basin, and are listed as a US Forest Service sensitive species (Region 2). The USFWS directs an interagency working group (Long-billed Curlew Working Group) that includes Regions 1, 2, and 6, as well as other federal and state agencies whose activities are focused on recovery of the species. Long-billed Curlews are also protected under the Migratory Bird Treaty Act.

## Breeding Studies

There have been limited research efforts on Long-billed Curlews. In Oregon, studies have been focused primarily on the national wildlife refuges along the Columbia River, where breeding birds spend time on refuge units of the Mid-Columbia River Refuge Complex. These include the Cold Springs, Columbia, Hanford Reach National Monument / Saddle Mountain, McNary, and Umatilla refuges, and the Department of Defense's Naval Weapons Systems Training Facility in Boardman (Blus et al. 1985, Gilligan et al. 1994, Dugger and Dugger 2002). Breeding curlews have also been studied at Malheur NWR and Summer Lake Wildlife Area (Blus et al. 1985, Warnock et al. 1998). The Columbia Basin of Oregon and Washington is a unique ecoregion because of the interspersion of remnant shrub-steppe habitats of varying quality with a high proportion of agricultural lands. Habitat on national wildlife refuges in the Columbia Basin is also diverse, with active farming on some refuges, native restoration programs, remnant sage-shrub patches, and large areas dominated by invasive cheatgrass.

Qualitatively, curlews seem to prefer to nest in invasive cheatgrass habitats over native bunchgrass and shrublands, but this has not been confirmed owing to small sample sizes in previous studies (Pampush and Anthony 1993, Stocking et al. 2010). Nests have also been found in agricultural fields, including buckwheat and timothy fields (same field, consecutive years) at Umatilla NWR and an alfalfa field at Columbia NWR. The nest in the alfalfa field was eventually abandoned, presumably because of dense vegetation that exceeded 20 cm in height at the time of abandonment. Three nests found in 2008 were located in irrigated cheatgrass at the corners of buckwheat fields. Mid-Columbia refuge staff created a 30 m buffer around these three nests that were not sprayed or disked, although the remainder of the field received those treatments. Two of the clutches hatched, and chicks used the non-disked areas for the first two weeks after hatching. This suggests that the 30 m buffer

Male Long-billed Curlew resting with Whimbrels during fall migration. Although not listed under the US Endangered Species Act, Long-billed Curlews are a species of high conservation concern because of ongoing population declines and degradation and destruction of nesting habitat due to irrigation-based agriculture and livestock grazing. Photo by Dan Roby

was sufficient for successful breeding, but further field and toxicological studies may be warranted to determine an appropriate buffer size for successful fledging (Stocking et al. 2010).

## Contaminants

In 1985, Blus et al. documented organochlorine pesticides, polychlorinated biphenyls (PCBs), and other contaminants in terrestrial bird eggs and blood on the Hanford Reach National Monument / Saddle Mountain NWR and throughout the Mid-Columbia River Refuge Complex in eastern Oregon and Washington. Similar ecotoxicants and heavy metals are still found in relatively high concentrations at several nonbreeding sites used by Long-billed Curlews (Sapozhnikova et al. 2004, Jiménez et al. 2005, García-Hernández et al. 2006, Schwarzbach et al. 2006). Blus et al. (1985) also found residue burden accumulated in adult fat tissue as a result of feeding on the wintering grounds and subsequently deposited into eggs during the breeding season. More recently, Long-billed Curlews breeding in Nevada and South Dakota have exhibited eggshell thinning, low levels of egg hatchability, and reproductive failure that are indicative of environmental pollutants (Hartman and Oring 2009). While loss or degradation of habitat is apparently the main threat to Long-billed Curlews in Oregon, the impact of contaminants needs more detailed examination. In addition, curlews are illegally shot throughout the annual cycle adding to their precarious status (Fellows and Jones 2009).

## Wind Energy Development in Arid Lands

SUSAN HAIG

As the human population grows, demands for energy increase and become more complicated as changing climates dictate the need for cleaner energy across the planet. Cleaner energy can come in the form of wind or solar power, but it often comes at a serious cost to the birds, bats, and butterflies living or migrating near the facilities (Erickson et al. 2014, Watson et al. 2018). Since the late twentieth century, Oregon has invested a great deal in wind and solar power in an effort to reduce greenhouse gases and slow the effects of climate change. With all the best of intentions, there are still negative consequences from these alternative energy sources to birds and other wildlife. There has also been an intense effort to mitigate these impacts (Katzner et al. 2019).

The US Fish and Wildlife Service estimates that each year 573,000 birds are killed in the United States by collisions with wind turbines (https://abcbirds.org/can-wind-energy-be-bird-safe/). Given the growth of wind power, this estimate is likely to grow to 1.4 million birds per year in the next 8–10 years in the United States (abcbirds.org). Given the current decline of birds (30% since 1970), already a devastating loss to North American avifauna, an increase is almost unthinkable in the future (Rosenberg et al. 2019). Turbine heights can range from 475 to 662 feet above ground level, which is an enormous reach and directly in the path of birds migrating, soaring, or flying locally. Most structures over 345 feet high are considered to be in the path of migrating birds. Thus wind turbines, power lines, and towers are among the fastest-growing threats to birds and bats in Canada, the United States, and beyond. In April 2020, there were more

than 70,800 operational commercial-scale wind turbines in the United States and many more currently under construction (Hoen et al. 2018).

Oregon has large wind energy resources, most of them located in rural eastern Oregon and near the Columbia River Gorge (see also the Offshore Renewable Energy Development section on p. 78). Oregon's wind generation capacity comes from large-scale wind projects that supply power directly to the electric grid. Onshore wind is the second-largest zero-carbon-emitting electricity resource in Oregon next to hydropower (Oregon Department of Energy 2020). As of 2020, wind power makes up 11.6% of Oregon's electricity generation and 4.69% of Oregon's energy consumption. Oregon's wind power capacity has grown substantially since construction of the state's first wind facility in 2001. With 3,415 megawatts of wind generation, Oregon is ninth nationally in terms of overall wind capacity and third among the 14 US states in the Western Electricity Coordinating Council. As of 1 October 2020, there were 46 existing wind farms and four state jurisdictional facilities under construction in Oregon. The state of Oregon has legislated that under Senate Bill 838, wind, solar, geothermal, and other types of renewable power must account for 25% of an electric utility's retail sales by 2025. Thus wind energy development will increase across the state.

Since development of these new energy sources, the race has been on to maintain cleaner energy but reduce the costs to wildlife. Much of the work to determine the impact of wind energy on wildlife worldwide has been led by OSU/USGS statistician Manuela Huso (Huso 2011, Huso and Dalthorp 2014, Huso et al. 2015). The obvious group of birds suffering mortality from running into wind turbines are raptors because they are large and

Golden Eagles, such as this adult, are especially vulnerable to injury or death from poorly situated wind farms in Oregon and other western states. Large numbers of Golden Eagles have been killed by wind turbines in the western states, and efforts to mitigate these losses need to be upscaled. Photo by Scott Carpenter

comparatively easy to locate when biologists survey for carcasses under a turbine (Watson et al. 2018). In the United States, eagle mortality has received the most attention because of their legal status under the Bald and Golden Eagle Protection Act. The negative effects of wind turbines on other raptor species are less well understood, and corresponding mitigation responses are less well developed. But Red-tailed Hawks and American Kestrels found at wind energy projects in the western United States make up the majority of known global fatalities for *Buteo* hawks and small falcons (Watson et al. 2018). More than 2,000 Golden Eagles have been killed by wind turbines in the Altamont Wind Resource Area in southern California (Allison 2017). Mortality of California Condors is another source of concern regarding wind turbines in California and now will be in Oregon as well (D'Elia et al. 2019).

A major challenge for biologists is to accurately estimate mortality for all taxa as a result of wind projects so better wind energy products can be developed (Huso 2011, Dalthorp and Huso 2014). For example, small birds are most likely to succumb to power structures but are hard to see and decay faster than large birds, and birds falling into water are not likely to be found. And when a turbine strikes a passing bird, it does not necessarily fall straight to the ground where it can be found by a biologist surveying directly under the wind turbine. And none of this considers the far smaller and more delicate bats, butterflies, and other insects that collide with turbines.

Solutions to the mortality problem are being developed on both sides of the discussion: the wind energy manufacturers and biologists. Biologists are coming up with better statistical survey methods, and wind power manufacturers are developing smaller, quieter, and bladeless wind turbines (Hsu et al. 2018). The

location or siting of a wind turbine can have a huge effect on faunal mortality. Even a simple mitigative measure like painting one turbine blade black reduces the "motion smear" that makes it difficult for birds to see the spinning blades. This resulted in a 71% decrease in collisions.

Other improvements are rapidly being developed. Professor of Mechanical Engineering Roberto Albertani of OSU leads a team that designed a sensor to detect a strike as well as provide taxonomic classification of the animal involved in the collision. This device could eliminate errors in not recording or identifying what hit a turbine and is a great step forward in the turbine mortality mystery, especially as it relates to offshore wind facilities, where it is nearly impossible to determine what has hit a tower. Most recently, McClure et al. (2021) devised a protocol such that wind turbines are slowed or stopped when wildlife are considered at risk. They found an 82% reduction in the fatality rate for eagles relative to a control site.

There is hope that mitigative measures will continue to be developed to enhance bird survival and use of a sustainable, carbon-free energy source.

## Burrowing Owls in Oregon
GREGORY GREEN

The presence of Burrowing Owls in Oregon was first noted in 1835 by John Kirk Townsend (1839), who served as a naturalist during the 1834–36 Wyeth Overland Expedition. He reported Burrowing Owls occurring in eastern Oregon during collection trips based out of Fort Vancouver. Burrowing Owls were also recorded in eastern Oregon during various railroad route surveys conducted along the Columbia River in the mid-1850s (Henry and Baird 1855, Newberry 1857, Suckley and Cooper

Adult Burrowing Owl, a species that was formerly common and widespread in much of Oregon but is now only locally common in parts of the arid eastern third of the state. In Oregon the species appears to be limited by the availability of suitable burrows in which to nest, usually dug by badgers, although badgers are also a major predator on Burrowing Owl eggs and nestlings. Photo by Scott Carpenter

1860). In 1860, James Graham Cooper and George Suckley described the species as abundant near The Dalles. Between 1874 and 1878, Charles Bendire, an army major and naturalist stationed at Camp Harney in southeastern Oregon, reported that Burrowing Owls were present but not common near the camp.

Arthur Roy Woodcock reported Burrowing Owls near Corvallis and considered them a year-round resident in Linn County in the late nineteenth century. A. C. Shelton also considered this owl an uncommon resident of Lane County, and Gabrielson and Jewett (1940) reported Burrowing Owls to be a regular inhabitant of the Jackson County prairie lands near Medford. After publication of Gabrielson and Jewett's *Birds of Oregon* (1940), many

publications and field reports focused on location records as opposed to ecological research (Lardy 1980).

Ecological research on Oregon's Burrowing Owls began in the 1970s when Maser et al. (1971) published a paper on the diet of central Oregon owls. From 1978 to 1981, Henny et al. (1984) investigated pesticide contamination in Columbia Basin raptors, including Burrowing Owls (see the Contaminants section on p. 50). The use of artificial burrows to facilitate collection of eggs for contaminants analyses also provided the opportunity to investigate the nesting ecology of this species.

In 1980, I began an M.S. program at OSU with Chuck Meslow, and subsequently Bob Anthony, to study habitat use and diet of

A brood of Burrowing Owl fledglings at the entrance to their nest burrow. Burrowing Owl nesting populations can be enhanced by providing artificial nest burrows in open grassland habitat with good prey resources. Photo by Mick Thompson

Burrowing Owls in the shrub-steppe between Arlington and Hat Rock State Park in the Columbia Basin. The study culminated in publications on nesting success and habitat use (Green and Anthony 1989), diet (Green et al. 1993), and management (Green and Anthony 1997). From 1995 to 1997, Aaron Holmes et al. (2003) returned to my study site at the Naval Weapons System Training Facility, Boardman, and further investigated nesting success and burrow longevity of Burrowing Owls.

My initial field effort, in spring of 1980, focused on locating active Burrowing Owl nests at the Naval Weapons System Training Facility, also called the Boardman Bombing Range, and the US Army Umatilla Storage and Supply Depot. This study was eventually expanded to include the Umatilla National Wildlife Refuge, Hat Rock State Park, and other locations of reported nesting owls. I found more than 100 nests during the initial search. OSU graduate student Geoff Pampush was in his second year studying habitat selection by nesting Long-billed Curlews at the Bombing Range, and both studies were part of a larger shrub-steppe ecosystem investigation.

The primary takeaway from the initial phase of the project was the dependency of Burrowing Owls on American badgers (*Taxidea taxus*) to provide suitable nesting burrows. Badgers were also a primary predator and, unlike prairie dogs (*Cynomys* spp.) in the Midwest, do not modify their vegetative habitat in a way that favors owls. Successful Burrowing Owl nests were surrounded by short vegetation, allowing a good horizontal view from the burrow entrance. Owls would also use nearby short perches to facilitate that view and generally nested in areas with low shrub coverage.

Collectively, these features provided owls with early detection of approaching predators, especially mammalian predators that could

excavate an owl trapped within a burrow. Burrowing Owls also selected available burrows surrounded by a high percentage of bare ground, indicative of a high population of ground-dwelling arthropod prey. Results also indicated that when owl pairs nested close to each other (e.g., less than 110 meters), one or both of the nests would fail, probably owing to intraspecific competition over local arthropod prey resources.

The large number of burrows found offered the opportunity for a comprehensive investigation into diet via analysis of owl pellets. A total of 5,559 pellets were collected and analyzed, providing insight into prey selection and seasonal variation in diet, habitat use by prey species, and a cross-sectional view of the small mammal and invertebrate prey base available. Owls preyed on more than 100 species, concentrating on small mammals in the early part of the nesting season and then arthropods as nesting progressed into summer. The results of the diet study were combined with another study conducted in the Columbia Basin at Hanford by Richard Fitzner and Lee Rogers and published in Green et al. (1993).

Recognizing that Burrowing Owl habitat use in particular and ecology in general are different in the shrub-steppe area from populations elsewhere in North America, Robert Anthony and I published a paper on management of Columbia Basin Burrowing Owls in 1997 (Green and Anthony 1997). In that publication, we noted that owl nesting habitat requirements included short vegetation, perches, and a high proportion of bare ground, and that if artificial burrows were offered, they should be spaced at least 110 m apart to avoid intraspecific competition.

In 1995, I returned to the Naval Weapons System Training Facility, Boardman (NWSTF Boardman) under contract with the US Navy to conduct various studies in preparation of an Integrated Natural Resource Management Plan, working with Kent Livezey from the US Navy and Russell Morgan with ODFW. Both groups combined to conduct a three-year study on Burrowing Owl nesting success and burrow longevity relative to soil type and cattle grazing. The crumbly nature of the soil along with trampling by cattle played a role in determining the longevity of an owl burrow. I was able to determine that burrow availability coinciding with other required features was highest in Columbia Basin areas that were backwater flood deposits from the Pleistocene Missoula Floods. These landscapes are flat and include enough sand to limit vegetation growth and allow ease in excavating by burrowing mammals yet contain enough loam to allow burrow structure to remain intact for at least a nesting season.

In 2009, David Johnson (with the Global Owl Project) followed up on my work at the Umatilla Army Storage and Supply Depot with a 12-year demographic study of Burrowing Owls using artificial burrows. Johnson's

research includes investigations into migration patterns using geolocators, longevity, lifetime reproductive success, nesting phenology, mate selection and fidelity, burrow decoration by males, morphology and wing characteristics, subspecific genetic comparison, male vocalizations and the role of heredity in transmission of male calls, juvenile dispersal, nest dispersal, and advancing capture techniques (D. H. Johnson et al. 2010).

Of note was the lack of owls and badger burrows on the depot at the beginning of Johnson's study in 2009 in comparison to when I began my study there in 1980, emphasizing the important role badgers play in the nesting success of Columbia Basin Burrowing Owls.

Currently, Burrowing Owls are absent as a nesting species from all areas in western Oregon where they once occurred. The wall-to-wall intensive agriculture in the valleys of western Oregon have pretty much eliminated Burrowing Owls in this entire region, except for the occasional overwintering birds.

# IV   Reflections

# Looking toward the Future of Avian Conservation in Oregon

SUSAN HAIG

The future of our world and its birds is crashing down on us far sooner than anyone expected. Horrific wildfires in Australia, Brazil, and Oregon only begin to demonstrate the devastation that human-induced climate change has wrought and will continue at an ever-accelerating pace. Our children will not see all that we have seen in the way of natural wonders. And there is no telling what our grandchildren will grow up to witness.

The urgency to act, devastation over what we have lost, and desperation to hold on to our precious natural resources in Oregon and beyond leads to the recognition that, like at no other time during the history of the earth, we all need to pitch in whatever skills and efforts we can to put a fist in a bursting dike. Our only hope is integration and implementation of research, theoretical and applied, into reasonable management approaches supported by governmental and nongovernmental organizations with field data collected by professionals, amateur birders, and others concerned about nature.

Among these efforts, our focus will have to be on conservation triage, a science that will now quickly evolve. For example, molecular markers currently used to describe mating systems, population differences, and taxonomy will sharply focus on identifying specific individuals and their pedigrees to properly rebuild populations that have shrunk to just a few individuals. CRISPR technology and beyond will be useful in editing species genomes to incapacitate invasive species, eliminate emergent infectious diseases, and perhaps even re-create some species that are going extinct. The role of zoos will become even more critical as we try to save species and genomes that have no other home. Conservation is not a perpetual battle that inevitably ends in extinction; the remarkable recovery of the Wood Duck in North America after being driven nearly to extinction in the early twentieth century and the implausible rescue of the California Condor from the very brink of extinction are but two of a large number of conservation success stories that should encourage us to continue the fight.

Tracking animals (individuals and populations) over short and long geographic and temporal distances will remain a priority as changing climates alter their phenologies over the annual cycle. Habitat shifts in response to climate change will alter much of what we have previously observed and management plans will need to be adapted accordingly. Large population shifts may be traceable via remote monitoring methods. For example, "listening devices" can be deployed in various ecosystems where bird vocalizations can be recorded. Environmental DNA (eDNA), CRISPR, and stable isotope techniques need to be further developed to more specifically identify species, populations, and individuals from soil, water, or even air samples. Yet perhaps most powerful are the millions of daily observations that birders enter into the eBird database at the Cornell Lab of Ornithology. The advanced techniques being developed cannot be verified unless there are eyes on the ground across the world on a daily basis.

Tracking individual movements throughout the year remains problematic. Although there is animal tracking technology (ICARUS) on the International Space Station, the constant need to further miniaturize animal tracking tags plagues us. The smaller and more powerful the tag, the longer and farther we can track (smaller) animals and understand how their

California Condor, a species that was extirpated from Oregon by early in the twentieth century. Condors were once uncommon but widespread in Oregon and appear to have succumbed to poisoning, ingestion of lead in shot carcasses, and persecution, but have been reintroduced to the state. Their fate rests on the need for hunters to use non-lead ammunition. Photo by Susan Haig

ranges and movements are changing. Further development of drones will also provide for more specific geographic and environmental data. They can help in desperate situations such as extensive climate-induced fires, floods, and droughts where their special heat-sensitive photographic capabilities can identify pockets of birds and other wildlife that might have survived the devastation.

Perhaps more than anything is the need to further develop better communication among scientists, managers, birders, agencies, legislators, the judiciary, and the public. We need to restore faith in science and gain a broader commitment to conservation with the understanding that, without global support, climate change will end our lives as we know them. Oregon has always been at the cutting edge of

environmental issues and solutions. Thus we need to further inspire young people from all ethnicities and socioeconomic backgrounds to become aware of their environmental surroundings, revel in the amazing beauty and diversity of birds, and care enough to safeguard their future.

As demonstrated in this book, Oregon ornithologists have always stepped up to address problems head-on and have clearly made a significant difference in resolving issues. We need only look back at the past few decades to see what we can accomplish. Ten years ago, who would have believed that there would be condors flying free in Oregon? In ten more years, will we have finally reestablished the Columbian Sharp-tailed Grouse, illustrated at the beginning of this section, as a viable

Wood Duck drake. Wood Ducks were hunted nearly to extinction soon after the turn of the twentieth century. Passage of the Migratory Bird Treaty Act in 1918 is credited with saving the species from extinction. Today, Wood Ducks are fairly common wherever suitable wooded wetlands occur in Oregon, despite being a harvested game species, and are an example of a major conservation success story. With similar conservation dedication, planning, and forethought, the imperiled Oregon bird species of today may experience a similar resurgence. Photo by Mick Thompson

Tufted Puffin adult in breeding plumage. The breeding population of Tufted Puffins along the Oregon coast has declined precipitously in the past 40 years for unknown reasons, but likely related to poor ocean conditions and associated declines in forage fish prey. Although proposed for listing under the US Endangered Species Act in Oregon, Washington, and California, the species is still doing well in Alaska. Photo by Dan Cushing

population in Oregon? Once extirpated, now reintroduced, the species is barely hanging on by a thread in eastern Oregon. And in ten more years, will we have the answers to the puzzling questions of why one of the most charismatic species in Oregon, the Tufted Puffin, is on the verge of being extirpated as a breeding species in Oregon, and how its dwindling population can be restored? We need to redouble efforts like these for the next few decades to save what we treasure most, the diverse and amazing earth we live in.

One of Aldo Leopold's many profound observations was that "One of the penalties of an ecological education is that one lives alone in a world of wounds." The challenge for those of us who care about birds and our environment is to not let ourselves become so overwhelmed by all the wounds that we give up trying. When you get discouraged by what we are leaving as a natural legacy for future generations, remember that nature is extraordinarily resilient. And we as scientists, conservationists, birders, and bird enthusiasts have demonstrated repeatedly that we can muster the vision, motivation, and technology to support nature's resilience. The birds are depending on us.

# List of Bird Species Regularly Occurring in Oregon

Key: *uncommon in Oregon; + introduced to Oregon; # extirpated from Oregon and reintroduced

| | |
|---|---|
| Order: WATERFOWL | *ANSERIFORMES* |
| Family: DUCKS AND GEESE | *ANATIDAE* |
| Emperor Goose | *Anser canagicus* |
| Snow Goose | *Anser caerulescens* |
| Ross's Goose | *Anser rossii* |
| Greater White-fronted Goose | *Anser albifrons* |
| Brant | *Branta bernicla* |
| Cackling Goose | *Branta hutchinsii* |
| Canada Goose | *Branta canadensis* |
| Trumpeter Swan | *Cygnus buccinator* |
| Tundra Swan | *Cygnus columbianus* |
| Wood Duck | *Aix sponsa* |
| Blue-winged Teal | *Spatula discors* |
| Cinnamon Teal | *Spatula cyanoptera* |
| Northern Shoveler | *Spatula clypeata* |
| Gadwall | *Mareca strepera* |
| Eurasian Wigeon | *Mareca penelope* |
| American Wigeon | *Mareca americana* |
| Mallard | *Anas platyrhynchos* |
| Northern Pintail | *Anas acuta* |
| Green-winged Teal | *Anas crecca* |
| Canvasback | *Aythya valisineria* |
| Redhead | *Aythya americana* |
| Ring-necked Duck | *Aythya collaris* |
| *Tufted Duck | *Aythya fuligula* |
| Greater Scaup | *Aythya marila* |
| Lesser Scaup | *Aythya affinis* |
| Harlequin Duck | *Histrionicus histrionicus* |
| Surf Scoter | *Melanitta perspicillata* |
| White-winged Scoter | *Melanitta deglandi* |
| Black Scoter | *Melanitta americana* |
| Long-tailed Duck | *Clangula hyemalis* |
| Bufflehead | *Bucephala albeola* |
| Common Goldeneye | *Bucephala clangula* |
| Barrow's Goldeneye | *Bucephala islandica* |
| Hooded Merganser | *Lophodytes cucullatus* |
| Common Merganser | *Mergus merganser* |
| Red-breasted Merganser | *Mergus serrator* |
| Ruddy Duck | *Oxyura jamaicensis* |

| | |
|---|---|
| Order: LANDFOWL | GALLIFORMES |
| Family: NEW WORLD QUAIL | ODONTOPHORIDAE |
| Mountain Quail | *Oreortyx pictus* |
| California Quail | *Callipepla californica* |
| Family: GROUSE AND PHEASANTS | PHASIANIDAE |
| +Wild Turkey | *Meleagris gallopavo* |
| Ruffed Grouse | *Bonasa umbellus* |
| Spruce Grouse | *Canachites canadensis* |
| Greater Sage-Grouse | *Centrocercus urophasianus* |
| Dusky Grouse | *Dendragapus obscurus* |
| Sooty Grouse | *Dendragapus fuliginosus* |
| #*Sharp-tailed Grouse | *Tympanuchus phasianellus* |
| +Gray Partridge | *Perdix perdix* |
| +Ring-necked Pheasant | *Phasianus colchicus* |
| +Chukar | *Alectoris chukar* |
| Order: GREBES | PODICIPEDIFORMES |
| Family: GREBES | PODICIPEDIDAE |
| Pied-billed Grebe | *Podilymbus podiceps* |
| Horned Grebe | *Podiceps auritus* |
| Red-necked Grebe | *Podiceps grisegena* |
| Eared Grebe | *Podiceps nigricollis* |
| Western Grebe | *Aechmophorus occidentalis* |
| Clark's Grebe | *Aechmophorus clarkii* |
| Order: DOVES AND PIGEONS | COLUMBIFORMES |
| Family: DOVES AND PIGEONS | COLUMBIDAE |
| +Rock Pigeon | *Columba livia* |
| Band-tailed Pigeon | *Patagioenas fasciata* |
| +Eurasian Collared-Dove | *Streptopelia decaocto* |
| *Common Ground-Dove | *Columbina passerina* |
| White-winged Dove | *Zenaida asiatica* |
| Mourning Dove | *Zenaida macroura* |
| Order: CUCKOOS AND ALLIES | CUCULIFORMES |
| Family: CUCKOOS | CUCULIDAE |
| *Yellow-billed Cuckoo | *Coccyzus americanus* |
| Order: NIGHTJARS | CAPRIMULGIFORMES |
| Family: NIGHTJARS | CAPRIMULGIDAE |
| Common Nighthawk | *Chordeiles minor* |
| Common Poorwill | *Phalaenoptilus nuttallii* |
| Order: SWIFTS AND HUMMINGBIRDS | APODIFORMES |
| Family: SWIFTS | APODIDAE |
| Black Swift | *Cypseloides niger* |
| Vaux's Swift | *Chaetura vauxi* |
| White-throated Swift | *Aeronautes saxatalis* |
| Family: HUMMINGBIRDS | TROCHILIDAE |
| Calliope Hummingbird | *Selasphorus calliope* |
| Rufous Hummingbird | *Selasphorus rufus* |
| Allen's Hummingbird | *Selasphorus sasin* |

| | |
|---|---|
| *Broad-tailed Hummingbird | *Selasphorus platycercus* |
| Black-chinned Hummingbird | *Archilochus alexandri* |
| Anna's Hummingbird | *Calypte anna* |
| *Costa's Hummingbird | *Calypte costae* |
| Order: CRANES, RAILS, AND ALLIES | *GRUIFORMES* |
| Family: RAILS | *RALLIDAE* |
| Virginia Rail | *Rallus limicola* |
| Sora | *Porzana carolina* |
| Yellow Rail | *Coturnicops noveboracensis* |
| American Coot | *Fulica americana* |
| Family: CRANES | *GRUIDAE* |
| Sandhill Crane | *Antigone canadensis* |
| Order: SHOREBIRDS | *CHARADRIIFORMES* |
| Family: STILTS AND AVOCETS | *RECURVIROSTRIDAE* |
| Black-necked Stilt | *Himantopus mexicanus* |
| American Avocet | *Recurvirostra americana* |
| Family: OYSTERCATCHERS | *HAEMATOPODIDAE* |
| Black Oystercatcher | *Haematopus bachmani* |
| Family: PLOVERS AND LAPWINGS | *CHARADRIIDAE* |
| Black-bellied Plover | *Pluvialis squatarola* |
| American Golden-Plover | *Pluvialis dominica* |
| Pacific Golden-Plover | *Pluvialis fulva* |
| Killdeer | *Charadrius vociferus* |
| Semipalmated Plover | *Charadrius semipalmatus* |
| *Mountain Plover | *Charadrius montanus* |
| Snowy Plover | *Charadrius nivosus* |
| Family: SANDPIPERS | *SCOLOPACIDAE* |
| *Upland Sandpiper | *Bartramia longicauda* |
| Whimbrel | *Numenius phaeopus* |
| Long-billed Curlew | *Numenius americanus* |
| Marbled Godwit | *Limosa fedoa* |
| *Bar-tailed Godwit | *Limosa lapponica* |
| *Hudsonian Godwit | *Limosa haemastica* |
| Ruddy Turnstone | *Arenaria interpres* |
| Black Turnstone | *Arenaria melanocephala* |
| Red Knot | *Calidris canutus* |
| Surfbird | *Calidris virgata* |
| Ruff | *Calidris pugnax* |
| *Sharp-tailed Sandpiper | *Calidris acuminata* |
| *Stilt Sandpiper | *Calidris himantopus* |
| Sanderling | *Calidris alba* |
| Dunlin | *Calidris alpina* |
| Rock Sandpiper | *Calidris ptilocnemis* |
| Baird's Sandpiper | *Calidris bairdii* |
| Least Sandpiper | *Calidris minutilla* |
| *Buff-breasted Sandpiper | *Calidris subruficollis* |
| Pectoral Sandpiper | *Calidris melanotos* |

| | |
|---|---|
| *Semipalmated Sandpiper | *Calidris pusilla* |
| Western Sandpiper | *Calidris mauri* |
| Short-billed Dowitcher | *Limnodromus griseus* |
| Long-billed Dowitcher | *Limnodromus scolopaceus* |
| Wilson's Snipe | *Gallinago delicata* |
| Spotted Sandpiper | *Actitis macularius* |
| Solitary Sandpiper | *Tringa solitaria* |
| Wandering Tattler | *Tringa incana* |
| Lesser Yellowlegs | *Tringa flavipes* |
| Willet | *Tringa semipalmata* |
| Greater Yellowlegs | *Tringa melanoleuca* |
| Wilson's Phalarope | *Phalaropus tricolor* |
| Red-necked Phalarope | *Phalaropus lobatus* |
| Red Phalarope | *Phalaropus fulicarius* |

Family: SKUAS — *STERCORARIIDAE*

| | |
|---|---|
| South Polar Skua | *Stercorarius maccormicki* |
| Pomarine Jaeger | *Stercorarius pomarinus* |
| Parasitic Jaeger | *Stercorarius parasiticus* |
| Long-tailed Jaeger | *Stercorarius longicaudus* |

Family: AUKS — *ALCIDAE*

| | |
|---|---|
| Common Murre | *Uria aalge* |
| Pigeon Guillemot | *Cepphus columba* |
| Marbled Murrelet | *Brachyramphus marmoratus* |
| *Scripps's Murrelet | *Synthliboramphus scrippsi* |
| Ancient Murrelet | *Synthliboramphus antiquus* |
| Cassin's Auklet | *Ptychoramphus aleuticus* |
| *Parakeet Auklet | *Aethia psittacula* |
| Rhinoceros Auklet | *Cerorhinca monocerata* |
| *Horned Puffin | *Fratercula corniculata* |
| Tufted Puffin | *Fratercula cirrhata* |

Family: GULLS AND TERNS — *LARIDAE*

| | |
|---|---|
| Black-legged Kittiwake | *Rissa tridactyla* |
| Sabine's Gull | *Xema sabini* |
| Bonaparte's Gull | *Chroicocephalus philadelphia* |
| Franklin's Gull | *Leucophaeus pipixcan* |
| Heermann's Gull | *Larus heermanni* |
| Short-billed Gull | *Larus brachyrhynchus* |
| Ring-billed Gull | *Larus delawarensis* |
| Western Gull | *Larus occidentalis* |
| California Gull | *Larus californicus* |
| Herring Gull | *Larus argentatus* |
| *Iceland Gull | *Larus glaucoides* |
| Glaucous-winged Gull | *Larus glaucescens* |
| Glaucous Gull | *Larus hyperboreus* |
| *Least Tern | *Sternula antillarum* |
| Caspian Tern | *Hydroprogne caspia* |
| Black Tern | *Chlidonias niger* |

| | |
|---|---|
| Common Tern | *Sterna hirundo* |
| Arctic Tern | *Sterna paradisaea* |
| Forster's Tern | *Sterna forsteri* |
| Elegant Tern | *Thalasseus elegans* |

Order: LOONS — GAVIIFORMES

Family: DIVERS — *GAVIIDAE*

| | |
|---|---|
| Red-throated Loon | *Gavia stellata* |
| Pacific Loon | *Gavia pacifica* |
| Common Loon | *Gavia immer* |
| *Yellow-billed Loon | *Gavia adamsii* |

Order: PETRELS — PROCELLARIIFORMES

Family: ALBATROSSES — *DIOMEDEIDAE*

| | |
|---|---|
| Laysan Albatross | *Phoebastria immutabilis* |
| Black-footed Albatross | *Phoebastria nigripes* |
| *Short-tailed Albatross | *Phoebastria albatrus* |

Family: SHEARWATERS AND ALLIES — PROCELLARIIDAE

| | |
|---|---|
| Northern Fulmar | *Fulmarus glacialis* |
| *Murphy's Petrel | *Pterodroma ultima* |
| *Mottled Petrel | *Pterodroma inexpectata* |
| *Hawaiian Petrel | *Pterodroma sandwichensis* |
| Buller's Shearwater | *Ardenna bulleri* |
| Short-tailed Shearwater | *Ardenna tenuirostris* |
| Sooty Shearwater | *Ardenna grisea* |
| Pink-footed Shearwater | *Ardenna creatopus* |
| Flesh-footed Shearwater | *Ardenna carneipes* |
| *Manx Shearwater | *Puffinus puffinus* |

Family: NORTHERN STORM-PETRELS — *HYDROBATIDAE*

| | |
|---|---|
| Fork-tailed Storm-Petrel | *Hydrobates furcatus* |
| Leach's Storm-Petrel | *Hydrobates leucorhous* |
| *Ashy Storm-Petrel | *Hydrobates homochroa* |

Order: GANNETS, CORMORANTS, AND ALLIES — SULIFORMES

Family: BOOBIES — *SULIDAE*

| | |
|---|---|
| *Brown Booby | *Sula leucogaster* |

Family: CORMORANTS — PHALACROCORACIDAE

| | |
|---|---|
| Brandt's Cormorant | *Urile penicillatus* |
| Pelagic Cormorant | *Urile pelagicus* |
| Double-crested Cormorant | *Nannopterum auritum* |

Order: PELICANS, HERONS, AND ALLIES — PELECANIFORMES

Family: PELICANS — PELECANIDAE

| | |
|---|---|
| American White Pelican | *Pelecanus erythrorhynchos* |
| Brown Pelican | *Pelecanus occidentalis* |

Family: HERONS — ARDEIDAE

| | |
|---|---|
| American Bittern | *Botaurus lentiginosus* |
| *Least Bittern | *Ixobrychus exilis* |
| Great Blue Heron | *Ardea herodias* |
| Great Egret | *Ardea alba* |
| Snowy Egret | *Egretta thula* |

| | |
|---|---|
| Cattle Egret | *Bubulcus ibis* |
| Green Heron | *Butorides virescens* |
| Black-crowned Night-Heron | *Nycticorax nycticorax* |
| Family: IBISES AND SPOONBILLS | THRESKIORNITHIDAE |
| White-faced Ibis | *Plegadis chihi* |
| Order: VULTURES | CATHARTIFORMES |
| Family: NEW WORLD VULTURES | CATHARTIDAE |
| #*California Condor | *Gymnogyps californianus* |
| Turkey Vulture | *Cathartes aura* |
| Order: DIURNAL RAPTORS | ACCIPITRIFORMES |
| Family: OSPREY | PANDIONIDAE |
| Osprey | *Pandion haliaetus* |
| Family: HAWKS, EAGLES, AND ALLIES | ACCIPITRIDAE |
| White-tailed Kite | *Elanus leucurus* |
| Golden Eagle | *Aquila chrysaetos* |
| Northern Harrier | *Circus hudsonius* |
| Sharp-shinned Hawk | *Accipiter striatus* |
| Cooper's Hawk | *Accipiter cooperii* |
| Northern Goshawk | *Accipiter gentilis* |
| Bald Eagle | *Haliaeetus leucocephalus* |
| Red-shouldered Hawk | *Buteo lineatus* |
| *Broad-winged Hawk | *Buteo platypterus* |
| Swainson's Hawk | *Buteo swainsoni* |
| Red-tailed Hawk | *Buteo jamaicensis* |
| Rough-legged Hawk | *Buteo lagopus* |
| Ferruginous Hawk | *Buteo regalis* |
| Order: OWLS | STRIGIFORMES |
| Family: BARN OWLS | TYTONIDAE |
| Barn Owl | *Tyto alba* |
| Family: TRUE OWLS | STRIGIDAE |
| Flammulated Owl | *Psiloscops flammeolus* |
| Western Screech-Owl | *Megascops kennicottii* |
| Great Horned Owl | *Bubo virginianus* |
| Snowy Owl | *Bubo scandiacus* |
| Northern Pygmy-Owl | *Glaucidium gnoma* |
| Burrowing Owl | *Athene cunicularia* |
| Spotted Owl | *Strix occidentalis* |
| Barred Owl | *Strix varia* |
| Great Gray Owl | *Strix nebulosa* |
| Long-eared Owl | *Asio otus* |
| Short-eared Owl | *Asio flammeus* |
| *Boreal Owl | *Aegolius funereus* |
| Northern Saw-whet Owl | *Aegolius acadicus* |
| Order: KINGFISHERS AND ALLIES | CORACIIFORMES |
| Family: KINGFISHERS | ALCEDINIDAE |
| Belted Kingfisher | *Megaceryle alcyon* |

Order: WOODPECKERS AND ALLIES

 Family: WOODPECKERS

  Lewis's Woodpecker

  Acorn Woodpecker

  Williamson's Sapsucker

  *Yellow-bellied Sapsucker

  Red-naped Sapsucker

  Red-breasted Sapsucker

  American Three-toed Woodpecker

  Black-backed Woodpecker

  Downy Woodpecker

  Hairy Woodpecker

  White-headed Woodpecker

  Northern Flicker

  Pileated Woodpecker

Order: FALCONS

 Family: FALCONS

  American Kestrel

  Merlin

  *Gyrfalcon

  Peregrine Falcon

  Prairie Falcon

Order: PASSERINES

 Family: NEW-WORLD FLYCATCHERS

  Ash-throated Flycatcher

  *Tropical Kingbird

  Western Kingbird

  Eastern Kingbird

  Olive-sided Flycatcher

  Western Wood-Pewee

  Willow Flycatcher

  *Least Flycatcher

  Hammond's Flycatcher

  Gray Flycatcher

  Dusky Flycatcher

  Pacific-slope Flycatcher

  Cordilleran Flycatcher

  Black Phoebe

  Say's Phoebe

  *Eastern Phoebe

 Family: SHRIKES

  Loggerhead Shrike

  Northern Shrike

 Family: VIREOS

  Hutton's Vireo

  Cassin's Vireo

*PICIFORMES*

 *PICIDAE*

  *Melanerpes lewis*

  *Melanerpes formicivorus*

  *Sphyrapicus thyroideus*

  *Sphyrapicus varius*

  *Sphyrapicus nuchalis*

  *Sphyrapicus ruber*

  *Picoides dorsalis*

  *Picoides arcticus*

  *Dryobates pubescens*

  *Dryobates villosus*

  *Dryobates albolarvatus*

  *Colaptes auratus*

  *Dryocopus pileatus*

*FALCONIFORMES*

 *FALCONIDAE*

  *Falco sparverius*

  *Falco columbarius*

  *Falco rusticolus*

  *Falco peregrinus*

  *Falco mexicanus*

*PASSERIFORMES*

 *TYRANNIDAE*

  *Myiarchus cinerascens*

  *Tyrannus melancholicus*

  *Tyrannus verticalis*

  *Tyrannus tyrannus*

  *Contopus cooperi*

  *Contopus sordidulus*

  *Empidonax traillii*

  *Empidonax minimus*

  *Empidonax hammondii*

  *Empidonax wrightii*

  *Empidonax oberholseri*

  *Empidonax difficilis*

  *Empidonax occidentalis*

  *Sayornis nigricans*

  *Sayornis saya*

  *Sayornis phoebe*

 *LANIIDAE*

  *Lanius ludovicianus*

  *Lanius borealis*

 *VIREONIDAE*

  *Vireo huttoni*

  *Vireo cassinii*

| | |
|---|---|
| Warbling Vireo | *Vireo gilvus* |
| Red-eyed Vireo | *Vireo olivaceus* |
| Family: CORVIDS | *CORVIDAE* |
| Canada Jay | *Perisoreus canadensis* |
| Pinyon Jay | *Gymnorhinus cyanocephalus* |
| Steller's Jay | *Cyanocitta stelleri* |
| *Blue Jay | *Cyanocitta cristata* |
| California Scrub-Jay | *Aphelocoma californica* |
| *Woodhouse's Scrub-Jay | *Aphelocoma woodhouseii* |
| Clark's Nutcracker | *Nucifraga columbiana* |
| Black-billed Magpie | *Pica hudsonia* |
| American Crow | *Corvus brachyrhynchos* |
| Common Raven | *Corvus corax* |
| Family: LARKS | *ALAUDIDAE* |
| Horned Lark | *Eremophila alpestris* |
| Family: SWALLOWS | *HIRUNDINIDAE* |
| Bank Swallow | *Riparia riparia* |
| Tree Swallow | *Tachycineta bicolor* |
| Violet-green Swallow | *Tachycineta thalassina* |
| Northern Rough-winged Swallow | *Stelgidopteryx serripennis* |
| Purple Martin | *Progne subis* |
| Barn Swallow | *Hirundo rustica* |
| Cliff Swallow | *Petrochelidon pyrrhonota* |
| Family: TITS | *PARIDAE* |
| Black-capped Chickadee | *Poecile atricapillus* |
| Mountain Chickadee | *Poecile gambeli* |
| Chestnut-backed Chickadee | *Poecile rufescens* |
| Oak Titmouse | *Baeolophus inornatus* |
| Juniper Titmouse | *Baeolophus ridgwayi* |
| Family: LONG-TAILED TITS | *AEGITHALIDAE* |
| Bushtit | *Psaltriparus minimus* |
| Family: NUTHATCHES | *SITTIDAE* |
| Red-breasted Nuthatch | *Sitta canadensis* |
| White-breasted Nuthatch | *Sitta carolinensis* |
| Pygmy Nuthatch | *Sitta pygmaea* |
| Family: TREECREEPERS | *CERTHIIDAE* |
| Brown Creeper | *Certhia americana* |
| Family: WRENS | *TROGLODYTIDAE* |
| Rock Wren | *Salpinctes obsoletus* |
| Canyon Wren | *Catherpes mexicanus* |
| House Wren | *Troglodytes aedon* |
| Pacific Wren | *Troglodytes pacificus* |
| Marsh Wren | *Cistothorus palustris* |
| Bewick's Wren | *Thryomanes bewickii* |
| Family: GNATCATCHERS | *POLIOPTILIDAE* |
| Blue-gray Gnatcatcher | *Polioptila caerulea* |

| Family: DIPPERS | CINCLIDAE |
|---|---|
|     American Dipper |     *Cinclus mexicanus* |
| Family: KINGLETS | *REGULIDAE* |
|     Golden-crowned Kinglet |     *Regulus satrapa* |
|     Ruby-crowned Kinglet |     *Corthylio calendula* |
| Family: SILVIID WARBLERS | *SYLVIIDAE* |
|     Wrentit |     *Chamaea fasciata* |
| Family: THRUSHES | *TURDIDAE* |
|     Western Bluebird |     *Sialia mexicana* |
|     Mountain Bluebird |     *Sialia currucoides* |
|     Townsend's Solitaire |     *Myadestes townsendi* |
|     Veery |     *Catharus fuscescens* |
|     Swainson's Thrush |     *Catharus ustulatus* |
|     Hermit Thrush |     *Catharus guttatus* |
|     American Robin |     *Turdus migratorius* |
|     Varied Thrush |     *Ixoreus naevius* |
| Family: MIMIDS | *MIMIDAE* |
|     Gray Catbird |     *Dumetella carolinensis* |
|     *Brown Thrasher |     *Toxostoma rufum* |
|     Sage Thrasher |     *Oreoscoptes montanus* |
|     Northern Mockingbird |     *Mimus polyglottos* |
| Family: STARLINGS | *STURNIDAE* |
|     +European Starling |     *Sturnus vulgaris* |
| Family: WAXWINGS | *BOMBYCILLIDAE* |
|     Bohemian Waxwing |     *Bombycilla garrulus* |
|     Cedar Waxwing |     *Bombycilla cedrorum* |
| Family: OLD WORLD SPARROWS | *PASSERIDAE* |
|     +House Sparrow |     *Passer domesticus* |
| Family: PIPITS | *MOTACILLIDAE* |
|     American Pipit |     *Anthus rubescens* |
| Family: FINCHES | *FRINGILLIDAE* |
|     Evening Grosbeak |     *Coccothraustes vespertinus* |
|     Pine Grosbeak |     *Pinicola enucleator* |
|     Gray-crowned Rosy-Finch |     *Leucosticte tephrocotis* |
|     Black Rosy-Finch |     *Leucosticte atrata* |
|     House Finch |     *Haemorhous mexicanus* |
|     Purple Finch |     *Haemorhous purpureus* |
|     Cassin's Finch |     *Haemorhous cassinii* |
|     Common Redpoll |     *Acanthis flammea* |
|     Red Crossbill |     *Loxia curvirostra* |
|     White-winged Crossbill |     *Loxia leucoptera* |
|     Pine Siskin |     *Spinus pinus* |
|     Lesser Goldfinch |     *Spinus psaltria* |
|     American Goldfinch |     *Spinus tristis* |
|     *Lawrence's Goldfinch |     *Spinus lawrencei* |

| | |
|---|---|
| Family: LONGSPURS AND ALLIES | *CALCARIIDAE* |
| Lapland Longspur | *Calcarius lapponicus* |
| *Chestnut-collared Longspur | *Calcarius ornatus* |
| Snow Bunting | *Plectrophenax nivalis* |
| Family: NEW WORLD SPARROWS | *PASSERELLIDAE* |
| Grasshopper Sparrow | *Ammodramus savannarum* |
| Black-throated Sparrow | *Amphispiza bilineata* |
| Lark Sparrow | *Chondestes grammacus* |
| *Lark Bunting | *Calamospiza melanocorys* |
| Chipping Sparrow | *Spizella passerina* |
| *Clay-colored Sparrow | *Spizella pallida* |
| Brewer's Sparrow | *Spizella breweri* |
| Fox Sparrow | *Passerella iliaca* |
| American Tree Sparrow | *Spizelloides arborea* |
| Dark-eyed Junco | *Junco hyemalis* |
| White-crowned Sparrow | *Zonotrichia leucophrys* |
| Golden-crowned Sparrow | *Zonotrichia atricapilla* |
| Harris's Sparrow | *Zonotrichia querula* |
| White-throated Sparrow | *Zonotrichia albicollis* |
| Sagebrush Sparrow | *Artemisiospiza nevadensis* |
| Vesper Sparrow | *Pooecetes gramineus* |
| Savannah Sparrow | *Passerculus sandwichensis* |
| Song Sparrow | *Melospiza melodia* |
| Lincoln's Sparrow | *Melospiza lincolnii* |
| Swamp Sparrow | *Melospiza georgiana* |
| California Towhee | *Melozone crissalis* |
| Green-tailed Towhee | *Pipilo chlorurus* |
| Spotted Towhee | *Pipilo maculatus* |
| Family: CHAT | *ICTERIIDAE* |
| Yellow-breasted Chat | *Icteria virens* |
| Family: GROSBEAKS, TANAGERS, AND ALLIES | *CARDINALIDAE* |
| Western Tanager | *Piranga ludoviciana* |
| *Rose-breasted Grosbeak | *Pheucticus ludovicianus* |
| Black-headed Grosbeak | *Pheucticus melanocephalus* |
| Lazuli Bunting | *Passerina amoena* |
| *Indigo Bunting | *Passerina cyanea* |
| Family: ICTERIDS | *ICTERIDAE* |
| Yellow-headed Blackbird | *Xanthocephalus xanthocephalus* |
| Bobolink | *Dolichonyx oryzivorus* |
| Western Meadowlark | *Sturnella neglecta* |
| *Hooded Oriole | *Icterus cucullatus* |
| Bullock's Oriole | *Icterus bullockii* |
| *Baltimore Oriole | *Icterus galbula* |
| Red-winged Blackbird | *Agelaius phoeniceus* |
| Tricolored Blackbird | *Agelaius tricolor* |
| Brown-headed Cowbird | *Molothrus ater* |

| | |
|---|---|
| *Rusty Blackbird | *Euphagus carolinus* |
| Brewer's Blackbird | *Euphagus cyanocephalus* |
| *Common Grackle | *Quiscalus quiscula* |
| *Great-tailed Grackle | *Quiscalus mexicanus* |
| Family: NEW WORLD WARBLERS | *PARULIDAE* |
| *Ovenbird | *Seiurus aurocapilla* |
| *Northern Waterthrush | *Parkesia noveboracensis* |
| *Black-and-white Warbler | *Mniotilta varia* |
| *Tennessee Warbler | *Leiothlypis peregrina* |
| *Virginia's Warbler | *Leiothlypis virginiae* |
| Orange-crowned Warbler | *Leiothlypis celata* |
| Nashville Warbler | *Leiothlypis ruficapilla* |
| MacGillivray's Warbler | *Geothlypis tolmiei* |
| Common Yellowthroat | *Geothlypis trichas* |
| American Redstart | *Setophaga ruticilla* |
| *Northern Parula | *Setophaga americana* |
| *Magnolia Warbler | *Setophaga magnolia* |
| Yellow Warbler | *Setophaga petechia* |
| *Chestnut-sided Warbler | *Setophaga pensylvanica* |
| *Blackpoll Warbler | *Setophaga striata* |
| *Black-throated Blue Warbler | *Setophaga caerulescens* |
| Palm Warbler | *Setophaga palmarum* |
| Yellow-rumped Warbler | *Setophaga coronata* |
| Black-throated Gray Warbler | *Setophaga nigrescens* |
| Townsend's Warbler | *Setophaga townsendi* |
| Hermit Warbler | *Setophaga occidentalis* |
| Wilson's Warbler | *Cardellina pusilla* |

# LITERATURE CITED

Adkins, J. Y., and D. D. Roby. 2010. *A Status Assessment of the Double-crested Cormorant* (Phalacrocorax auritus) *in Western North America: 1998–2009.* Final report to the US Army Corps of Engineers, Portland, Oregon.

Adkins, J. Y., D. D. Roby, D. E. Lyons, K. N. Courtot, K. Collis, H. R. Carter, W. D. Shuford, and P. J. Capitolo. 2014. Recent population size, trends, and limiting factors for the double-crested cormorant in western North America. *Journal of Wildlife Management* 78:1131–42. doi:10.1002/jwmg.737.

Agee, J. K. 1991. Fire history along an elevational gradient in the Siskiyou Mountains, Oregon. *Northwest Science* 65:188–99.

Alagona, P. S. 2004. Biography of a "feathered pig": the California Condor conservation controversy. *Journal of the History of Biology* 37:557–83.

Alexander, J. D. 1999. Bird-habitat relationships in the Klamath/Siskiyou Mountains. Master's thesis, Southern Oregon University, Ashland.

Alexander, J. D. 2011. Advancing landbird conservation on western federally managed lands with management- and policy-relevant science. Ph.D. dissertation, Prescott College, Prescott, Arizona.

Alexander, J. D., C. J. Ralph, K. Hollinger, and K. Hogoboom. 2004a. Using a wide-scale landbird monitoring network to determine landbird distribution and productivity in the Klamath Bioregion. Pages 33–41 in K. L. Mergenthaler, J. E. Williams, and J. Jules, eds. *Proceedings of the Second Conference on Klamath-Siskiyou Ecology.* Siskiyou Field Institute, Cave Junction, Oregon.

Alexander, J. D., C. J. Ralph, K. Hollinger, B. Hogoboom, N. E. Seavy, and S. Janes. 2004b. Understanding effects of fire suppression, fuels treatment, and wildfire on bird communities in the Klamath-Siskiyou Ecoregion. Pages 42–46 in K. L. Mergenthaler, J. E. Williams, and J. Jules, eds. *Proceedings of the Second Conference on Klamath-Siskiyou Ecology.* Siskiyou Field Institute, Cave Junction, Oregon.

Alexander, J. D., N. E. Seavy, and P. Hosten. 2007. Using bird conservation plans to evaluate ecological effects of fuels reduction in southwest Oregon oak woodland and chaparral. *Forest Ecology Management* 238(1):375–83.

Alexander, J. D., J. L. Stephens, and N. E. Seavy. 2008. Livestock utilization and bird community composition in mixed-conifer forest and oak woodland in southern Oregon. *Northwest Science* 82:7–17.

Alexander, J. D., J. L. Stephens, G. R. Geupel, and T. C. Will. 2009. Decision support tools: bridging the gap between science and management. Pages 283-91 in *Tundra to Tropics: Connecting Birds, Habitats, and People. Proceedings of the 4th International Partners in Flight Conference.* Partners in Flight, McAllen, Texas.

Alexander, J. D., J. L. Stephens, S. Veloz, L. Salas, J. S. Rousseau, C. J. Ralph, and D. A. Sarr. 2017. Using regional bird density distribution models to evaluate protected area networks and inform conservation planning. *Ecosphere* 8:e01799.

Alexander, J. D., C. R. Gillespie, S. Evans-Peters, and B. Brown. 2020a. *Klamath-Siskiyou Oak Network Strategic Conservation Action Plan Version 1.0.* KBO-2020-0013. Klamath-Siskiyou Oak Network and Klamath Bird Observatory, Ashland, Oregon.

Alexander, J. D., C. C. Macias, S. M. D. Younkman, T. Luszcs, R. V. Maria, B. Smith, and D. Casey.

2020b. An integrated conservation strategy for western temperate, Mexican pine-oak, and tropical cloud-forest birds: North America to Central America. Partners in Flight and Klamath Bird Observatory, Ashland, Oregon.,

Allen, J. N. 1980. The ecology and behavior of the Long-billed Curlew in southeastern Washington. *Wildlife Monographs* 73:1–67.

Allison, I. S. 1966. *Fossil Lake, Oregon—Its Geology and Fossil Faunas*. Studies in Geology 9. Oregon State University Press, Corvallis.

Allison, T. D. 2017. A review of options for mitigating take of Golden Eagles at wind energy facilities. *Journal of Raptor Research* 51:319–33.

Altman, B. 1999. Status and conservation of state sensitive grassland bird species in the Willamette Valley. Avifauna Northwest unpublished report to Oregon Department of Fish and Wildlife, Salem.

Altman, B. 2008. Ground-truthing habitat-based population estimates for birds in oak habitats in the Northern Pacific Rainforest Bird Conservation Region. American Bird Conservancy unpublished report to US Fish and Wildlife Service, Washington, DC.

Altman, B. 2011. Historical and current distribution and populations of bird species in prairie-oak habitats in the Pacific Northwest. *Northwest Science* 85:194–222.

Altman, B. 2015a. Oregon Vesper Sparrow range-wide inventory and habitat assessment: final report. State Wildlife Grant G1024-06. American Bird Conservancy unpublished report to Center for Natural Lands Management, Olympia, Washington.

Altman, B. 2015b. Grassland birds and prairie restoration in the West Eugene Wetlands Conservation Area, 2014. American Bird Conservancy unpublished report to Bureau of Land Management, Eugene, Oregon.

Altman, B. 2020. Artificial conspecific attraction to establish populations of Oregon Vesper Sparrow on prairie and savannah conservation lands. American Bird Conservancy unpublished report to Oregon Wildlife Foundation, Portland.

Altman, B. 2021. An assessment of limiting factors and conservation actions for a metapopulation of Oregon Vesper Sparrows in the Willamette Valley, Oregon, 2016-2020. Unpublished report, American Bird Conservancy, The Plains, Virginia.

Altman, B., and J. D. Alexander. 2012. *Habitat Conservation for Landbirds in Coniferous Forests of Western Oregon and Washington*. Version 2.0. Oregon-Washington Partners in Flight, American Bird Conservancy and Klamath Bird Observatory, Ashland, Oregon.

Altman, B., and M. Blakeley-Smith. 2011. Western Meadowlarks and prairie habitat in the Willamette Valley: population enhancement through habitat enhancement. Unpublished report, Avifauna Northwest and Institute for Applied Ecology, Corvallis, Oregon.

Altman, B., and J. Lloyd. 2012. Bird-habitat relationships in oak habitats of western Oregon. Unpublished report, American Bird Conservancy and Ecostudies Institute, Olympia, Washington.

Altman, B., and J. L. Stephens. 2012. *Land Manager's Guide to Bird Habitat and Populations in Oak Ecosystems of the Pacific Northwest*. American Bird Conservancy and Klamath Bird Observatory, Ashland, Oregon.

Altman, B., C. R. Gillespie, and J. L. Stephens. 2017a. OakBirdPop: an online interactive supplement to the *Land Manager's Guide to Bird Habitat and Populations in Oak Ecosystems of the Pacific Northwest*. Version 1.0. American Bird Conservancy, Corvallis, Oregon, and Klamath Bird Observatory, Ashland, Oregon.

Altman, B., S. Evans-Peters, E. Hilton-Kim, N. Maness, J. Stephens, and B. Taylor. 2017b. *Prairie, Oaks, and People: A Conservation Business Plan to Revitalize the Prairie-Oak Habitats of the Pacific Northwest*. American Bird Conservancy, Center for Natural Lands Management, Klamath Bird Observatory, Pacific Birds Habitat Joint Venture, and Willamette Partnership, Corvallis, Oregon.

Alverson, E. R. 2005. Recovering prairies and savannas in a sea of forest: a conservation challenge in the Pacific Northwest. *Plant Talk* 40:23–27.

Anderson, C. D., D. D. Roby, and K. Collis. 2004. Conservation implications of the large colony of Double-crested Cormorants on East Sand Island, Columbia River estuary, Oregon, U.S.A. *Waterbirds* 27:155–60.

Anderson, D. R., and K. P. Burnham. 1994. Demographic analysis of Northern Spotted Owl populations. Pages 66–75 in *USDI Fish and Wildlife Service: Final Draft Recovery Plan for the Northern Spotted Owl*, vol. 2. US Fish and Wildlife Service Directory, Region 1, Portland, Oregon.

Anderson, H. E., A. Wolf, and R. A. Martin. 2013. Streaked Horned Lark conspecific attraction feasibility study. Center for Natural Lands Management unpublished report submitted to US Fish and Wildlife Service, Port of Portland, Joint Base Lewis-McChord, The Nature Conservancy, and Metro, Portland, Oregon.

Anderson, J. 1971. Eagle habitat requirements and forest management. Pages 56–69 in *Proceedings of Fish and Wildlife Habitat Management Training Conference*. USDA Forest Service, Region 6, Eugene, Oregon.

Anderson, R. J. 1985. Bald Eagles and forest management. Pages 189–93 in *Forestry and Wildlife Management in Canada: A Symposium*. University of British Columbia, Vancouver.

Anderson, S. H. 1970a. Ecological relationships of birds in forests of western Oregon. Ph.D. dissertation, Oregon State University, Corvallis.

Anderson, S. H. 1970b. The avifaunal composition of Oregon white oak stands. *Condor* 72:417–23.

Anthony, C. R. 2020. Thermal ecology and populations dynamics of female Greater Sage-Grouse following wildfire in Trout Creek Mountains of Oregon and Nevada. Ph.D. dissertation, Oregon State University, Corvallis.

Anthony, R. G., M. G. Garrett, and C. A. Schuler. 1993. Environmental contaminants in Bald Eagles in the Columbia River estuary. *Journal of Wildlife Management* 57:10–19.

Anthony, R. G., G. A. Green, E. D. Forsman, and S. K. Nelson. 1996. Avian abundance in riparian zones of three forest types in the Cascade Mountains, Oregon. *Wilson Bulletin* 108:280–91.

Anthony, R. G., E. D. Forsman, A. B. Franklin, D. R. Anderson, K. P. Burnham, et al. 2006. Status and trends in demography of Northern Spotted Owls, 1985–2003. *Wildlife Monographs* 163.

Arbib, R. S., ed. 1970. Audubon Field Notes discovery award. *Audubon Field Notes* 24:654.

Bailey, O. A. 2018. Extrinsic and intrinsic factors associated with reproductive success of Caspian Terns (*Hydroprogne caspia*) at East Sand Island, Columbia River estuary. M.S. thesis, Oregon State University, Corvallis.

Baker, L. M., M. Z. Peery, E. E. Burkett, S. W. Singer, D. L. Suddjian, and S. R. Beissinger. 2006. Nesting habitat characteristics of the Marbled Murrelet in central California redwood forests. *Journal of Wildlife Management* 70:939–46.

Banet, N. V., A. F. Evans, K. Collis, D. D. Roby, A. Turecek, Q. Payton, B. Cramer, and T. J. Lawes. In review. Predation on juvenile salmonids by double-crested cormorants nesting on the Astoria-Megler Bridge. Submitted to the *North American Journal of Fisheries Management*.

Barbaree, B. A., S. K. Nelson, B. D. Dugger, D. D. Roby, H. R. Carter, D. L. Whitworth, and S. H. Newman. 2014. Nesting ecology of Marbled Murrelets at a remote mainland fjord in southeast Alaska. *Condor: Ornithological Applications* 116:173–84.

Barber, O. 1941. Juvenile Marbled Murrelet found on Coos River. *Murrelet* 22:38–39.

Barker, F. K., K. J. Burns, J. Klicka, S. M. Lanyon, and I. J. Lovette. 2015. New insights into New World biogeography: an integrated view from the phylogeny of blackbirds, cardinals, sparrows, tanagers, warblers, and allies. *Auk: Ornithological Advances* 132:333–48.

Barnosky, C. W. 1985. Late Quaternary vegetation near Battle Ground Lake, southern Puget Trough, Washington. *Geological Society of America Bulletin* 96:263–71.

Baron, H. M., P. Ruggiero, N. J. Wood, E. L. Harris, J. Allan, P. D. Komar, and P. Corcoran. 2015. Incorporating climate change and morphological uncertainty into coastal change hazard assessments. *Natural Hazards* 75:2081–102.

Barr, B. R., M. E. Koopman, C. D. Williams, S. J. Vynne, R. Hamilton, and B. Doppelt. 2010. *Preparing for Climate Change in the Klamath Basin: Special Report*. Climate Leadership Initiative, National Center for Conservation Science and Policy, Eugene, Oregon.

Barrowclough, G. F., J. G. Groth, W. M. Mauck, and M. E. Blair. 2019. Phylogeography and species limits in the Red-shouldered Hawk (*Buteo lineatus*):

Characterization of the Northern Florida Suture Zone in birds. *Ecology and Evolution* 9:6245–58.

Bartos Smith, S., J. E. McKay, J. K. Richardson, A. A. Shipley, and M. T. Murphy. 2016a. Demography of a ground nesting bird in an urban system: are populations self-sustaining? *Urban Ecosystems* 19:577–98.

Bartos Smith, S., J. E. McKay, M. T. Murphy, and D. A. Duffield. 2016b. Spatial patterns of extra-pair paternity for Spotted Towhees (*Pipilo maculatus*) in urban parks. *Journal of Avian Biology* 47:815–23.

Bartuszevige, A. M., P. L. Kennedy, and R. V. Taylor. 2012. Sixty-seven years of landscape changes in the last, large remnant of the Pacific Northwest Bunchgrass Prairie. *Natural Areas Journal* 32:166–70.

Bate, L. J. 1995. Monitoring woodpecker abundance and habitat in the central Oregon Cascades. M.S. thesis, University of Idaho, Moscow.

Batterson, W. M., and W. B. Morse. 1948. *Oregon Sage Grouse*. Oregon Fauna Series 1. Oregon State Game Commission, Portland.

Bayer, R. D. 1986. Breeding success of seabirds along the mid-Oregon coast concurrent with the 1983 El Niño. *Murrelet* 67:23-26.

Bayer, R. D. 1989. *The Cormorant/Fisherman Conflict in Tillamook County, Oregon*. Studies in Oregon Ornithology 6. Gahmken Press, Newport, Oregon.

Bayer, R. D., and R. W. Ferris. 1987. *Reed Ferris' 1930–1934 Bird Banding Records and Bird Observations for Tillamook County, Oregon*. Studies in Oregon Ornithology 3. Gahmken Press, Newport, Oregon.

Baz, A. 2018. Woodpeckers in the city: habitat use and minimum area requirements of woodpeckers in urban parks and natural areas in Portland, Oregon. M.S. thesis, Portland State University, Portland, Oregon.

Beasley, W. J., and S. J. Dundas. 2021. Hold the line: modeling private coastal adaptation through shoreline armoring decisions. *Journal of Environmental Economics and Management* 105:102397.

Beason, R. C. 2020. Horned Lark (*Eremophila alpestris*), version 1.0. In S. M. Billerman, ed. *Birds of the World*. Cornell Lab of Ornithology, Ithaca, New York.

Becker, C. G., C. R. Fonseca, C. F. Baptista Haddad, R. Fernandes Batista, and P. I. Prado. 2007. Habitat split and the global decline of amphibians. *Science* 318:1775–77. https://doi.org/10.1126/science.1149374.

Beissinger, S. R., and M. Z. Peery. 2007. Reconstructing the historical demography of an endangered seabird. *Ecology* 88:296–305.

Benke, A. C., and C. E. Cushing. 2005. *Rivers of North America: The Natural History*. Academic Press, San Diego, California.

Bent, A. C. 1919. Life histories of North American diving birds. *US National Museum Bulletin* 107:1–239.

Berlanga, H., J. A. Kennedy, T. D. Rich, M. C. Arizmendi, C. J. Beardmore, et al. 2010. *Saving Our Shared Birds: Partners in Flight Tri-National Vision for Landbird Conservation*. Cornell Lab of Ornithology, Ithaca, New York.

Betts, M. G., J. C. Hagar, J. W. Rivers, J. D. Alexander, K. McGarigal, and B. C. McComb. 2010. Thresholds in forest bird occurrence as a function of the amount of early-seral broadleaf forest at landscape scales. *Ecological Applications* 20:2116–30.

Betts, M. G., J. Verschuyl, J. Giovanini, T. Stokely, and A. J. Kroll. 2013. Initial experimental effects of intensive forest management on avian abundance. *Forest Ecology and Management* 310:1036–44.

Betts, M. G., L. Fahrig, A. S. Hadley, K. E. Halstead, J. Bowman, W. D. Robinson, J. A. Wiens, and D. B. Lindenmayer. 2014. A species-centered approach for uncovering generalities in organism responses to habitat loss and fragmentation. *Ecography* 7:517–27.

Betts, M. G., B. Phalan, S. J. Frey, J. S. Rousseau, and Z. Yang. 2017. Old-growth forests buffer climate-sensitive bird populations from warming. *Diversity and Distributions* 24:439 47.

Betts, M. G., J. Gutiérrez Illán, Z. Yang, S. M. Shirley, and C. D. Thomas. 2019. Synergistic effects of climate and land-cover change on long-term bird population trends of the Western USA: a test of modeled predictions. *Frontiers of Ecological Evolution* 7:1–11.

Betts, M. G., J. M. Northrup, J. A. Bailey Guerrero, L. J. Adrean, S. K. Nelson, et al. 2020. Squeezed by a habitat split: warm ocean conditions and old-forest loss interact to reduce long-term occupancy of a threatened seabird. *Conservation Letters* 13:e12745.

Betts, M. G., B. T. Phalan, C. Wolf, S. C. Baker, C. Messier, K. J. Puettmann, R. Green, S. H. Harris, D. P. Edwards, D. B. Lindenmayer, and A. Balmford. 2021. Producing wood at least cost to biodiversity: integrating Triad and sharing–sparing approaches to inform forest landscape management. *Biological Reviews* brv.12703.

Binford, L. C., B. G. Elliott, and S. W. Singer. 1975. Discovery of a nest and the downy young of the Marbled Murrelet. *Wilson Bulletin* 87:303–19.

BirdLife International. 2018. Brachyramphus marmoratus. *IUCN Red List of Threatened Species* 2010:e.T22694870A24483089.

Bloxton, T. D., Jr., and M. G. Raphael. 2008. *Breeding Ecology of the Marbled Murrelet in Washington State: Five-Year Project Summary (2004–2008).* USDA Forest Service, Pacific Northwest Research Station, Olympia, Washington.

Blus, L. J., C. J. Henny, and A. J. Krynitsky. 1985. Organochlorine-induced mortality and residues in Long-billed Curlews from Oregon. *Condor* 87:563–65.

Boehlert, G., and A. Gill. 2010. Environmental and ecological effects of ocean renewable energy development—a current synthesis. *Oceanography* 23:68-81.

Bond, N. A., M. F. Cronin, H. Freeland, and N. Mantua. 2015. Causes and impacts of the 2014 warm anomaly in the NE Pacific. *Geophysical Research Letters* 42:3414–20.

Bottorff, J. 1989. *Concept Plan for Waterfowl Habitat Protection, Klamath Basin.* North American Waterfowl Management Plan Category 28. US Fish and Wildlife Service, Portland, Oregon.

Boyd, R. T. 1999. Strategies of Indian burning in the Willamette Valley. Pages 94–138 in R. T. Boyd, ed. *Indians, Fire and the Land in the Pacific Northwest.* Oregon State University Press, Corvallis.

Brabant R., N. Vanermen, E. W. N. Stienen, and S. Degraer, 2015. Towards a cumulative collision risk assessment of local and migrating birds in North Sea offshore wind farms. *Hydrobiologia* 756:63–74.

Bradley, R. W., and F. Cooke. 2001. Cliff and deciduous tree nests of Marbled Murrelets in southwestern British Columbia. *Northwestern Naturalist* 82:52–57.

Bradley, R. W., F. Cooke, L. W. Lougheed, and W. S. Boyd. 2004. Inferring breeding success through radio telemetry in the Marbled Murrelet. *Journal of Wildlife Management* 68:318–31.

Braun, C. E., M. F. Baker, R. L. Eng, J. S. Gashwiler, and M. H. Schroeder. 1976. Conservation committee report on effects of alteration of sagebrush communities on the associated avifauna. *Wilson Bulletin* 88:165–71.

Brooks, A. 1926. The mystery of the Marbled Murrelet. *Murrelet* 7:1–2.

Brooks, A. 1928. Does the Marbled Murrelet nest inland? *Murrelet* 9:68.

Brophy, L. S. 2019. *Comparing Historical Losses of Forested, Scrub-Shrub, and Emergent Tidal Wetlands on the Oregon Coast, USA: A Paradigm Shift for Estuary Restoration and Conservation.* Prepared for the Pacific States Marine Fisheries Commission and the Pacific Marine and Estuarine Fish Habitat Partnership. Estuary Technical Group, Institute for Applied Ecology, Corvallis, Oregon.

Broshot, N. E. 1999. The effects of urbanization and human disturbance upon plant community structure and bird species richness, diversity, and abundance in a natural forested area (Forest Park) in Portland, Oregon. Ph.D. dissertation, Portland State University, Oregon.

Brown, E. R., tech. ed. 1985. *Management of Wildlife and Fish Habitats in Forests of Western Oregon and Washington.* USDA Forest Service R6-F&WL-192-1985. Pacific Northwest Region, Portland, Oregon.

Brown, S., C. Hickey, B. Harrington, and R. Gill, eds. 2001. *The U.S. Shorebird Conservation Plan,* 2nd ed. Manomet Center for Conservation Sciences, Manomet, Massachusetts.

Browning, M. R. 1973. Bendire's records of Red-shouldered Hawk (*Buteo lineatus*) and Yellow-bellied Sapsucker (*Sphyrapicus varius nuchalis*) in Oregon. *Murrelet* 54:34–35.

Browning, M. R. 1975. The distribution and occurrence of the birds of Jackson County, Oregon, and surrounding areas. Pages 1–69 in *North American Fauna*. US Fish and Wildlife Service, Washington, DC.

Browning, M. R. 1977. The types and type-localities of *Oreortyx pictus* (Douglas) and *Ortyx plumiferus* Gould. *Proceedings of the Biology Society of Washington* 90:808–12.

Browning, M. R. 1979. Type specimens of birds collected in Oregon. *Northwest Science* 53:132–40.

Browning, M. R. 2019. A review of the subspecific and species status of Warbling Vireo (*Vireo gilvus*). *Oregon Birds* 45:89–99.

Buck, J. 2017. *Assessment of Water Quality at Ankeny and Baskett Slough National Wildlife Refuges*. US Fish and Wildlife Service, Washington, DC.

Buck, J. A., R. G. Anthony, C. A. Schuler, F. B. Isaacs, and D. E. Tillitt. 2005. Changes in productivity and contaminants in Bald Eagles nesting along the lower Columbia River, USA. *Environmental Toxicology and Chemistry* 24:1779-92.

Buehler, D. A. 2000. Bald Eagle (*Haliaeetus leucocephalus*). In A. Poole and F. Gill, eds. *The Birds of North America*, No. 506. Birds of North America, Inc., Philadelphia, Pennsylvania.

Bull, E. L. 1980. Resource partitioning among woodpeckers in northeastern Oregon. Ph.D. dissertation, University of Idaho, Moscow.

Bull, E. L., and E. C. Meslow. 1977. Habitat requirements of the Pileated Woodpecker in northeastern Oregon. *Journal of Forestry* 75:335–37.

Bull, E. L., S. R. Peterson, and J. W. Thomas. 1986. *Resource Partitioning among Woodpeckers in Northeastern Oregon*. Note PNW-444. USDA, US Forest Service Research, Portland, Oregon.

Bull, E. L., J. E. Hohmann, and M. G. Henjum. 1987. Northern Pygmy-Owl nests in northeastern Oregon. *Journal of Raptor Research* 21:77–78.

Bureau of Land Management. 2016. *Northwestern and Coastal Oregon Record of Decision and Approved Resource Management Plan*. Bureau of Land Management, Portland, Oregon.

Burnham, K. P., D. R. Anderson, and G. C. White. 1996. Meta-analysis of vital rates of the Northern Spotted Owl. *Studies in Avian Biology* 17:92–101.

Cahall, R. E., and J. P. Hayes. 2009. Influences of postfire salvage logging on forest birds in the Eastern Cascades, Oregon, USA. *Forest Ecology and Management* 257(3):1119–28.

Cahall, R. E., J. P. Hayes, and M. G. Betts. 2013. Will they come? Long-term response by forest birds to experimental thinning supports the "Field of Dreams" hypothesis. *Forest Ecology and Management* 304:137–49.

Cancellieri, S., and M. T. Murphy. 2013. Experimental examination of nest reuse by an open-cup nesting passerine: savings of time/energy or nest site shortage? *Animal Behaviour* 85:1287–94.

Carson, R. 1962. *Silent Spring*. Houghton Mifflin, Boston, Massachusetts.

Carter, H. R., and M. L. Morrison, eds. 1992. Status and conservation of the Marbled Murrelet in North America. *Proceedings of the Western Foundation of Vertebrate Zoology* 5.

Carter, H. R., and S. G. Sealy. 1987. Inland records of downy young and fledgling Marbled Murrelets in North America. *Murrelet* 68:58–63.

Carter, H. R., and S. G. Sealy. 2005. Who solved the mystery of the Marbled Murrelet? *Northwestern Naturalist* 86:2–11.

Carter, H. R., A. L. Sowls, M. S. Rodway, U. W. Wilson, R. W. Lowe, G. J. McChesney, F. Gress, and D. W. Anderson. 1995. Population size, trends, and conservation problems of the Double-crested Cormorant on the Pacific coast of North America. *Colonial Waterbirds* 18:189–215.

Cederholm, C. J., D. H. Johnson, R. E. Bilby, L. G. Dominguez, A. M. Garrett, et al. 2001. Pacific salmon and wildlife-ecological contexts, relationships, and implications for management. Pages 628–84 in D. H. Johnson and T. A. O'Neil, dis. *Wildlife-Habitat Relationships in Oregon and Washington*. Oregon State University Press, Corvallis.

Center for Biological Diversity, Oregon Natural Resources Council, Friends of the San Juans, and Northwest Ecosystem Alliance. 2002. *Petition to List Streaked Horned Lark (Eremophila alpestris strigata) as a Federally Endangered Species*. Center for Biological Diversity, Tucson, Arizona. https://ecos.fws.gov/docs/tess/petition/646.pdf.

Chambers, C. L., W. C. McComb, and J. C. Tappeiner. 1999. Breeding bird responses to three silvicultural treatments in the Oregon Coast Range.

*Ecological Applications* 9:171–85.

Chan, F., J. A. Barth, C. A. Blanchette, R. H. Byrne, F. Chavez, et al. 2017. Persistent spatial structuring of coastal ocean acidification in the California Current System. *Scientific Reports* 7:2526. https://doi.org/10.1038/s41598-017-02777-y.

Chase, M., and G. R. Geupel. 2005. The use of avian focal species for conservation planning in California. Pages 130–42 in C. J. Ralph and T. D. Rich, eds. *Bird Conservation Implementation and Integration in the Americas: Proceedings of the Third International Partners in Flight Conference.* General Technical Report PSW-GTR-191. Pacific Southwest Research Station, USDA Forest Service, Albany, California.

Chatwin, T. A., M. H. Mather, and T. D. Giesbrecht. 2002. Changes in pelagic and double-crested cormorant nesting populations in the Strait of Georgia, British Columbia. *Northwestern Naturalist* 83:109–17.

Chen, D., and R. C. Hale. 2010. A global review of polybrominated diphenyl ether flame retardant contamination in birds. *Environment International* 36:800–811.

Church, M. E., R. Gwiazda, R. W. Riseborough, C. P. Chamberlain, S. Farry, W. Heinrich, B. A. Rideout, and D. R. Smith. 2006. Ammunition is the principal source of lead accumulated by California Condors re-introduced to the wild. *Environmental Science and Technology* 40:6143–50.

Chutter, C. M., L. J. Redmond, N. W. Cooper, A. C. Dolan, D. A. Duffield, and M. T. Murphy. 2016. Paternal behavior in a socially monogamous but sexually promiscuous passerine bird. *Behaviour* 153:443–66.

Clifton, G. T., J. A. Carr, and A. A. Biewener. 2018. Comparative hindlimb myology of foot-propelled swimming birds. *Journal of Anatomy* 232(1):105–23.

Cline, B. B., and S. M. Haig. 2011. Seasonal movements of Wilson's Snipe (*Gallinago delicata*) across multiple phases of the annual cycle in Oregon. *Auk* 128:543–55.

Clinton, W. J. 2000. Establishment of the Cascade-Siskiyou National Monument. Proclamation No. 7318, 65 Fed. Reg. 37249–52.

Coates, P. S., M. A. Ricca, B. G. Prochazka, M. L.

Brooks, K. E. Doherty, T. Kroger, E. J. Blomberg, C. A. Hagen, and M. L. Casazza. 2016. Wildfire, climate, and invasive grass interactions negatively impact an indicator species by reshaping sagebrush ecosystems. *Proceedings of the National Academy of Sciences* 113(45):12745–50.

Coleman, R. G., and A. R. Kruckeberg. 1999. Geology and plant life of the Klamath-Siskiyou Mountain region. *Natural Areas Journal* 19:320–40.

Collar, S., D. D. Roby, and D. E. Lyons. 2017. Top-down and bottom-up interactions influence fledging success at North America's largest colony of Caspian Terns (*Hydroprogne caspia*). *Estuaries and Coasts* 40:1808–18. doi:10.1007/s12237-017-0238-x.

Collis, K., D. D. Roby, D. P. Craig, S. K. Adamany, J. Y. Adkins, and D. E. Lyons. 2002. Colony size and diet composition of piscivorous waterbirds on the Lower Columbia River: implications for losses of juvenile salmonids to avian predation. *Transactions of the American Fisheries Society* 131:537–50.

Collis, K., D. D. Roby, A. F. Evans, D. E. Lyons, T. J. Lawes, Q. Payton, B. Cramer, A. Turecek, and A. G. Patterson. 2021. Caspian Tern management in the Columbia Plateau region. Pages 114–210 in D. D. Roby, A. F. Evans, and K. Collis, eds. *Avian Predation on Salmonids in the Columbia River Basin: A Synopsis of Ecology and Management.* Synthesis report submitted to the US Army Corps of Engineers, Walla Walla, Washington; Bonneville Power Administration, Portland, Oregon; Grant County Public Utility District / Priest Rapids Coordinating Committee, Ephrata, Washington; and Oregon Department of Fish and Wildlife, Salem.

Contreras, A. L., et al. 2022. *A History of Oregon Ornithology: From Territorial Days to the Rise of Birding.* Oregon State University Press, Corvallis.

Cooper, N. W., M. T. Murphy, and L. J. Redmond. 2009. Age- and sex-dependent arrival date in the Eastern Kingbird. *Journal of Field Ornithology* 80:35–41.

Cooper, N. W., M. T. Murphy, L. J. Redmond, and A. C. Dolan. 2011. Reproductive consequences of spring arrival date in the Eastern Kingbird. *Journal of Ornithology* 152:143–52.

Cornelius, J. M., G. Perreau, V. R. Bishop, J. S. Krause, R. Smith, T. P. Hahn, and S. L. Meddle. 2018. Social information changes stress hormone receptor expression in the songbird brain. *Hormones and Behavior* 97:31–38.

Cornely, J. E., S. P. Thompson, E. L. McLaury, and L. D. Napier. 1985. A summary of Trumpeter Swan production on Malheur National Wildlife Refuge, Oregon. *Murrelet* 66:50–55.

Courtot, K. N., D. D. Roby, J. Y. Adkins, D. E. Lyons, D. T. King, and R. S. Larsen. 2012. Colony connectivity of Pacific coast double-crested cormorants based on post-breeding dispersal from the region's largest colony. *Journal of Wildlife Management* 76:1462–71. doi:10.1002/jwmg.403.

Cramer, B., A. F. Evans, Q. Payton, K. Collis, and D. D. Roby. 2021a. Relative impacts of Double-crested Cormorants and Caspian Terns on juvenile salmonids in the Columbia River estuary: a retrospective analysis of the PIT tag data. Pages 418–45 in D. D. Roby, A. F. Evans, and K. Collis, eds. *Avian Predation on Salmonids in the Columbia River Basin: A Synopsis of Ecology and Management.* Synthesis report submitted to the US Army Corps of Engineers, Walla Walla, Washington; Bonneville Power Administration, Portland, Oregon; Grant County Public Utility District / Priest Rapids Coordinating Committee, Ephrata, Washington; and Oregon Department of Fish and Wildlife, Salem.

Cramer, B., K. Collis, A. F. Evans, D. D. Roby, D. E. Lyons, T. J. Lawes, Q. Payton, and A. Turecek. 2021b. Predation on juvenile salmonids by colonial waterbirds nesting at unmanaged colonies in the Columbia River basin. Pages 446–535 in D. D. Roby, A. F. Evans, and K. Collis, eds. *Avian Predation on Salmonids in the Columbia River Basin: A Synopsis of Ecology and Management.* Synthesis report submitted to the US Army Corps of Engineers, Walla Walla, Washington; Bonneville Power Administration, Portland, Oregon; Grant County Public Utility District / Priest Rapids Coordinating Committee, Ephrata, Washington; and Oregon Department of Fish and Wildlife, Salem.

Crawford, J. A. 1982. *History of Sage Grouse in Oregon.* Oregon Wildlife. Oregon Department of Fish and Wildlife, Portland.

Cross, S. P., and J. K. Simmons. 1983. *Bird Populations of the Mixed-Hardwood Forests near Roseburg, Oregon.* Technical Report 82-2-05. Oregon Department of Fish and Wildlife Nongame Wildlife Program, Salem.

Curtis, J. 2014. Sixty years of avian community change in the Willamette Valley, Oregon. M.S. thesis, Oregon State University, Corvallis.

Cushman, S. A., and K. McGarigal. 2003. Hierarchical, multi-scale decomposition of species-environment relationships. *Landscape Ecology* 17:637–46.

Cuthbert, F. J., and L. Wires. 1999. Caspian Tern (*Sterna caspia*). In A. Poole and F. Gill, eds. *The Birds of North America*, No. 403. Birds of North America, Inc., Philadelphia, Pennsylvania.

Dahl, T. E. 1990. *Wetlands Losses in the United States, 1780s to 1980s.* US Department of the Interior, Fish and Wildlife Service, Washington, DC.

Davies, K. W., C. S. Boyd, J. L. Beck, J. D. Bates, T. J. Svejcar, and M. A. Gregg. 2011. Saving the sagebrush sea: an ecosystem conservation plan for big sagebrush plant communities. *Biological Conservation* 144:2573–84.

Davis, R. J., K. M. Dugger, S. Mohoric, L. Evers, and W. C. Aney. 2011. *Northwest Forest Plan—The First 15 Years (1994–2008): Status and Trends of Northern Spotted Owl Populations and Habitats.* General Technical Report PNW-GTR-850. USDA Forest Service, Pacific Northwest Research Station, Portland, Oregon.

Davis, R. J., B. Hollen, J. Hobson, J. E. Gower, and D. Keenum. 2016. *Northwest Forest Plan—The First 20 Years (1994–2013): Status and Trends of Northern Spotted Owl Habitats.* General Technical Report PNW-GTR-929. USDA Forest Service, Pacific Northwest Research Station, Portland, Oregon.

Dawson, W. L., and J. H. Bowles. 1909. *The Birds of Washington*, vol. 2. Occidental Publishing, Seattle, Washington.

Dedrick, K. 1989. San Francisco Bay tidal marsh-land acreages: recent and historic values. Pages 383–98 in O. T. Magoon, ed. *Proceedings of the 6th Symposium in Coastal Ocean and Management*

(*Coastal Zone 1989*). American Society of Engineers, New York.

Deignan, H. G. 1961. Type specimens of birds in the United States National Museum. *Bulletin of the United States National Museum* 221:1–718. https://doi.org/10.5479/si.03629236.221.

D'Elia, J., and S. M. Haig. 2013. *The California Condor in the Pacific Northwest*. Oregon State University Press, Corvallis.

D'Elia, J., S. M. Haig, J. M. Johnson, B. Marcot, and R. Young. 2015. Activity-specific ecological niche models for planning reintroductions of California Condors (*Gymnogyps californianus*). *Biological Conservation* 184:90–99.

D'Elia, J., S. M. Haig, M. Miller, and T. Mullins. 2016. Ancient DNA reveals substantial genetic diversity in the California Condors (*Gymnogyps californianus*) prior to a population bottleneck. *Condor: Ornithological Applications* 118:703–14.

D'Elia, J., J. Brandt, L. J. Burnett, S. M. Haig, J. Hollenbeck, et al. 2019. Applying circuit theory and landscape linkage maps to reintroduction planning for California Condors. *PLoS ONE* 14(12):e0226491.

DellaSala, D. A., S. B. Reid, T. J. Frest, J. R. Strittholt, and D. M. Olson. 1999. A global perspective on the biodiversity of the Klamath-Siskiyou Ecoregion. *Natural Areas Journal* 19:300–319.

DellaSala, D. A., R. Baker, D. Heiken, C. A. Frissell, J. R. Karr, S. K. Nelson, B. R. Noon, D. Olson, and J. Strittholt. 2015. Building on two decades of ecosystem management and biodiversity conservation under the Northwest Forest Plan, USA. *Forests* 6:3326–52.

DeMars, C. A., D. K. Rosenberg, and J. B. Fontaine. 2010. Multi-scale factors affecting bird use of isolated remnant oak trees in agro-ecosystems. *Biological Conservation* 143:1485–92.

Department of Land Conservation and Development. 2021a. *Oregon Land Use Planning Goals Program*. Oregon Department of Land Conservation and Development, Salem.

Department of Land Conservation and Development. 2021b. *Oregon Coastal Management Program*. Oregon Department of Land Conservation and Development, Salem.

Department of Land Conservation and Development. 2021c. *Territorial Sea Plan*. Oregon Department of Land Conservation and Development, Salem.

Deshler, J. D., and M. T. Murphy. 2012. The breeding biology of the Northern Pygmy Owl: do the smallest of the small have an advantage? *Condor* 114:314–22.

Desholm, M., A. D. Fox, P. D. L. Beasley, and J. Kahlert. 2006. Remote techniques for counting and estimating the number of bird wind turbine collisions at sea: a review. *Ibis* 148(s1):76–89.

de Wit, C. A. 2002. An overview of brominated flame retardants in the environment. *Chemosphere* 46:583–624.

Dickens, S. N. 1985. Klamath Falls and Altamount, Oregon. *Yearbook of the Association of Pacific Coast Geographers* 47:27–38.

Diller, L., K. Hamm, D. Early, D. Lamphear, K. Dugger, C. Yackulic, C. Schwarz, P. Carlson, and T. McDonald. 2016. Demographic response of Northern Spotted Owls to Barred Owl removal in coastal northern California. *Journal of Wildlife Management* 80:691–707. doi:10.1002/jwmg.1046.

Dixon, R. D. 1995a. Ecology of White-headed Woodpeckers in the central Oregon Cascades. M.S. thesis, University of Idaho, Moscow.

Dixon, R. D. 1995b. *Density, Nest-Site Selection and Roost Characteristics, Home Range, and Habitat Use, and Behavior of White-headed Woodpeckers: Deschutes and Winema National Forests, Oregon*. Nongame Report 93-3-01. Oregon Department of Fish and Wildlife, Salem.

Dixon, R. D. 1998. *An Assessment of White-headed Woodpeckers in a Regional Landscape*. Contract number 101-98. US Forest Service Region 6, Portland, Oregon.

Dobkin, D. S., and J. D. Sauder. 2004. *Shrubsteppe Landscapes in Jeopardy: Distributions, Abundances, and the Uncertain Future of Birds and Small Mammals in the Intermountain West*. High Desert Ecological Research Institute, Bend, Oregon.

Doerge, K. F. 1978. Aspects of the geographic ecology of the Acorn Woodpecker (*Melanerpes formicivorus*). M.S. thesis, Oregon State University, Corvallis.

Dolan, A., K. Sexton, L. Redmond, and M. T. Murphy. 2007. Dawn song of Eastern Kingbirds: intrapopulation variability and sociobiological correlates. *Behaviour* 144(10):1273–95.

Dolan, A. C., M. T. Murphy, L. J. Redmond, and D. Duffield. 2009. Maternal characteristics and the production and recruitment of sons in the eastern kingbird (*Tyrannus tyrannus*). *Behavioral Ecology and Sociobiology* 63(10):1527–37.

Donnelly, J. P., D. E. Naugle, D. P. Collins, B. D. Dugger, B. W. Allred, J. D. Tack, and V. J. Dreitz. 2019. Synchronizing conservation to seasonal wetland hydrology and waterbird migration in semi-arid landscapes. *Ecosphere* 10(6):e02758. doi:10.1002/ecs2.2758.

Dorr, B. S., J. J. Hatch, and D. V. Weseloh. 2014. Double-crested Cormorant (*Phalacrocorax auritus*), version 2.0. In A. F. Poole, ed. *The Birds of North America*. Cornell Lab of Ornithology, Ithaca, New York. https://doi.org/10.2173/bna.441.

Douglas, D. 1829. Observations on the *Vultur californianus* of Shaw. *Zoology Journal* 4:328–30.

Douglas, D. 1959. *Journal Kept by David Douglas during His Travels in North America, 1823–1827.* Wesley and Son's, Antiquarian Press, New York.

Draheim, H. M., P. Baird, and S. M. Haig. 2012. Temporal analysis of mtDNA variation reveals decreased genetic diversity in Least Terns. *Condor* 114:145–54.

Drovetski, S. V., S. F. Pearson, and S. Rohwer. 2006. Streaked Horned Lark *Eremophila alpestris strigata* has distinct mitochondrial DNA. *Conservation Genetics* 6:875–83.

Dubois, A., and A. Nemesio. 2007. Does nomenclatural availability of nomina of new species or subspecies require the deposition of vouchers in collections? *Zootaxa* 1409(1):1–22.

Duchac, L. S., D. B. Lesmeister, K. M. Dugger, Z. J. Ruff, and R. J. Davis. 2020. Passive acoustic monitoring effectively detects Northern Spotted Owls and Barred Owls over a range of forest conditions. *Condor* 122:1–22.

Duerr, A. E., T. M. Donovan, and D. E. Capen. 2009. Management-induced reproductive failure and breeding dispersal in Double-crested Cormorants on Lake Champlain. *Journal of Wildlife Management* 71:2565–74.

Dugger, B. D., and K. M. Dugger, 2002. Long-billed Curlew: *Numenius americanus*. In A. Poole and F. Gill, eds. *The Birds of North America*, No. 628. Birds of North America, Philadelphia, Pennsylvania.

Dugger, K. M., F. F. Wagner, R. G. Anthony, and G. S. Olson. 2005. The relationship between habitat characteristics and demographic performance of Northern Spotted Owls in southern Oregon. *Condor* 107:865–80.

Dugger, K. M., R. G. Anthony, and L. S. Andrews. 2011. Transient dynamics of invasive competition: Barred Owls, Spotted Owls, and the demons of competition present. *Ecological Applications* 21:2459–68.

Dugger, K. M., E. D. Forsman, A. B. Franklin, R. J. Davis, G. C. White, et al. 2016. The effects of habitat, climate, and Barred Owls on long-term demography of Northern Spotted Owls. *Condor* 118:57–116.

Dunk, J. R, W. J. Zielinski, and H. H. Welsh. 2006. Evaluating reserves for species richness and representation in northern California. *Diversity and Distributions* 12:434–42.

eBird. 2021. eBird: an online database of bird distribution and abundance [web application]. eBird, Cornell Lab of Ornithology, Ithaca, New York.

Eddleman, W. R., F. L. Knopf, B. Meanley, F. A. Reid, and R. Zembal. 1988. Conservation of North American rallids. *Wilson Bulletin* 100:458–75.

Eddy, R. H. 1953. Summer bird habitats in the Corvallis area, Willamette Valley, Oregon. M.S. thesis, Oregon State University, Corvallis.

Eliot, W. A. 1923. *Birds of the Pacific Coast.* G. P. Putnam and Sons, New York.

Ellis, M. S., P. L. Kennedy, W. D. Edge, and T. A. Sanders. 2019. Twenty-year changes in riparian bird communities of east-central Oregon. *Wilson Bulletin* 131:43–61.

Eltzroth, E. K. 1983. Breeding biology and mortality of Western Bluebirds near Corvallis, Oregon. *Sialia* 5:83–87.

Environment Canada. 2014. *Recovery Strategy for the Marbled Murrelet* (Brachyramphus marmoratus) *in Canada.* Species at Risk Act Recovery Strategy Series. Environment Canada, Ottawa.

Erickson, W. P., G. D. Johnson, D. M. Strickland, D. P. Young Jr., K. J. Sernka, and R. E. Good. 2001. *Avian Collisions with Wind Turbines: A Summary of Existing Studies and Comparisons to Other Sources of Avian Collision Mortality in the United States*. No. DOE-00SF22100-. Western EcoSystems Technology, Cheyenne, Wyoming; RESOLVE, Washington, DC.

Erickson, W. P., M. M. Wolfe, K. J. Bay, D. H. Johnson, and J. L. Gehring. 2014. A comprehensive analysis of small-passerine fatalities from collision with turbines at wind energy facilities. *PLoS ONE* 9(9):e107491.

Evans, A. F., N. J. Hostetter, D. D. Roby, K. Collis, D. E. Lyons, B. P. Sandford, R. D. Ledgerwood, and S. Sebring. 2012. Systemwide evaluation of avian predation on juvenile salmonids from the Columbia River based on recoveries of passive integrated transponder tags. *Transactions of the American Fisheries Society* 141:975–89.

Evans, A., Q. Payton, B. Cramer, K. Collis, N. J. Hostetter, D. D. Roby, and C. Dotson. 2019. Cumulative effects of avian predation on Upper Columbia River steelhead. *Transactions of the American Fisheries Society* 148:896–913.

Evans Mack, D., W. P. Ritchie, S. K. Nelson, E. Kuo-Harrison, P. Harrison, and T. E. Hamer. 2003. *Methods for Surveying Marbled Murrelets in Forests: A Revised Protocol for Land Management and Research*. Technical Publication 2. Pacific Seabird Group, Corvallis, Oregon.

Evans-Peters, G. R., B. D. Dugger, and M. J. Petrie. 2012. Plant community composition and waterfowl food production on Wetland Reserve Program easements compared to those on managed public lands in western Oregon and Washington. *Wetlands* 32:391–99.

Evenden, F. G. 1949. Habitat relations of typical austral and boreal avifauna in the Willamette Valley, Oregon. Ph.D. dissertation, Oregon State College, Corvallis.

Farner, D. S. 1952. *The Birds of Crater Lake National Park*. University of Kansas Press, Lawrence.

Fellows, S. D., and S. L. Jones. 2009. *Status Assessment and Conservation Action Plan for the Long-billed Curlew* (Numenius americanus). Biological Technical Publication FWS/BTP-R6012–2009. Department of the Interior, Fish and Wildlife Service, Washington, DC.

Fickas, K. C., W. B. Cohen, and Z. Yang. 2016. Landsat-based monitoring of annual wetland change in the Willamette Valley of Oregon, USA from 1972 to 2012. *Wetlands Ecology and Management* 24(1):73–92.

Figueira, L., P. Martins, C. J. Ralph, J. L. Stephens, J. D. Alexander, and J. D. Wolfe. 2020. Effects of breeding and molt activity on songbird site fidelity. *Auk* 137(4):1–15.

Fijn, R. C., K. L. Krijgsveld, M. J. M. Poot, and S. Dirksen. 2015. Bird movements at rotor heights measured continuously with vertical radar at a Dutch offshore wind farm. *Ibis* 157:558–66.

Finkelstein, M. E., D. F. Doak, D. George, J. Burnett, J. Brandt, M. Church, J. Grantham, and D. R. Smith. 2012. Lead poisoning and the deceptive recovery of the critically endangered California Condor. *Proceedings of the National Academy of Sciences* 109:11,449–54.

Finley, W. L. 1907a. Among the gulls on Klamath Lake. *Condor* 9:12–16.

Finley, W. L. 1907b. The grebes of southern Oregon. *Condor* 9:97–101.

Finley, W. L. 1908. Life history of the California Condor part II—historical data and range of the condor. *Condor* 10:5–10.

Fleischer, R. C., G. Fuller, and D. B. Ledig. 1995. Genetic structure of endangered Clapper Rail (*Rallus longirostris*) populations in southern California. *Conservation Biology* 9:1234–43.

Fleskes, J. P. 2012. Wetlands of the Central Valley of California and Klamath Basin. Pages 357–70 in D. P. Batzer and A. H. Baldwin, eds. *Wetland Habitats of North America: Ecology and Conservation Concerns*. University of California Press, Berkeley.

Flowers, J., R. Albertani, T. Harrison, B. Polagye, and R. Suryan. 2014. Design and initial component tests of an integrated avian and bat collision detection system for offshore wind turbines. Presented at the 2nd Annual Marine Energy Technology Symposium, Seattle, Washington, April 15–18, 2014.

Fontaine, J. B., D. C. Donato, W. D. Robinson, B. E. Law, and J. B. Kauffman. 2009. Bird communities following high-severity fire: response to single and repeat fires in a mixed-evergreen forest, Oregon, USA. *Forest Ecology and Management* 257:1496–504.

Forest Ecosystem Management Assessment Team. 1993. *Forest Ecosystem Management: An Ecological, Economic, and Social Assessment.* USDA, US Department of the Interior, Washington, DC.

Forsman, E. D., E. C. Meslow, and M. J. Strub. 1977. Spotted Owl abundance in young versus old-growth forests, Oregon. *Wildlife Society Bulletin* 5:43–47.

Forsman, E. D., E. C. Meslow, and H. M. Wight. 1984. Distribution and biology of the Spotted Owl in Oregon. *Wildlife Monographs* 87:1–64.

Forsman, E. D., R. G. Anthony, J. A. Reid, P. J. Loschl, S. G. Sovern, et al. 2002. Natal and breeding dispersal of Northern Spotted Owls. *Wildlife Monographs* 149:1–35.

Forsman, E. D., R. G. Anthony, E. C. Meslow, and C. J. Zabel. 2004. Diets and foraging behavior of Northern Spotted Owls in Oregon. *Journal of Raptor Research* 38:214–30.

Forsman, E. D., R. G. Anthony, K. M. Dugger, E. M. Glenn, A. B. Franklin, et al. 2011. *Population Demography of Northern Spotted Owls.* Studies in Avian Biology 40. Cooper Ornithological Society, University of California Press, Berkeley.

Foster, D. 2002. Refuges and reclamation: conflicts in the Klamath Basin, 1904–1964. *Oregon Historical Quarterly* 103(2):150–87.

Foster, L. J., K. M. Dugger, C. A. Hagen, and D. A. Budeau. 2018. Greater Sage-Grouse vital rates after wildfire. *Journal of Wildlife Management* 83:121–34.

Franklin, A. B., K. P. Burnham, G. C. White, R. G. Anthony, E. D. Forsman, C. Schwarz, J. D. Nichols, and J. Hines. 1999. *Rangewide Status and Trends in Northern Spotted Owl Populations.* USGS Colorado Cooperative Fish and Wildlife Research Unit, Colorado State University, Fort Collins.

Franklin, A. B., D. R. Anderson, R. J. Gutiérrez, and K. P. Burnham. 2000. Climate, habitat quality, and fitness in Northern Spotted Owl populations in northwestern California. *Ecological Monographs* 70:539–90.

Franklin, A. B., K. M. Dugger, D. B. Lesmeister, R. J. Davis, J. D. Wiens, et al. 2021. Range-wide declines of Northern Spotted Owls populations in the Pacific Northwest: a meta-analysis. *Biological Conservation* 259. doi:10.1016/j.biocon.2021.109168.

Franklin, J. F., and C. T. Dyrness. 1973. *Natural Vegetation of Oregon and Washington.* General Technical Report PNW-GTR-008. USDA Forest Service, Pacific Northwest Research Station, Portland, Oregon.

Franklin, J. F., and C. T. Dyrness. 1988. *Natural Vegetation of Oregon and Washington.* Oregon State University Press, Corvallis.

Franklin, J. F., T. A. Spies, R. Van Pelt, A. B. Carey, D. A. Thornburgh, et al. 2002. Disturbances and structural development of natural forest ecosystems with silvicultural implications, using Douglas-fir forests as an example. *Forest Ecological Management* 155:399–423.

Franklin, J. F., R. J. Mitchell, and B. J. Palik. 2007. *Natural Disturbance and Stand Development Principles for Ecological Forestry.* General Technical Report NRS-19. USDA Forest Service, Northern Research Station, Portland, Oregon.

Franklin, J. F., K. N. Johnson, D. J. Churchill, K. Hagmann, D. Johnson, and J. Johnston. 2013. *Restoration of Dry Forests in Eastern Oregon: A Field Guide.* The Nature Conservancy, Portland, Oregon.

Frenzel, R. W. 1998. Nest sites and nesting success of White-headed Woodpeckers on the Winema and Deschutes National Forests in 1997. Unpublished report to Oregon Natural Heritage Program, The Nature Conservancy of Oregon, Portland.

Frenzel, R. W. 2000. Nest sites, nesting success, and turn-over rates of White-headed Woodpeckers on the Deschutes and Winema National Forests in 2000. Unpublished report to Oregon Natural Heritage Program, The Nature Conservancy of Oregon, Portland.

Frenzel, R. W. 2004. Nest-site occupancy, nesting success, and turnover rates of White-headed Woodpeckers in the Oregon Cascade Mountains in 2004. Unpublished report to Portland Audubon Society, Oregon Department of Fish and Wildlife, Bureau of Land Management, and US Forest Service, Portland, Oregon.

Frenzel, R. W., and R. G. Anthony. 1989. Relationship of diets and environmental contaminants in wintering bald eagles. *Journal of Wildlife Management* 53(3):792–802.

Frenzel, R. W., and K. J. Popper. 1998. Densities of White-headed Woodpeckers and other woodpeckers in study areas on the Winema and Deschutes National Forests, Oregon in 1997. Unpublished report to Oregon Natural Heritage Program, The Nature Conservancy of Oregon, Portland.

Frey, S. J. K., A. S. Hadley, S. L. Johnson, M. Schulze, J. A. Jones, and M. G. Betts. 2016. Spatial models reveal the microclimatic buffering capacity of old-growth forests. *Science Advances* 2:e1501392.

Friends of Tryon Creek. 1994. *A Forest in the City: Your Guide to Tryon Creek State Park*. Far West Book Service, Portland, Oregon.

Funk, W. C., T. D. Mullins, and S. M. Haig. 2007. Conservation genetics of Snowy Plovers (*Charadrius alexandrinus*) in the Western Hemisphere: population genetic structure and delineation of subspecies. *Conservation Genetics* 8:1287–309.

Funk, W. C., E. D. Forsman, T. D. Mullins, and S. M. Haig. 2008. Introgression and dispersal among Spotted Owl (*Strix occidentalis*) subspecies. *Ecological Applications* 1:161–71.

Funk, W. C., E. D. Forsman, J. M. Johnson, T. D. Mullins, and S. M. Haig. 2010. Evidence for recent population bottlenecks in Northern Spotted Owls (*Strix occidentalis caurina*). *Conservation Genetics* 11:1013–21.

Furness, R. W. 1978. Energy requirements of seabird communities: a bioenergetics model. *Journal of Animal Ecology* 47:39–53.

Gabrielson, I. N. 1931. The Birds of the Rogue River Valley, Oregon. *Condor* 33:110–21.

Gabrielson, I. N., and S. G. Jewett. 1940. *Birds of Oregon*. Oregon State College, Corvallis.

Gabrielson, I. N., and S. G. Jewett. 1970. *Birds of the Pacific Northwest: With Special Reference to Oregon*. Dover, New York.

Galen, C. 1989. *A Preliminary Assessment of the Status of Lewis's Woodpecker in Wasco, County,*

*Oregon*. Nongame Report 88-3-01. Oregon Department of Fish and Wildlife, Salem.

García-Hernández, J., Y. V. Sapozhnikova, D. Schlenk, A. Z. Mason, O. Hinojosa-Huerta, J. J. Rivera-Díaz, N. A. Ramos-Delgado, and G. Sánchez-Bon. 2006. Concentration of contaminants in breeding bird eggs from the Colorado River Delta, Mexico. *Environmental Toxicology and Chemistry: An International Journal* 25(6):1640–47.

Garcia-Reyes, M., W. J. Sydeman, D. S. Schoeman, R. R. Rykaczewski, B. A. Black, A. J. Smit, and S. J. Bograd. 2015. Under pressure: climate change, upwelling, and Eastern Boundary Current Upwelling Systems. *Frontiers in Marine Science* 2:109. http://dx.doi.org/10.3389/fmars.2015.00109.

Gardner, M. 2015. *Analysis of Shoreline Armoring and Erosion Policies along the Oregon Coast*. Oregon Department of Land Conservation and Development, Salem.

Gashwiler, J. S. 1977. *Bird Populations in Four Vegetation Types in Central Oregon*. Special Scientific Report—Wildlife Number 205. US Fish and Wildlife Service, Portland, Oregon.

Gashwiler, J., and A. Ward. 1966. Western redcedar seed, a food of Pine Siskins. *Murrelet* 47:73–75.

Gaunt, A. S., and L. W. Oring, eds. 2010. *Guidelines to the Use of Wild Birds in Research*, 3rd ed. Ornithological Council, Washington, DC.

Gill, R. E., and L. R. Mewaldt. 1983. Pacific coast Caspian Terns: dynamics of an expanding population. *Auk* 100:369–81.

Gillespie, C. R., and J. L. Stephens. 2020. *Breeding Density of Six Chaparral-Associated Songbirds in Large (<2 ha) Chaparral Retention Patches at Table Rocks Oak Restoration Sites*. Report KBO-2020-0005. Klamath Bird Observatory, Ashland, Oregon.

Gillespie, C. R., J. L. Stephens, K. E. Halstead, and J. D. Alexander. 2017. *Amount of Chaparral Habitat at the Landscape Scale Influences Site Level Occupancy for Three Chaparral-Associated Bird Species*. Report KBO-2017-0003. Klamath Bird Observatory, Ashland, Oregon.

Gillespie, C. R., J. D. Alexander, J. L. Stephens, and S. M. Rockwell. 2018. *Using Science to Improve*

Habitat and Preserve Biological Diversity in the Cascade-Siskiyou National Monument. Report KBO-2018-0005. Klamath Bird Observatory, Ashland, Oregon.

Gilligan, J., M. Smith, D. Rogers, and A. Contreras. 1994. *Birds of Oregon: Status and Distribution.* Cinclus Publications, McMinnville, Oregon.

Gilmer, D. S., J. L. Yee, D. M. Mauser, and J. L. Hainline. 2004. *Waterfowl Migration on Klamath Basin National Wildlife Refuges, 1953–2001.* Biological Science Report USGS/BRD/BSR-2003-2004. US Geological Survey, Biological Resources Division, Reston, Virginia.

Girard, P., J. Y. Takekawa, and S. R. Beissinger. 2010. Uncloaking a cryptic, threatened rail with molecular markers: origins, connectivity, and demography of a recently-discovered population. *Conservation Genetics* 11:2409–18.

Goggans, R. 1985. Habitat use by Flammulated Owls in northeastern Oregon. M.S. thesis, Oregon State University, Corvallis.

Goldade, C. M., J. A. Dechant, D. H. Johnson, A. L. Zimmerman, B. E. Jamison, J. O. Church, and B. R. Euliss. 2002. *Effects of Management Practices on Wetland Birds: Yellow Rail.* Northern Prairie Wildlife Research Center, Jamestown, North Dakota.

Grecian, W. J., R. Inger, M. J. Attrill, S. Bearhop, B. J. Godley, M. J. Witt, and S. C. Votier. 2010. Potential impacts of wave-powered marine renewable energy installations on marine birds. *Ibis* 152:683–97. https://doi.org/10.1111/J.1474-919x.2010.01048.X.

Green, D. L. 1965. Developmental history of European beachgrass (*Ammophilia arenaria* (L.) Link) plantings on the Oregon coastal sand dunes. M.S. thesis, Oregon State University, Corvallis.

Green, G. A. 1983. Ecology of breeding burrowing owls in the Columbia Basin, Oregon. M.S. thesis, Oregon State University, Corvallis.

Green, G. A., and R. G. Anthony. 1989. Nesting success and habitat relationships of Burrowing Owls in the Columbia Basin, Oregon. *Condor* 91(2):347–54.

Green, G. A., and R. G. Anthony. 1997. Ecological considerations for management of breeding Burrowing Owls in the Columbia Basin. Pages 117–21 in J. L. Lincer and K. Steenhof, eds. *The Burrowing Owl, Its Biology and Management: Including the Proceedings of the First International Symposium.* Raptor Research Report 9. Raptor Research Foundation, Topeka, Kansas.

Green, G. A., R. E. Fitzner, R. G. Anthony, and L. E. Rogers. 1993. Comparative diets of Burrowing Owls in Oregon and Washington. *Northwest Science* 67:88–93.

Green, R. E., and J. P. Scharlemann. 2003. Egg and skin collections as a resource for long-term ecological studies. *Bulletin-British Ornithologists Club* 123:165–76.

Gremillet, D., G. Wright, A. Lauder, D. N. Carss, and S. Wanless. 2003. Modelling the daily food requirements of wintering great cormorants: a bioenergetics tool for wildlife management. *Journal of Applied Ecology* 40:266–77.

Griffee, W. E. 1944. First Oregon nest of Yellow Rail (*Coturnicops noveboracensis*). *Murrelet* 25:29.

Grinnell, J. 1897. Notes on the Marbled Murrelet. *Osprey* 1:115–17.

Grove, R. A., C. J. Henny, and J. L. Kaiser. 2009. Ospreys: worldwide sentinel species for assessing and monitoring environmental contamination in rivers, lakes, reservoirs and estuaries. *Journal of Toxicology and Environmental Health B* 12:25–44.

Grover, J. C., and J. M. Grover. 2000. *Oral History of David B. Marshall.* US Fish and Wildlife Service, Portland, Oregon. https://digitalmedia.fws.gov/digital/collection/document/id/1151/.

Guiguet, C. J. 1956. Enigma of the Pacific. *Audubon Magazine* 58:164–67, 174.

Gullion, G. W. 1951. Birds of the southern Willamette Valley, Oregon. *Condor* 53:129–49.

Gunderson, D. T., D. A. Duffield, T. Randall, N. Wintle, D. N. D'Alessandro, J. M. Rice, and D. Shepardson. 2012. Organochlorine contaminants in blubber from stranded marine mammals collected from the northern Oregon and southern Washington coasts: implications for re-introducing California Condors, *Gymnogyps californianus*, in Oregon. *Bulletin of Environmental Contamination and Toxicology* 90(3):269–73.

Gutiérrez, R. J., A. B. Franklin, and W. S. Lahaye. 2020. Spotted Owl (*Strix occidentalis*). In A. F. Poole and F. B. Gill, eds. *Birds of the World.* Cornell Lab of

Ornithology, Ithaca, New York. https://doi-org .ezproxy.proxy.library.oregonstate.edu/10.2173/ bow.spoowl.01.

Hagar, D. C. 1960. The interrelationships of logging, birds and timber regeneration in the Douglas-fir region of northwestern California. *Ecology* 41:116–25.

Hagar, J. C., and L. M. Sherman. 2018. *Distribution and Habitat Use of Snag-Nesting Purple Martins in Western Oregon.* Interagency Special Status / Sensitive Species Program Administrative Report. USDA Forest Service Region 6 and US Department of the Interior Bureau of Land Management, Portland, Oregon.

Hagar, J. C., and M. A. Stern. 2001. Avifauna in oak woodlands of the Willamette Valley, Oregon. *Northwestern Naturalist* 82:12–25.

Hagar, J. C., W. McComb, and W. Emmingham. 1996. Bird communities in commercially thinned and unthinned Douglas-fir stands of western Oregon. *Wildlife Society Bulletin* 24:353–66.

Hagar, J. C., S. Howlin, and L. Ganio. 2004. Short-term response of songbirds to experimental thin-ning of young Douglas-fir forests in the Oregon Cascades. *Forest Ecology and Management* 199:333–47.

Hagen, C. A. 2011. *Greater Sage-Grouse Conservation Assessment and Strategy for Oregon: A Plan to Maintain and Enhance Populations and Habitat.* Oregon Department of Fish and Wildlife, Salem.

Haig, S. M., and F. W. Allendorf. 2006. Hybrids and pol-icy. Pages 150–63 in J. M. Scott, D. D. Goble, and F. W. Davis, eds. *The Endangered Species Act at Thirty,* vol. 2. *Conserving Biodiversity in Human-Dominated Landscape.* Island Press, Washington, DC.

Haig, S. M., and J. D'Elia. 2010. Defining avian subspecies and endangered species laws. *Ornithological Monographs* 67:24–34.

Haig, S. M., C. L. Gratto-Trevor, T. D. Mullins, and M. A. Colwell. 1997. Population identification of Western Hemisphere shorebirds throughout the annual cycle. *Molecular Ecology* 6:413–27.

Haig, S. M., D. W. Mehlman, and L. W. Oring. 1998. Avian movements and wetland connectivity in landscape conservation. *Conservation Biology* 12:749–58.

Haig, S. M., R. S. Wagner, E. D. Forsman, and T. D. Mullins. 2001. Geographic variation and genetic structure in Spotted Owls. *Conservation Genetics* 2:25–40.

Haig, S. M., L. W. Oring, P. M. Sanzenbacher, and O. W. Taft. 2002. Space use, migratory connectiv-ity, and population segregation among Willets breeding in the western Great Basin. *Condor* 104:620–30.

Haig, S. M., E. D. Forsman, and T. D. Mullins. 2004a. Subspecies relationships and genetic structure in the Spotted Owl. *Conservation Genetics* 5:683–705.

Haig, S. M., T. D. Mullins, E. D. Forsman, P. Trail, and L. Wennerberg. 2004b. Genetic identification of Spotted Owls, Barred Owls, and their hybrids: evolutionary and legal implications. *Conservation Biology* 18:1347–57.

Haig, S. M., E. A. Beever, S. M. Chambers, H. M. Draheim, B. D. Dugger, et al. 2006. Taxonomic considerations in listing subspecies under the U.S. Endangered Species Act. *Conservation Biology* 20:1584–94.

Haig, S. M., J. D'Elia, G. Herring, C. Eagles-Smith, J. Fair, J. Gervais, J. Rivers, and J. Schulz. 2014. Perspectives in ornithology: the persistent prob-lem of lead poisoning in birds from ammunition and fishing tackle. *Condor* 116:408–28.

Haig, S. M., S. P. Murphy, J. H. Matthews, I. Arismendi, and M. Safeeq. 2019. Climate-altered Great Basin wetlands and deterioration of a mi-gratory flyway. *Nature Scientific Reports* 9:4666.

Hallman, T. A. 2018. Modeling fine-scale avian dis-tributions and densities with multi-scale models: predicting the past and present. Ph.D. dissertation, Oregon State University, Corvallis.

Halofsky, J. E., D. L. Peterson, and B. J. Harvey. 2020. Changing wildfire, changing forests: the effects of climate change on fire regimes and vegetation in the Pacific Northwest, USA. *Fire Ecology* 16:1–4.

Halstead, K. E., J. L. Stephens, and J. D. Alexander. 2012. *Bird Abundance and Composition across a Gradient of Oak Forest Types in the Rogue Basin of Southwestern Oregon.* Report KBO-2012-0011. Klamath Bird Observatory, Ashland, Oregon.

Halstead, K. E., J. D. Alexander, A. S. Hadley, J. L. Stephens, Z. Yang, and M. G. Betts. 2019. Using a species-centered approach to predict bird community responses to habitat fragmentation. *Landscape Ecology* 34:1919–35.

Hamer, T. E., S. K. Nelson, J. E. Jones, and J. Verschuyl. 2021. Marbled Murrelet nest site selection at three fine spatial scales. *Avian Conservation and Ecology* 16:4.

Hansel-Kuehn, V. J. 2003. The Dalles Roadcut (Fivemile Rapids) avifauna: evidence for a cultural origin. M.S. thesis, Washington State University, Pullman.

Hansen, A. J., W. C. McComb, R. Vega, M. G. Raphael, and M. Hunter. 1995. Bird habitat relationships in natural and managed forests in the west Cascades of Oregon. *Ecological Applications* 5:555–69.

Hansen, H. P. 1947. Postglacial forest succession, climate, and chronology in the Pacific Northwest. *Transactions of the American Philosophical Society* 37:1–126.

Hapke, C. J., P. N. Adams, J. Allan, A. Ashton, G. B. Griggs, M. A. Hampton, J. Kelly, and A. P. Young. 2014. The rock coast of the USA. Pages 137–54 in D. M. Kennedy, W. J. Stephenson, and L. A. Naylor, eds. *Rock Coast Geomorphology: A Global Synthesis*. Geological Society, London.

Harcombe, P. A., S. E. Greene, K. A. Kramer, S. A. Acker, T. A. Spies, and T. E. Valentine. 2004. The influence of fire and windthrow dynamics on a coastal spruce-hemlock forest in Oregon, USA, based on aerial photographs spanning 40 years. *Forest Ecology and Management* 194:71–82.

Harrington-Tweit, B., P. W. Mattocks, and E. S. Hunn. 1979. North Pacific Coast region. *American Birds* 33:892.

Harris, L. D. 1984. *The Fragmented Forest: Island Biogeography Theory and the Preservation of Biotic Diversity*. University of Chicago Press, Chicago, Illinois.

Harris, S. H., and M. G. Betts. 2021. Bird abundance is highly dynamic across succession in early seral tree plantations. *Forest Ecology and Management* 483:118902.

Hartlaub, G. 1852. R. Titian Peale's Vögel der "United States Exploring Expedition." *Archiv für Naturgeschichte* 18:93–138.

Hartman, C. A., and L. W. Oring. 2009. Reproductive success of Long-Billed Curlews (*Numenius americanus*) in northeastern Nevada hayfields. *Auk* 126:420-30.

Hartman, C. A., J. T. Ackerman, M. P. Herzog, C. Strong, and D. Trachtenbarg. 2019. Social attraction used to establish Caspian Tern nesting colonies in San Francisco Bay. *Global Ecology and Conservation* 20:e00757. https://doi.org/10.1016/j.gecco.2019.e00757.

Harvey, C. N., G. Garfield, N. Williams, K. Tolimieri, K. Andrews, et al. 2020. *Ecosystem Status Report of the California Current for 2019–20: A Summary of Ecosystem Indicators Compiled by the California Current Integrated Ecosystem Assessment Team (CCIEA)*. Technical Memorandum NMFS-NWFSC-160. US Department of Commerce, National Oceanic and Atmospheric Administration, Seattle, Washington.

Hennings, L. A., and D. W. Edge. 2003. Riparian bird community structure in Portland, Oregon: habitat, urbanization, and spatial scale patterns. *Condor* 105:288–302.

Henny, C. J. 1977. Birds of prey, DDT, and Tussock Moths in Pacific Northwest. *Transactions North American Wildlife and Natural Resources Conference* 42:397–411.

Henny, C. J., and M. B. Naughton. 1998. Wintering Canada Geese in the Willamette Valley. Pages 676–77 in M. J. Mac, P. A. Opler, C. E. Puckett Haecker, and P. D. Doran, eds. *Status and Trends of the Nation's Biological Resources*, 2 vols. Biological Resources Division, US Geological Survey, Reston, Virginia.

Henny, C. J., and M. W. Nelson. 1981. Decline and present status of breeding Peregrine Falcons in Oregon. *Murrelet* 62:43–53.

Henny, C. J., and H. M. Wight. 1969. An endangered Osprey population: estimates of mortality and production. *Auk* 86:188–98.

Henny, C. J., W. S. Overton, and H. M. Wight. 1970. Determining parameters for populations by structural models. *Journal of Wildlife Management* 34:690–703.

Henny, C. J., L. J. Blus, and C. J. Stafford. 1984. Effects of heptachlor on American kestrels in the Columbia Basin, Oregon. *Journal of Wildlife Management* 47:1080–87.

Henny, C. J., L. J. Blus, E. J. Kolbe, and R. E. Fitzner. 1985. Organophosphate insecticide (Famphur) topically applied to cattle kills magpies and hawks. *Journal of Wildlife Management* 49:648–58.

Henny, C. J., R. A. Grove, J. L. Kaiser, and B. L. Johnson. 2010. North American Osprey populations and contaminants: historic and contemporary perspectives. *Journal of Toxicology and Environmental Health, Part B* 13:579–603.

Henny, C. J., R. A. Grove, J. L. Kaiser, B. L. Johnson, C. Furl, and R. J. Letcher. 2011. Wastewater dilution index partially explains observed polybrominated diphenyl ether flame retardant concentrations in Osprey eggs from Columbia River Basin, 2008–2009. *Ecotoxicology* 20:682–97.

Henry, J., and S. F. Baird. 1855. *Reports of Explorations and Surveys: To Ascertain the Most Practicable and Economical Route for a Railroad from the Mississippi River to the Pacific Ocean*, vol. 1. T. Ford, Printer.

Herring, G., C. A. Eagles-Smith, and M. T. Wagner. 2016. Ground squirrel shooting and potential lead exposure in breeding avian scavengers. *PLoS ONE* 11:e0167926.

Herring, G., C. A. Eagles-Smith, and J. A. Buck. 2017. Characterizing Golden Eagle risk to lead and anticoagulant rodenticide exposure. *Journal of Raptor Research* 51:273–92.

Herring, G., C. A. Eagles-Smith, and D. E. Varland. 2018. Mercury and lead exposure in avian scavengers from the Pacific Northwest suggest risks to California Condors—implications for reintroduction and recovery. *Environmental Pollution* 243, Part A:610–19.

Herring, G., C. A. Eagles-Smith, J. A. Buck, A. E. Shiel, C. R. Vennum, et al. 2020. The lead (Pb) lining of agriculture-related subsidies: enhanced Golden Eagle growth rates tempered by Pb exposure. *Ecosphere* 11(1):e03006. doi:10.1002/ecs2.3006.

Hessburg, P. F., T. A. Spies, D. A. Perry, C. N. Skinner, A. H. Taylor, et al. 2016. Tamm review: management of mixed-severity fire regime forests in Oregon, Washington, and northern California.

*Forest Ecology and Management* 366:221–50.

Hilderbrand, G. V., S. D. Farley, C. C. Schwartz, and C. T. Robbins. 2004. Importance of salmon to wildlife: implications for integrated management. *Ursus* 15:1–9.

Hipfner, J. M., L. K. Blight, R. W. Lowe, S. I. Wilhelm, G. J. Robertson, R. T. Barrett, T. Anker-Nilssen, and T. P. Good. 2012. Unintended consequences: how the recovery of sea eagle *Haliaeetus* spp. populations in the Northern Hemisphere is affecting seabirds. *Marine Ornithology* 40:39–52.

Hodder, J., and M. R. Graybill. 1985. Reproduction and survival of seabirds in Oregon during the 1982–1983 El Niño. *Condor* 87:535–41.

Hoen, B. D., J. E. Diffendorfer, J. T. Rand, L.A. Kramer, C. P. Garrity, and H. E. Hunt. 2018. United States Wind Turbine Database (v4.3, 14 January 2022): data release. US Geological Survey, American Clean Power Association, and Lawrence Berkeley National Laboratory. https://doi.org/10.5066/F7TX3DN0.

Hofstadter, D. F., N. F. Kryshak, C. M. Wood, B. P. Dotters, K. N. Roberts, et al. 2022. Arresting the spread of invasive species in continental systems. *Frontiers in Ecology and the Environment*. https://doi.org/10.1002/fee.2458.

Hollenbeck, J. P., V. A. Saab, and R. Frenzel. 2011. Habitat suitability and survival of nesting White-headed Woodpeckers in unburned forests of central Oregon. *Journal of Wildlife Management* 75:1061–71.

Hollenbeck, J. P., J. M. Johnson, and S. M. Haig. 2014. Using terrestrial laser scanning to support ecological research in the rocky intertidal zone. *Journal of Coastal Conservation Planning and Management* 18:701–14. doi:10.1007/s11852-014-0346-8.

Hollenbeck, J. P., S. M. Haig, E. D. Forsman, J. M. Johnson, and J. D. Wiens. 2018. Latitudinal variation in dispersal by juvenile Northern Spotted Owls over 30 years. *Condor* 120:530–42.

Holm, S. R., B. R. Noon, J. D. Wiens, and W. J. Ripple. 2016. Potential trophic cascades triggered by the Barred Owl range expansion. *Wildlife Society Bulletin* 40:615–24.

Holmes, A. L., G. A. Green, R. L. Morgan, and K. B. Livezey. 2003. Burrowing Owl nest success and burrow longevity in north central Oregon. *Western North American Naturalist* 63(2):244–50.

Holmes, M. W., T. T. Hammond, G. O. Wogan, R. E. Walsh, K. LaBarbera, et al. 2016. Natural history collections as windows on evolutionary processes. *Molecular Ecology* 25(4):864–81.

Holmes, A., J. D. Maestas, and D. E. Naugle. 2017. Bird responses to western juniper removal in sagebrush steppe. *Rangeland Ecology and Management* 70:87–94.

Hornaday, W. T. 1916. *Save the Sagebrush Grouse from Extinction*. Bulletin No. 5. Permanent Wild Life Protection Fund, New York Zoological Society, New York.

Horton, C. A. 2014. Top-down influences of Bald Eagles on Common Murre populations in Oregon. M.S. thesis, Oregon State University, Corvallis.

Horvath, E. 1999. *Distribution, Abundance, and Nest-Site Characteristics of Purple Martins in Oregon*. Technical Report 99-1-01. Oregon Department of Fish and Wildlife Diversity Program, Salem.

Hostetter, N. J., A. F. Evans, Q. Payton, D. D. Roby, D. E. Lyons, and K. Collis. 2021. Factors affecting the susceptibility of juvenile salmonids to avian predation. Pages 665–711 in D. D. Roby, A. F. Evans, and K. Collis, eds. *Avian Predation on Salmonids in the Columbia River Basin: A Synopsis of Ecology and Management*. Synthesis report submitted to the US Army Corps of Engineers, Walla Walla, Washington; Bonneville Power Administration, Portland, Oregon; Grant County Public Utility District / Priest Rapids Coordinating Committee, Ephrata, Washington; and Oregon Department of Fish and Wildlife, Salem.

Houck, M. C., and M. J. Cody. 2009. *Wild in the City: A Guide to Portland's Natural Areas*. Audubon Society of Portland, Portland, Oregon.

Houle, M. 2007. *The Prairie Keepers: Secrets of the Zumwalt*, 2nd ed. Oregon State University Press, Corvallis.

Houle, M. 2010. *One City's Wilderness*. Oregon State University Press, Corvallis.

Howell, S. N. G. 2010. *Molt in North American Birds*. Houghton Mifflin Harcourt, New York.

Hsu, C., and R. Albertani. 2021. Wind turbine event detection by support vector machine. *Wind Energy* 24(7):672–85. https://doi.org/10.1002/we.2596.

Hsu, C., R. Albertani, and R. M. Suryan. 2018. Wind turbine sensor array for monitoring avian and bat collisions. *Wind Energy* 21:255–63. https://doi.org/10.1002/we.2160.

Huff, M. H., N. E. Seavy, J. D. Alexander, and C. J. Ralph. 2005. Fire and birds in maritime Pacific Northwest. *Studies in Avian Biology* 30:46–62.

Hulse, D., A. Branscomb, J. Giocochea-Duclos, S. Gregory, L. Ashkenas, et al. 1998. *Willamette River Basin Planning Atlas*, 1st ed. Pacific Northwest Ecosystem Research Consortium, Institute for a Sustainable Environment, University of Oregon, Eugene.

Hulse, D. W., J. M. Eilers, K. Freemark, C. Hummons, and D. White. 2000. Planning alternative future landscapes in Oregon: evaluating effects on water quality and biodiversity. *Landscape Journal* 19:1–19.

Huso, M. M. 2011. An estimator of wildlife fatality from observed carcasses. *Environmetrics* 22(3):318–29.

Huso, M., and D. Dalthorp. 2014. Accounting for unsearched areas in estimating wind turbine-caused fatality. *Journal of Wildlife Management* 78:347–58.

Huso, M., D. Dalthorp, D. Dail, and L. Madsen. 2015. Estimating turbine-caused bird and bat fatality when zero carcasses are observed. *Ecological Applications* 25:1213–25.

ICF International. 2010. *Habitat Conservation Plan for the Western Snowy Plover*. Prepared for Oregon Parks and Recreation Department, Portland.

Impara, P. C. 1997. Spatial and temporal patterns of fire in the forests of the central Oregon Coast Range. Ph.D. dissertation, Oregon State University, Corvallis.

IPCC. Intergovernmental Panel on Climate Change. 2022. *Climate Change 2022: Impacts, Adaptation,* and *Vulnerability*. Contribution of Working Group II to the Sixth Assessment Report of the Intergovernmental Panel on Climate Change. Edited by H.-O. Pörtner, D. C. Roberts, M. Tignor, E. S. Poloczanska, K. Mintenbeck, et al. Cambridge University Press, Cambridge, Massachusetts.

Isaacs, F. B., and R. G. Anthony. 2008. Bald Eagle nest locations and history of use in Oregon and the Washington portion of the Columbia River Recovery Zone, 1971 through 2007. Oregon Cooperative Fish and Wildlife Research Unit, Oregon State University, Corvallis.

Isaacs, F. B., and R. G. Anthony. 2011. *Bald Eagles* (Haliaeetus leucocephalus) *Nesting in Oregon and along the Lower Columbia River, 1978–2007.* Final Report. Oregon Cooperative Fish and Wildlife Research Unit, Department of Fisheries and Wildlife, Oregon State University, Corvallis.

Ivey, G. L. 1990. Population status of Trumpeter Swans in southeast Oregon. Pages 118–22 in D. Compton, ed. *Proceedings and Papers of the Eleventh Trumpeter Swan Society Conference.* Trumpeter Swan Society, Maple Plain, Minnesota.

Ivey, G. L., and C. P. Herziger. 2000. *Distribution of Greater Sandhill Crane Pairs in Oregon, 1999/00.* Nongame Technical Report 03-01-00. Oregon Department of Fish and Wildlife, Portland.

Ivey, G. L., M. A. Stern, and C. G. Carey. 1988. An increasing White-faced Ibis population in Oregon. *Western Birds* 19:105–8.

Ivey, G. L., C. P. Herziger, and T. J. Hoffmann. 2005. Annual movements of Pacific Coast Sandhill Cranes. *Proceedings of the North American Crane Workshop* 9:25–35.

Jackson, A., C. Eagles-Smith, and C. Emery. 2020. Spatial variation in aquatic invertebrate and riparian songbird mercury exposure across a river-reservoir system with a legacy of mercury contamination. *Ecotoxicology* 29(8):1195–204.

Jacobs, M. 1945. *Kalapuya Texts.* Publications in Anthropology 11. University of Washington, Seattle.

Jacox, M. G., E. L. Hazen, K. D. Zaba, D. L. Rudnick, C. A. Edwards, A. M. Moore, and S. J. Bogard. 2016. Impacts of the 2015–2016 El Niño on the California Current System: early assessment and comparison to past events. *Geophysical Research Letters* 43:7072–80.

James, H. F. 2017. Getting under the skin: a call for specimen-based research on the internal anatomy of birds. Pages 11–22 in The Extended Specimen: Emerging Frontiers in Collections-Based Ornithological Research, ed. Michael S. Webster. *Studies in Avian Biology* 50. CRC Press, Boca Raton, Florida.

Janes, S. W., and L. Ryker. 2011. Geographic variation in Type I songs of Black-throated Gray Warblers. *Wilson Journal of Ornithology* 123(2):339–46.

Jehl, J. R., and N. K. Johnson. 1994. A century of avifaunal change in western North America. *Studies in Avian Biology* 15:1–3.

Jehl, J. R., Jr. 1988. Biology of the Eared Grebe and Wilson's Phalarope in the nonbreeding season: a study of adaptations to saline lakes. *Studies in Avian Biology* 12:1–74.

Jenkins, J. M. A., D. B. Lesmeister, E. D. Forsman, K. M. Dugger, S. H. Ackers, et al. 2019. Social status, forest disturbance, and Barred Owls shape long-term trends in breeding dispersal distance of Northern Spotted Owls. *Condor: Ornithological Applications* 121:1–17. doi:10.1093/condor/duz055.

Jenkins, J. M. A., D. B. Lesmeister, E. D. Forsman, K. M. Dugger, S. H. Ackers, et al. 2021. Conspecific and congeneric interactions shape increasing rates of breeding dispersal of Northern Spotted Owls. *Ecological Applications* 31(7):e02398. doi:10.1002/eap.2398.

Jewett, S. G. 1930. Notes on the Dowell bird collection. *Condor* 32:123–24.

Jewett, S. G. 1934. The mystery of the Marbled Murrelet deepens. *Murrelet* 15:24.

Jiménez, B., R. Rodríguez-Estrella, R. Merino, G. Gómez, L. Rivera, M. J. González, E. Abad, and J. Rivera. 2005. Results and evaluation of the first study of organochlorine contaminants (PCDDs, PCDFs, PCBs, and DDTs), heavy metals and metalloids in birds from Baja California, México. *Environmental Pollution* 133(1):139–46.

Jimerson, T. M., E. A. McGee, D. W. Jones, R. J. Svilich, E. Hotalen, et al. 1996. *A Field Guide to the Tanoak and Douglas-Fir Plant Associations in Northwestern California.* Technical Publication R5-ECOL-TP-009. USDA Forest Service, Pacific Southwest Region, San Francisco.

Jobanek, G. A., and D. B. Marshall. 1992. John K. Townsend's 1836 report of the birds of the Lower Columbia River region, Oregon and Washington. *Northwestern Naturalist* 73:1–14.

Johnson, D. H., D. C. Gillis, M. A. Gregg, J. L. Rebholz, J. L. Lincer, and J. R. Belthoff. 2010. *Users Guide to Installation of Artificial Burrows for Burrowing Owls.* Tree Top, Selah, Washington.

Johnson, E. M., and D. K. Rosenberg. 2006. Granary-site selection by Acorn Woodpeckers in the Willamette Valley, Oregon. *Northwest Science* 80:177–83.

Johnson, M., P. Clarkson, M. I. Goldstein, S. M. Haig, R. B. Lanctot, D. F. Tessler, and D. Zwiefelhofer. 2010. Seasonal movements, winter range use, and migratory connectivity of the Black Oystercatcher. *Condor* 112:731–43.

Johnson, T. N., P. L. Kennedy, T. DelCurto, and R. V. Taylor. 2011. Bird community responses to cattle stocking rates in a Pacific Northwest bunchgrass prairie. *Agriculture, Ecosystems and Environment* 144:338–46.

Kahle, L. Q., M. E. Flannery, and J. P. Dumbacher. 2016. Bird-window collisions at a west-coast urban park museum: analyses of bird biology and window attributes from Golden Gate Park, San Francisco. *PLoS ONE* 11(1):e0144600.

Katzner, T. E., D. M. Nelson, J. E. Diffendorfer, A. E. Duerr, C. J. Campbell, et al. 2019. Wind energy: an ecological challenge. *Science* 366(6470):1206–7.

Keister, G. P., Jr., and R. G. Anthony. 1983. Characteristics of bald eagle communal roosts in the Klamath Basin, Oregon, and California. *Journal of Wildlife Management* 42:1072–79.

Keister, G. P., Jr., R. G. Anthony, and E. J. O'Neill. 1987. Use of communal roosts and foraging areas by bald eagles wintering in the Klamath Basin. *Journal of Wildlife Management* 51:415–20.

Kelly, E. G., E. D. Forsman, and R. G. Anthony. 2003. Are Barred Owls displacing Spotted Owls? *Condor* 105:45–53.

Kennedy, P. L., S. J. DeBano, A. M. Bartuszevige, and A. S. Lueders. 2009. Effects of native and non-native grassland plant communities on breeding passerine birds: implications for restoration of northwest bunchgrass prairie. *Restoration Ecology* 17:515–25.

Kennedy, P. L., T. DelCurto, S. J. DeBano, R. V. Taylor, T. N. Johnson, S. Wyffels, C. Kimoto, H. Schmalz, and R. Limb. 2012. Responses of a Pacific Northwest bunchgrass food web to experimental manipulations of stocking rate. Pages 17-19 in A. Glaser, ed. *Proceedings of the 1st Biennial Conference on the Conservation of America's Grasslands. America's Grasslands Conference: Status, Threats, and Opportunities.* National Wildlife Federation, Washington, DC, and South Dakota State University, Brookings.

Kennedy, P. L., A. M. Bartuszevige, M. Houle, A. B. Humphrey, K. M. Dugger, and J. Williams. 2014. Stable occupancy by breeding hawks (*Buteo* spp.) over 25 years on a privately managed bunchgrass prairie in northeastern Oregon, USA. *Condor: Ornithological Applications* 116:435–45.

Kennedy, P. L., T. DelCurto, S. J. DeBano, T. N. Johnson, S. Wyffels, C. Kimoto, R. V. Taylor, and H. Schmalz. 2017. Can cattle, birds and bees co-exist? Experimental results from the Zumwalt Prairie in northeastern Oregon. *Oregon Cattlemen Association Magazine* (July/August):42–46.

Keyser A. J., M. T. Keyser, and D. E. L. Promislow. 2004. Life-history variation and demography in Western Bluebirds (*Sialia mexicana*) in Oregon. *Auk* 2004:118–33.

Kiat, Y., Y. Vortman, and N. Sapir. 2019. Feather moult and bird appearance are correlated with global warming over the last 200 years. *Nature Communications* 10(1):1–7.

Kiff, L. F. 2005. History, present status, and future prospects of avian eggshell collections in North America. *Auk* 122:994–99.

Kimoto, C., S. J. DeBano, R. W. Thorp, R. V. Taylor, H. J. Schmalz, T. DelCurto, T. N. Johnson, P. L.

Kennedy, and S. Rao. 2012. Short-term responses of native bees to livestock and implications for managing ecosystem services in grasslands. *Ecosphere* 3(10):88. http://dx.doi.org/10.1890/ES12–00118.1.

King, R. J. 2013. *The Devil's Cormorant: A Natural History*. University of New Hampshire Press, Lebanon.

Kline, J. D., M. E. Harmon, T. A. Spies, A. T. Morzillo, R. J. Pabst, B. C. McComb, F. Schnekenburger, K. A. Olsen, B. Csuti, and J. C. Vogeler. 2016. Evaluating carbon storage, timber harvest, and habitat possibilities for a western Cascades (USA) forest landscape. *Ecological Applications* 26:2044–59.

Klomp, J. E., M. T. Murphy, S. Bartos Smith, J. E. McKay, I. Ferrera, and A-L. Reysenbach. 2008. Cloacal microbial communities of female Spotted Towhees: microgeographic variation and individual sources of variability. *Journal of Avian Biology* 39:530–38.

Knick, S. T., D. S. Dobkin, J. T. Rotenberry, M. A. Schroeder, W. M. Vander Haegen, and C. van Riper III. 2003. Teetering on the edge or too late? Conservation and research issues for avifauna of sagebrush habitats. *Condor* 105(4):611–34.

Koford, C. B. 1953. *The California Condor*. Research Report 4. National Audubon Society, New York.

Kress, S. W. 1983. The use of decoys, sound recordings, and gull control for re-establishing a tern colony in Maine. *Colonial Waterbirds* 6:185–96.

Kress, S. W. 1998. Applying research for effective management: case studies in seabird restoration. Pages 141–54 in J. M. Marzluff and R. Sallabanks, eds. *Avian Conservation*. Island Press, Washington, DC.

Kroll, A. J., J. D. Johnston, T. D. Stokely, and G. W. Meigs. 2020. From the ground up: managing young forests for a range of ecosystem outcomes. *Forest Ecology and Management* 464:118055.

Kroodsma, D. E. 2008a. *The Backyard Birdsong Guide: Western North America*. Chronicle Books, San Francisco, California.

Kroodsma, D. E. 2008b. *The Backyard Birdsong Guide: Eastern and Central North America*. Chronicle Books, San Francisco, California.

Kroodsma, D. E. 2015. *The Singing Life of Birds: The Art and Science of Listening to Birdsong*. Houghton Mifflin Harcourt, Boston, Massachusetts.

Kunz, T., E. B. Arnett, B. M. Cooper, W. P. Erickson, R. P. Larkin, T. Mabee, M. L. Morrison, M. D. Strickland, and J. M. Szewczak. 2007. Assessing impacts of wind-energy development on nocturnally active birds and bats: a document. *Journal of Wildlife Management* 71:2449–86.

Lackey, R. T., D. H. Lach, and S. L. Duncan, eds. 2006. *Salmon 2100: The Future of Wild Pacific Salmon*. American Fisheries Society, Bethesda, Maryland.

Landys, M. M., M. Ramenofsky, and J. C. Wingfield. 2006. Actions of glucocorticoids at a seasonal baseline as compared to stress-related levels in the regulation of periodic life processes. *General and Comparative Endocrinology* 48:132–49.

Langham, G. M., J. G. Schuetz, T. Distler, C. U Soykan, and C. B. Wilsey. 2015. Conservation status of North American birds in the face of future climate change. *PLoS ONE* 10(9):e0135350.

Lardy, M. E. 1980. Raptor inventory and Ferruginous Hawk breeding biology in southeastern Oregon. M.S. thesis, University of Idaho, Moscow.

Larson, R., J. Eilers, K. Kruez, W. T. Pecher, S. DasSarma, and S. Dougill. 2016. Recent desiccation-related ecosystem changes at Lake Abert, Oregon: a terminal alkaline salt lake. *Western North American Naturalist* 76:389–404.

Lawes, T. J., D. D. Roby, K. S. Bixler, D. E. Lyons, K. Collis, A. F. Evans, and A. G. Patterson. 2021a. Caspian Tern management at alternative colony sites outside the Columbia River basin. Pages 211–78 in D. D. Roby, A. F. Evans, and K. Collis, eds. *Avian Predation on Salmonids in the Columbia River Basin: A Synopsis of Ecology and Management*. Synthesis report submitted to the US Army Corps of Engineers, Walla Walla, Washington; Bonneville Power Administration, Portland, Oregon; Grant County Public Utility District / Priest Rapids Coordinating Committee, Ephrata, Washington; and Oregon Department of Fish and Wildlife, Salem.

Lawes, T. J., K. S. Bixler, D. D. Roby, D. E. Lyons, K. Collis, et al. 2021b. Double-crested cormorant management in the Columbia River estuary.

Pages 279–417 in D. D. Roby, A. F. Evans, and K. Collis, eds. *Avian Predation on Salmonids in the Columbia River Basin: A Synopsis of Ecology and Management*. Synthesis report submitted to the US Army Corps of Engineers, Walla Walla, Washington; Bonneville Power Administration, Portland, Oregon; Grant County Public Utility District / Priest Rapids Coordinating Committee, Ephrata, Washington; and Oregon Department of Fish and Wildlife, Salem.

Lawes, T. J., D. D. Roby, and D. E. Lyons. 2022. *2021 Pacific Flyway Caspian Tern Population Monitoring*. Final Annual Report to the US Fish and Wildlife Service, Portland, Oregon.

Lawonn, M. J. In press. *Status Assessment of the Double-crested Cormorant (*Nannopterum auritum*) in the Columbia River Estuary and Implications for Salmonid Recovery*. Oregon Department of Fish and Wildlife, Salem.

Laybourne, R. C., and C. J. Dove. 1994. Preparation of birdstrike remains for identification. Pages 531–34 in *Proceedings and Working Papers of the Bird Strike Committee Meeting: Europe 22, Vienna*. US Air Force, Colorado Springs.

Leopold, A. 1933. *Game Management*. University of Wisconsin Press, Madison.

Leopold, A. 1949. *A Sand County Almanac*. Oxford University Press, New York.

Lesmeister, D. B., C. L. Appel, R. J. Davis, C. B. Yackulic, and Z. J. Ruff. 2021. *Simulating the Effort Necessary to Detect Changes in Northern Spotted Owl (*Strix occidentalis caurina*) Populations Using Passive Acoustic Monitoring*. Research Paper PNW-RP-618. USDA Forest Service, Pacific Northwest Research Station, Portland, Oregon.

Leston, L., and T. A. Bookhout. 2020. Yellow Rail (*Coturnicops noveboracensis*). In A. F. Poole, ed. *Birds of the World*. Cornell Lab of Ornithology, Ithaca, New York.

Lewis, M., W. Clark, Members of the Corps of Discovery, and G. Moulton, eds. 2002. *Journals of the Lewis and Clark Expedition*. Lincoln Libraries Electronic Text Center, University of Nebraska Press, Lincoln.

Lichatowich, J. A. 1999. *Salmon without Rivers: A History of the Pacific Salmon Crisis*. Island Press, Washington, DC.

Liebezeit, J., A. O'Connor, J. E. Lyons, C. Shannon, S. Stephensen, and E. Elliott-Smith. 2020. Black Oystercatcher (*Haematopus bachmani*) population size, use of marine reserve complexes, and spatial distribution in Oregon. *Northwestern Naturalist* 101(1):14–26.

Linck, E., E. Bridge, J. Duckles, A. Navarro-Sigüenza, and S. Rohwer. 2016. Assessing migration patterns in *Passerina ciris* using the world's bird collections as an aggregated resource. *PeerJ* 4:e1871. doi:10.7717/peerj.1871.

Lindstrand, L., III, and M. Humes. 2009. White-headed Woodpecker occurrences in Sun Pass State Forest, south-central Oregon. *Northwestern Naturalist* 90:212–16.

Lint, J., B. Noon, R. Anthony, E. Forsman, M. Raphael, M. Collopy, and E. Starkey. 1999. *Northern Spotted Owl Effectiveness Monitoring Plan for the Northwest Forest Plan*. General Technical Report PNW-GTR-440. USDA Forest Service, Washington, DC.

Littlefield, C. D., and S. P. Thompson. 1979. Distribution and status of the Central Valley population of Greater Sandhill Cranes. Pages 113–20 in J. C. Lewis, ed. *Proceedings of the 1978 Crane Workshop*. Colorado State University, Fort Collins.

Littlefield, C. D., M. A. Stern, and R. W. Schlorff. 1994. Summer distribution, status, and trends of Greater Sandhill Crane populations in Oregon and California. *Northwestern Naturalist* 75:1–10.

Livezey, K. B. 2009a. Range expansion of Barred Owls, part I: chronology and distribution. *American Midland Naturalist* 161:49–56.

Livezey, K. B. 2009b. Range expansion of Barred Owls, part II: facilitating ecological changes. *American Midland Naturalist* 161:323 49.

Long, L. L., and J. D. Wolfe. 2019. Review of the effects of barred owls on spotted owls. *Journal of Wildlife Management* 83:1281–96. doi:10.1002/jwmg.21715.

Lorenz, T. J., M. G. Raphael, R. D. Young, D. Lynch, S. K. Nelson, and W. R. McIver. 2021. *Status and Trend of Nesting Habitat for the Marbled Murrelet under the Northwest Forest Plan, 1993 to 2017*.

General Technical Report PNW-GTR-998, USDA Forest Service, Portland, Oregon.

Loy, W. G., S. Allan, A. R. Buckley, and J. E. Meacham. 2001. *Atlas of Oregon*, 2nd ed. University of Oregon Press, Eugene.

Lundsten, S., and K. J. Popper. 2002. Breeding ecology of Yellow Rails at Fourmile Creek, Wood River Wetland, Mares Egg Spring, and Klamath additional areas in southern Oregon, 2001. Unpublished report to Lakeview District Bureau of Land Management, US Fish and Wildlife Service, US Forest Service, The Nature Conservancy, Portland, Oregon.

Lutz, H. L., V. V. Tkach, and J. D. Weckstein. 2017. Methods for specimen-based studies of avian symbionts: 1. Pages 157–84 in *The Extended Specimen*. CRC Press, Boca Raton, Florida.

Lyon, M. W. 1918. Biological Society of Washington, report of the secretary, October 20, 1917. *Journal of the Washington Academy of Sciences* 8:25–28.

Lyons, D. E. 2010. Bioenergetics-based predator-prey relationships between piscivorous birds and juvenile salmonids in the Columbia River estuary. Ph.D. dissertation, Oregon State University, Corvallis.

Lyons, D. E., D. D. Roby, and K. Collis. 2005. Foraging ecology of Caspian Terns in the Columbia River estuary, USA. *Waterbirds* 28:280–91.

Lyons, D. E., D. D. Roby, A. F. Evans, N. J. Hostetter, and K. Collis. 2014. *Benefits to Columbia River Anadromous Salmonids from Potential Reductions in Predation by Double-crested Cormorants Nesting at the East Sand Island Colony in the Columbia River Estuary*. Final Report. US Army Corps of Engineers Portland District, Portland, Oregon.

Lyons, J. E., J. A. Royle, S. M. Thomas, E. Elliott-Smith, J. R. Evenson, E. G. Kelly, R. L. Milner, D. R. Nysewander, and B. A. Andres. 2012. Large-scale monitoring of shorebird populations using count data and *N*-mixture models: Black Oystercatchers (*Haematopus bachmani*) surveys by land and sea. *Auk* 129:645–52.

Madsen, S., D. Evans, T. Hamer, P. Hanson, S. Miller, S. K. Nelson, D. Roby, and M. Stapanian. 1999. *Marbled Murrelet Effectiveness Monitoring Plan for the Northwest Forest Plan*. General Technical Report PNW-GTR-439, USDA Forest Service, Portland, Oregon.

Mannan, R., and E. C. Meslow. 1984. Bird populations and vegetation characteristics in managed and old-growth forests, northeastern Oregon. *Journal of Wildlife Management* 48:1219–38.

Marcot, B. G., and J. W. Thomas. 1997. *Of Spotted Owls, Old Growth, and New Policies: A History since the Interagency Scientific Committee Report*. General Technical Report PNW-GTR-408. US Department of Agriculture, Forest Service, Pacific Northwest Research Station, Portland, Oregon.

Marks, J. S., and J. H. Doremus. 2000. Are Northern Saw-whet Owls nomadic? *Journal of Raptor Research* 34:299–304.

Marks, J. S., A. Nightingale, and J. M. McCullough. 2015. On the breeding biology of Northern Saw-whet Owls (*Aegolius acadicus*). *Journal of Raptor Research* 49:486–97.

Marques, A. T., H. Batalha, S. Rodrigues, H. Costa, M. J. R Pereira, C. Fonseca, M. Mascarenhas, and J. Bernardino. 2014. Understanding bird collisions at wind farms: an updated review on the causes and possible mitigation strategies. *Biological Conservation* 179:40–52.

Marshall, D. B. 1969. *Endangered Plants and Animals of Oregon: III. Birds*. Special Report 278. Agricultural Experiment Station, Oregon State University, Corvallis.

Marshall, D. B. 1988a. The Marbled Murrelet joins the old-growth forest conflict. *American Birds* 42:202–12.

Marshall, D. B. 1988b. *Status of the Marbled Murrelet in North America: With Special Emphasis on Populations in California, Oregon, and Washington*. Biological Report 88(30). US Fish and Wildlife Service, Washington, DC.

Marshall, D. B. 1997. *Status of the White-headed Woodpecker in Oregon and Washington*. Audubon Society of Portland, Oregon.

Marshall, D. B. 2008. *Memoirs of a Wildlife Biologist*. Audubon Society of Portland, Oregon.

Marshall, D. B., M. W. Chilcote, and H. Weeks. 1996. *Species at Risk: Sensitive, Threatened and Endangered Vertebrates of Oregon*, 2nd ed. Oregon Department of Fish and Wildlife, Portland.

Marshall, D. B., M. G. Hunter, and A. L. Contreras, eds. 2003. *Birds of Oregon: A General Reference*. Oregon State University Press, Corvallis.

Maser, C., E. W. Hammer, and S. H. Anderson. 1971. Food habits of the burrowing owl in central Oregon. *Northwest Science* 45(1):19–26.

Mason, N. A., P. O. Title, C. Cicero, K. J. Burns, and R. C. K. Bowie. 2014. Genetic variation among western populations of the Horned Lark (*Eremophila alpestris*) indicates recent colonization of the Channel Islands off southern California, mainland-bound dispersal, and postglacial range shifts. *Auk* 131:162–74.

Materna, W. J., and J. Buck. 2007. *Assessment of Impacts to Aquatic Organisms from Pesticide Use on the Willamette Valley National Wildlife Refuge Complex*. US Fish and Wildlife Publications 187. US Fish and Wildlife Service, Washington, DC.

Mathewson, W. 1986. *William L. Finley: Pioneer Wildlife Photographer*. Oregon State University Press, Corvallis.

Matthews, S. E., D. P. Craig, K. Collis, and D. D. Roby. 2003. Double-crested Cormorant (*Phalacrocorax auritus*). Pages 56–58 in D. B. Marshall, M. G. Hunter, and A. L. Contreras, eds. *Birds of Oregon: A General Reference*. Oregon State University Press, Corvallis.

Mauser, D. M., R. L. Jarvis, and D. S. Gilmer. 1994. Movements and habitat use of mallard broods in northeastern California. *Journal of Wildlife Management* 58:88–94.

May, C. L. 2002. Debris flows through different forest age classes in the central Oregon Coast Range. *Journal of the American Water Resources Association* 38:1–17.

McClure, C. J. W., B. W. Rolek, L. Dunn, J. D. McCabe, L. Martinson, and T. Katzner. 2021. Eagle fatalities are reduced by automated curtailment of wind turbines. *Journal of Applied Ecology* 58:446–52.

McComb, B. C., and S. A. Cushman. 2020. Editorial: synergistic effects of pervasive stressors on ecosystems and biodiversity. *Frontiers in Ecology and Evolution* 8:398.

McComb, B. C., T. Spies, and K. Olsen. 2007. Sustaining biodiversity in the Oregon Coast Range: potential effects of forest policies in a multi-ownership province. *Ecology and Society* 12:29.

McGarigal, K., and W. C. McComb. 1995. Relationships between landscape structure and breeding birds in the Oregon Coast Range. *Ecological Monographs* 65:235–60.

McIver, W., J. Baldwin, M. M. Lance, S. F. Pearson, C. Strong, D. Lynch, M. G. Raphael, R. Young, and N. Johnson. 2020. *Marbled Murrelet Effectiveness Monitoring, Northwest Forest Plan: At-Sea Monitoring—2019 Summary Report*. US Fish and Wildlife Service, Arcata, California.

McShane, C., T. Hamer, H. Carter, G. Swartzman, V. Friesen, et al. 2004. *Evaluation Report for the 5-Year Status Review of the Marbled Murrelet in Washington, Oregon, and California*. US Fish and Wildlife Service, Portland, Oregon.

Mercer, B. 2005. *People of the River: Native Arts of the Oregon Territory*. University of Washington Press, Seattle.

Mercer, D. M., S. M. Haig, D. D. Roby, and T. D. Mullins. 2013. Phylogeography and population structure in Double-crested Cormorants. *Conservation Genetics* 14:823–36.

Mellen-McLean, K., B. G. Marcot, J. L. Ohmann, K. Waddell, E. A. Willhite, S. A. Acker, S. A. Livingston, B. B. Hostetler, B. S. Webb, and B. A. Garcia. 2017. *DecAID: The Decayed Wood Advisor for Managing Snags, Partially Dead Trees, and Down Wood for Biodiversity in Forests of Washington and Oregon*. Version 3.0. USDA Forest Service, Pacific Northwest Region and Pacific Northwest Research Station, US Department of the Interior Fish and Wildlife Service, Oregon State Office, Portland.

Meslow, E. C. 1993. Spotted Owl protection: unintentional evolution toward ecosystem management. *Endangered Species Update* 10(3-4):34–38.

Michel, M. L., S. B. Saunders, T. D. Meehan, and C. D. Wilsey. 2021. Effects of stewardship on protected area effectiveness for coastal birds. *Conservation Biology* 35:1484–95.

Millar, J. G. 2002. The protection of eagles and the Bald and Golden Eagle Protection Act. *Journal of Raptor Research* 36(1 suppl.):29–31.

Miller, A. H. 1942. A California Condor bone from the coast of southern Oregon. *Murrelet* 23:77.

Miller, L. 1957. Bird remains from an Oregon Indian midden. *Condor* 59:59–63.

Miller, M. P., S. M. Haig, D. B. Ledig, M. F. Vander Heyden, and G. Bennett. 2011. Will an "island" population of voles be re-colonized if eradicated? Insights from molecular genetic analyses. *Journal of Wildlife Management* 75:1812–18.

Miller, M. P., J. Parrish, J. Walters, and S. M. Haig. 2012a. Variation in migratory behavior influences regional genetic structure among American kestrel (*Falco sparverius*) populations in North America. *Journal of Heredity* 103:503–14.

Miller, M. P., S. M. Haig, T. D. Mullins, K. J. Popper, and M. Green. 2012b. Evidence for population bottlenecks and subtle genetic structure in the Yellow Rail. *Condor* 114:100–112.

Miller, M. P., S. M. Haig, T. D. Mullins, L. Ruan, B. Casler, et al. 2015. Intercontinental genetic structure and gene flow in dunlin (*Calidris alpina*), a potential vector of avian influenza. *Evolutionary Applications* 8:149–71.

Miller, M. P., S. M. Haig, T. D. Mullins, and E. D. Forsman. 2017. Genetic differentiation and inferred dynamics of a hybrid zone between northern spotted owls (*Strix occidentalis caurina*) and California spotted owls (*S. o. occidentalis*) in northern California. *Ecology and Evolution* 7:6871–83.

Miller, M. P., R. J. Davis, E. D. Forsman, T. D. Mullins, and S. M. Haig. 2018a. Isolation by distance versus landscape resistance: understanding dominant patterns of genetic structure in northern spotted owls (*Strix occidentalis caurina*). *PloS ONE* 13(8):e0201720.

Miller, M. P., S. M. Haig, E. D. Forsman, R. G. Anthony, L. Diller, K. Dugger, A. Franklin, T. Fleming, and S. Gremel. 2018b. Variation in inbreeding rates across the range of Northern Spotted Owls (*Strix occidentalis caurina*): insights from over 30 years of monitoring data. *Auk* 135:821–33.

Miller, M. R., J. Y. Takekawa, J. P. Fleskes, D. L. Orthmeyer, M. L. Casazza, and W. M. Perry. 2005. Spring migration of northern pintails from California's Central Valley wintering area tracked with satellite telemetry: routes, timing, and destinations. *Canadian Journal of Zoology* 83(10):1314–32.

Miller, R. F., and J. A. Rose. 1995. Historic expansion of *Juniperus occidentalis* (western juniper) in southeastern Oregon. *Great Basin Naturalist* 55:37–45.

Miller, R. F., S. T. Knick, D. A. Pyke, C. W. Meinke, S. E. Hanser, M. J. Wisdom, A. L. Hild, S. Knick, and J. Connelly. 2011. Characteristics of sagebrush habitats and limitations to long-term conservation. Greater Sage-Grouse: ecology and conservation of a landscape species and its habitats. *Studies in Avian Biology* 38:145–84.

Miller, S. L., M. G. Raphael, G. A. Falxa, C. Strong, J. Baldwin, et al. 2012. Recent population decline of the Marbled Murrelet in the Pacific Northwest. *Condor* 114:771–81.

Mitchell, C. D., and M. W. Eichholz. 2020. Trumpeter Swan (*Cygnus buccinator*), version 1.0. In P. G. Rodewald, ed. *Birds of the World*. Cornell Lab of Ornithology, Ithaca, New York.

Monahan, W. B., and R. J. Hijmans. 2007. *Distributional Dynamics of Invasion and Hybridization by* Strix *spp. in Western North America*. Ornithological Monographs 63. American Ornithological Society, Chicago, Illinois.

Moore, J. N. 2016. Recent desiccation of western Great Basin saline lakes: lessons from Lake Abert, Oregon, USA. *Science of the Total Environment* 554:142–54.

Moore, R. P., and A. A. Kotaich. 2010. *Reproductive Success of Streaked Horned Larks (*Eremophila alpestris strigata*) in Oregon's Varied Agricultural Landscape, Mid- and Southern Willamette Valley, 2009*. US Fish and Wildlife Service, Oregon State Office, Portland.

Moran, D., and K. Kanemoto. 2017. Identifying species threat hotspots from global supply chains. *Nature Ecology and Evolution* 1:1–5.

Morgan, G. 1952. Florence boasts big sand project. *Eugene Register Guard*, 8 July 1952.

Morrison, M. L., and E. C. Meslow. 1983. Bird community structure on early-growth clearcuts in western Oregon. *American Midland Naturalist* 110:129–37.

Morrison, M. L., and E. C. Meslow. 1984. Effects of the herbicide glyphosate on bird community structure, western Oregon. *Forest Science* 30:95–106.

Morse, J. A., A. N. Powell, and M. D. Tetreau. 2006. Productivity of Black Oystercatchers: effects of recreational disturbance in a national park. *Condor* 108:623–33.

Mote, P. W., J. Abatzoglou, K. D. Dello, K. Hegewisch, and D. E. Rupp. 2019. *Fourth Oregon Climate Assessment Report*. Oregon Climate Change Research Institute, Corvallis.

Murphy, M. T., L. J. Redmond, A. C. Dolan, N. W. Cooper, C. M. Chutter, and S. Cancellieri. 2020. Population decline of a long-distance migratory passerine at the edge of its range: nest predation, nest replacement and immigration. *Journal of Avian Biology* 51(6):e02286. doi:10.1111/jav.02286.

Myers, B. M., D. T. Rankin, K. J. Burns, and C. J. Clark. 2019. Behavioral and morphological evidence of an Allen's x Rufous hummingbird (*Selasphorus sasin* x *S. rufus*) hybrid zone in southern Oregon and northern California. *Auk* 136(4):ukz049.

National Marine Fisheries Service. 2008. *Endangered Species Act Section 7(a)(2) Consultation Biological Opinion and Magnuson-Stevens Fishery Conservation and Management Act Essential Fish Habitat Consultation: Consultation on Remand for Operation of the Federal Columbia River Power System, 11 Bureau of Reclamation Projects in the Columbia Basin and ESA Section 10(a)(I)(A) Permit for Juvenile Fish Transportation Program*. National Marine Fisheries Service, Seattle, Washington.

National Marine Fisheries Service. 2010. *Supplemental Consultation on Remand for Operation of the Federal Columbia River Power System, 11 Bureau of Reclamation Projects in the Columbia Basin and ESA Section 10(a)(I)(A) Permit for Juvenile Fish Transportation Program*. National Marine Fisheries Service, Northwest Region, Seattle, Washington.

National Marine Fisheries Service. 2019a. *Status of Federally Listed West Coast Salmon and Steelhead*. 84 FR 53117-53119. National Marine Fisheries Service, Northwest Region, Seattle, Washington.

National Marine Fisheries Service. 2019b. *Endangered Species Act Section 7(a)(2) Biological Opinion and Magnuson-Stevens Fishery Conservation and Management Act Essential Fish Habitat Consultation for the Continued Operation and Maintenance of the Columbia River System*. National Marine Fisheries Service, Portland, Oregon.

National Park Service, US Fish and Wildlife Service, and the Yurok Tribe. 2019. *Northern California Condor Restoration Program Environmental Assessment* [Draft]. Redwood National Park, Crescent City, California.

National Park Service, Redwood National Park, US Fish and Wildlife Service, and the Yurok Tribe. 2020. *Northern California Condor Restoration Program Final Environmental Assessment*. Redwood National Park, Crescent City, California.

Naughton, M. B., D. S. Pitkin, R. W. Lowe, K. J. So, and C. S. Strong. 2007. *Catalog of Oregon Seabird Colonies*. Biological Technical Publication FWS/BTP-R1009 2007. US Department of Interior, Fish and Wildlife Service, Region 1, Washington, DC.

Nehls, H. 1998. *Records of the Oregon Bird Records Committee*. Oregon Bird Records Committee, Portland.

Nelson, S. K. 1986. *Observations of Marbled Murrelets in Inland, Older-Aged Forests of Western Oregon*. Contribution 54. USDA Forest Service, Pacific Northwest Experiment Station, Olympia, Washington.

Nelson, S. K. 1990. *Distribution of the Marbled Murrelet in Western Oregon*. Contribution 89-9-02. Oregon Department of Fish and Wildlife, Portland.

Nelson, S. K. 1997. Marbled Murrelet (*Brachyramphus marmoratus*). In A. F. Poole and F. B. Gill, eds. *The Birds of North America.* Cornell Lab of Ornithology, Ithaca, New York. https://doi.org/10.2173/bna.276.

Nelson, S. K. 2003. Marbled Murrelet. In D. B. Marshall, M. G. Hunter, and A. L. Contreras, eds. *Birds of Oregon: A General Reference*. Oregon State University Press, Corvallis.

Nelson, S. K. 2020. Marbled Murrelet (*Brachyramphus marmoratus*). In A. F. Poole and

F. B. Gill, eds. *Birds of the World*. Cornell Lab of Ornithology, Ithaca, New York. https://doi .org/10.2173/bow.marmur.01.

Nelson, S. K., and T. E. Hamer 1995a. Nest success and the effects of predation on Marbled Murrelets. In C. J. Ralph, G. L. Hunt, and J. F. Piatt, eds. *Ecology and Conservation of the Marbled Murrelet*. General Technical Report PSW-GTR-152. US Forest Service, Albany, California.

Nelson, S. K., and T. E. Hamer. 1995b. Nesting biology and behavior of the Marbled Murrelet. Pages 57–67 in C. J. Ralph, G. L. Hunt Jr., M. G. Raphael, and J. F. Piatt, eds. *Ecology and Conservation of the Marbled Murrelet*. General Technical Report PSW-GTR-152. US Forest Service, Albany, California.

Nelson, S. K., and R. W. Peck. 1995. Behavior of the Marbled Murrelet at nine nest sites in Oregon. *Northwestern Naturalist* 76:43–53.

Nelson, S. K., and S. G. Sealy, eds. 1995. Biology of the Marbled Murrelet: inland and at-sea. *Northwestern Naturalist* 76:1–119.

Nelson, S. K., and A. K. Wilson. 2002. Marbled Murrelet habitat characteristics of state lands in western Oregon. Unpublished final report. Oregon Cooperative Fish and Wildlife Research Unit, Oregon State University, Department of Fisheries and Wildlife, Corvallis.

Nelson, S. K., M. L. C. McAllister, M. A. Stern, D. H. Varoujean, and J. M. Scott. 1992. The Marbled Murrelet in Oregon, 1899–1987. *Proceedings of the Western Foundation of Vertebrate Zoology* 5:61–91.

Nelson, S. K., B. A. Barbaree, B. D. Dugger, and S. H. Newman. 2010. *Marbled Murrelet Breeding Ecology at Port Snettisham, Southeast Alaska, in 2005–2008*. Alaska Department of Fish and Game, Juneau.

Nettleship, D. N., and D. C. Duffy, eds. 1995. The Double-crested Cormorant: biology, conservation and management. *Colonial Waterbirds* 18:1–256.

Newberry, J. S. 1857. Report upon the birds. Pages 73–106 in *Report upon the Zoology of the Route*, vol. 2. US War Department, Washington, DC.

Nielsen-Pincus, N. 2005. Nest site selection, nest success, and density of selected cavity-nesting birds in northeastern Oregon with a method for improving accuracy of density estimates. M.S. thesis. University of Idaho, Moscow.

Nonaka, E., and T. A. Spies. 2005. Historical range of variability in landscape structure: a simulation study in Oregon, USA. *Ecological Applications* 15:1727–46.

North American Bird Conservation Initiative, US Committee. 2010. *The State of the Birds 2010 Report on Climate Change*. US Department of the Interior, Washington, DC.

North American Bird Conservation Initiative, US Committee. 2011. *The State of the Birds 2011 Report on Public Lands and Water*. US Department of the Interior, Washington, DC.

North American Bird Conservation Initiative, US Committee. 2014. *The State of the Birds, United States*. US Department of the Interior, Washington, DC.

North American Bird Conservation Initiative, US Committee. 2016. *The State of North America's Birds 2016*. US Department of the Interior, Washington, DC.

Northrup, J. M., J. W. Rivers, S. K. Nelson, D. D. Roby, and M. G. Betts. 2018. Assessing the utility of satellite transmitters for identifying nest locations and foraging behavior of the threatened Marbled Murrelet *Brachyramphus marmoratus*. *Marine Ornithology* 46:47–55.

Northrup, J., Y. Zhiqiang, J. R. Rivers, and M. G. Betts. 2019. Synergistic effects of climate and land-use change influence broad-scale population declines. *Global Change Biology* 25:1561–75.

Novitch, N. R., M. Westberg, and R. M. Zink. 2015. Migration of Alder Flycatchers (*Empidonax alnorum*) and Willow Flycatchers (*Empidonax traillii*) through the Tuxtla Mountains, Veracruz, Mexico, and the identification of migrant flycatchers in collections. *Wilson Journal of Ornithology* 127(1):142–45.

Nowak, M. C., K. J. Mougey, D. P. Collins, and B. A. Grisham. 2018. Mixing of two greater sandhill crane populations in northeast Oregon. *Proceedings of the North American Crane Workshop* 14:110–14.

Nuechterlein, G. L., and D. P. Buitron. 1989. Diving differences between Western and Clark's Grebes. *Auk* 106:467–70.

Obama, B. H. 2017. Boundary Enlargement of the Cascade-Siskiyou National Monument. Proclamation No. 9564, 82 Fed. Reg. 6145:6250.

Olsen, A. C., K. Yates, D. E. Naugle, J. D. Maestas, J. D. Tack, J. Smith, and C. A. Hagen. 2021. Reversing tree expansion in sagebrush steppe yields population-level benefit for imperiled grouse. *Ecosphere* 12(6):e03551. doi:10.1002/ecs2.3551.

Olson, G. S., E. M. Glenn, R. G. Anthony, E. D. Forsman, J. A. Reid, P. J. Loschl, and W. J. Ripple. 2004. Modeling demographic performance of Northern Spotted Owls relative to forest habitat in Oregon. *Journal of Wildlife Management* 68:1039–53.

Olson, S. L. 1989. David Douglas and the original description of the Hawaiian Goose. *Elepaio* 49:49–51.

Oregon Department of Energy. 2020. *2020 Biennial Energy Report*. Oregon Department of Energy, Salem.

Oregon Department of Fish and Wildlife. 1994. *Oregon Conservation Program for the Western Snowy Plover* (Charadrius alexandrinus nivosus). Oregon Department of Fish and Wildlife, Portland.

Oregon Department of Fish and Wildlife. 2012. *Biological Status Assessment of the Bald Eagle* (Haliaeetus leucocephalus) *and Findings to Remove the Bald Eagle from the List of Threatened Species under the Oregon Endangered Species Act*. Oregon Department of Fish and Wildlife, Salem.

Oregon Department of Fish and Wildlife. 2016a. *Oregon Nearshore Strategy*. Oregon Department of Fish and Wildlife, Salem.

Oregon Department of Fish and Wildlife. 2016b. *Oregon Conservation Strategy*. Oregon Department of Fish and Wildlife, Salem.

Oregon Department of Fish and Wildlife. 2017. *Draft Status Review of the Marbled Murrelet* (Brachyramphus marmoratus) *in Oregon and Evaluation of Criteria to Reclassify the Species from Threatened to Endangered under the Oregon Endangered Species Act*. Oregon Fish and Wildlife Commission, Oregon Department of Fish and Wildlife, Salem.

Oregon Department of Fish and Wildlife. 2020. *Climate and Ocean Change Policy*. Oregon Department of Fish and Wildlife, Salem.

Oregon Department of Forestry. 2021. Specified resource site protection plans, Chapter 629, Division 665. Oregon State Archives, Salem.

Oregon Parks and Recreation Department. 2005. *Ocean Shore Management Plan*. Oregon Parks and Recreation Department, Salem.

Oregon State Parks and Recreation Branch. 1977. *Oregon's Beaches: A Birthright Preserved*. Oregon State Parks and Recreation Branch, Salem.

O'Reilly, K., and J. C. Wingfield. 2003. Seasonal, age, and sex differences in weight, fat reserves, and plasma corticosterone in Western Sandpipers. *Condor* 105:13–26.

Oring, L. W., L. Neel, and K. E. Oring. 2013. *Intermountain West Regional Shorebird Plan*. US Shorebird Conservation Plan.

Osterback, A. M. K., D. M. Frechette, A. O. Shelton, S. A. Hayes, M. H. Bond, S. A. Shaffer, and J. W. Moore. 2013. High predation on small populations: avian predation on imperiled salmonids. *Ecosphere* 4:1–21.

Pacific Flyway Council. 2013. A monitoring strategy for the western population of Double-crested Cormorants within the Pacific Flyway. Pacific Flyway Council, US Fish and Wildlife Service, Portland, Oregon.

Pacific Seabird Group. 1982. Consideration of Marbled Murrelets in old-growth forest management: a resolution of the Pacific Seabird Group. *Pacific Seabird Group Bulletin* 9:62–63.

Pacific Seabird Group. 1986. Management of the Marbled Murrelet: a resolution of the Pacific Seabird Group. Drafted at the Marbled Murrelet Management Workshop, Pacific Seabird Group Annual Meeting, 9 December 1986. Pacific Seabird Group, Corvallis, Oregon.

Page, G. W., M. A. Stern, and P. W. C. Paton. 1995. Differences in wintering areas of Snowy Plovers from inland breeding sites in western North America. *Condor* 97:258–62.

Pain, D. J., R. Mateo, and R. W. Green. 2019. Effects of lead from ammunition on birds and other wildlife: a review and update. *Ambio* 48:935–53.

Pampush, G. J., and R. J. Anthony. 1993. Nest success, habitat utilization, and nest site selection of Long-billed Curlews in the Columbia Basin, Oregon. *Condor* 95:957–67.

Pank, L. 1976. Effects of seed and background colors on seed acceptance by birds. *Journal of Wildlife Management* 40:769-74.

Paton, P. W. C., C. J. Ralph, and H. R. Carter. 1988. *The Pacific Seabird Group's Marbled Murrelet Survey and Intensive Inventory Handbook*. US Department of Agriculture Forest Service Redwood Sciences Lab, Arcata, California.

Paton, P. W. C., C. J. Ralph, H. R. Carter, and S. K. Nelson 1990. *Surveying Marbled Murrelets at Inland Forested Sites: A Guide*. General Technical Report PSW-120, Pacific Southwest Research Station, USDA Forest Service, Berkeley, California.

Patten, M. A., and C. L. Pruett. 2009. The Song Sparrow, *Melospiza melodia*, as a ring species: patterns of geographic variation, a revision of subspecies, and implications for speciation. *Systematics and Biodiversity* 7:33–62.

Patterson, R. L. 1952. *Sage Grouse of Wyoming*. Wyoming Game and Fish Commission, Sage Books, Denver, Colorado.

Payton, Q., A. F. Evans, N. J. Hostetter, D. D. Roby, B. Cramer, and K. Collis. 2020. Measuring the additive effects of predation on prey survival across spatial scales. *Ecological Applications* e02193. doi:10.1002/EAP.2193.

Payton, Q., A. F. Evans, N. J. Hostetter, B. Cramer, K. Collis, and D. D. Roby. 2021. Additive effects of avian predation on the survival of juvenile salmonids in the Columbia River basin. Pages 581–618 in D. D. Roby, A. F. Evans, and K. Collis, eds. *Avian Predation on Salmonids in the Columbia River Basin: A Synopsis of Ecology and Management*. Synthesis report submitted to the US Army Corps of Engineers, Walla Walla, Washington; Bonneville Power Administration, Portland, Oregon; Grant County Public Utility District / Priest Rapids Coordinating Committee, Ephrata, Washington; and Oregon Department of Fish and Wildlife, Salem.

Pearcy, W. G., J. Fisher, R. Brodeur, and S. Johnson. 1985. Effects of the 1983 El Niño on coastal nekton off Oregon and Washington. Pages 188–204 in W. S. Wooster and D. L. Fluharty, eds. *El Niño North: Niño Effects in the Eastern Subarctic Pacific Ocean*. Washington Sea Grant, Seattle.

Pearson, S. F., R. Moore, and S. M. Knapp. 2012. Nest exclosures do not improve Streaked Horned Lark nest success. *Journal of Field Ornithology* 83:315–22.

Peck, R. W., S. K. Nelson, and T. L. DeSanto 1994. Tree climbing as a technique for locating Marbled Murrelet nests. *Northwest Science* 68:143.

Peck-Richardson, A. G. 2017. Double-crested Cormorants (*Phalacrocorax auritus*) and Brandt's Cormorants (*P. penicillatus*) breeding on East Sand Island in the Columbia River estuary: foraging ecology, colony connectivity, and overwinter dispersal. M.S. thesis, Oregon State University, Corvallis.

Peery, M. Z., S. R. Beissinger, S. H. Newman, E. Burkett, and T. D. Williams. 2004. Applying the declining population paradigm: diagnosing causes of poor reproduction in the Marbled Murrelet. *Conservation Biology* 18:1088–98.

Peterson, W. T., J. L. Fisher, P. T. Strub, X. Du, C. Risien, J. Peterson, and C. T. Shaw. 2017. The pelagic ecosystem in the Northern California Current off Oregon during the 2014–2016 warm anomalies within the context of the past 20 years. *Journal of Geophysical Research: Oceans* 122(9):7267–90. doi:10.1002/2017jc012952.

Peterson, W. T., J. L. Fisher, C. A. Morgan, S. M. Zeman, B. J. Burke, and K. C. Jacobson. 2019. *Ocean Ecosystem Indicators of Salmon Marine Survival in the Northern California Current*. Fish Ecology Division, Northwest Fisheries Science Center, National Marine Fisheries Service, Newport, Oregon.

Phalan, B. T., J. M. Northrup, Z. Yang, R. L. Deal, J. S. Rousseau, T. A. Spies, and M. G. Betts. 2019. Impacts of the Northwest Forest Plan on forest composition and bird populations. *Proceedings of the National Academy of Sciences of the United States of America* 116:3322–27.

Phillips, D. 1986. Science centres: a lesson for art galleries? *International Journal of Museum Management and Curation* 5:259–66.

Piatt, J. F., K. J. Kuletz, A. E. Burger, S. A. Hatch, V. L. Friesen, T. P. Birt, M. L. Arimitsu, G. S. Drew, A. M. A. Harding, and K. S. Bixler. 2007. *Status Review of the Marbled Murrelet* (Brachyramphus marmoratus) *in Alaska and British Columbia*. Open-File Report 2006-1387. US Geological Survey, Reston, Virginia.

Piatt, J. F., J. K. Parrish, H. M. Renner, S. K. Schoen, T. T. Jones, et al. 2020. Extreme mortality and reproductive failure of common murres resulting from the northeast Pacific marine heatwave of 2014–2016. *PLoS ONE* 15(1):e0226087. https://doi.org/10.1371/journal.pone.0226087.

Pitelka, F. 1981. The condor case: an uphill struggle in a downhill crush. *Auk* 98:634–35.

Pitman, R. L., J. Hodder, M. R. Graybill, and D. H. Varoujean. 1985. Catalog of Oregon Seabird Colonies. Unpublished report, US Fish and Wildlife Service, Portland, Oregon.

Plissner, J. H., S. M. Haig, and L. W. Oring. 1999. Within and among year movements of American Avocets in the western Great Basin. *Wilson Bulletin* 111:314–20.

Plissner, J. H., S. M. Haig, and L. W. Oring. 2000a. Post-breeding movements of American Avocets and implications for wetland connectivity in the western Great Basin. *Auk* 117:290–98.

Plissner, J. H., L. W. Oring, and S. M. Haig. 2000b. Space use of killdeer at a Great Basin breeding area. *Journal of Wildlife Management* 64:421–29.

Popper, K. J., and M. A. Stern. 1996. Breeding ecology of Yellow Rails at Jack Spring and Mare's Egg Spring, Klamath Co., OR. Unpublished report to Winema National Forest, Bureau of Land Management at the Klamath Falls Resource Area, and Oregon Department of Fish and Wildlife, Bend. The Nature Conservancy, Portland, Oregon.

Popper, K. J., and M. A. Stern. 2000. Nesting ecology of Yellow Rails in southcentral Oregon. *Journal of Field Ornithology* 71:460–66.

Prill, A. G. 1924. Nesting birds of Lake County, Oregon. *Wilson Bulletin* 36:24–25.

Pritchard, K. R. 2015. Bird abundance and microhabitat associations with oak mistletoe in Willamette Valley oak woodlands. M.S. thesis, Oregon State University, Corvallis.

Pumphrey, V., T. DelCurto, and M. Vavra. 2001. *100 years of agricultural research*. Agricultural Experiment Station Special Report 1033. Eastern Oregon Agricultural Research Center Union Station, Oregon State University, Corvallis.

Ralph, C. J., and S. K. Nelson. 1992. *Methods of surveying Marbled Murrelets at inland forest sites*. Marbled Murrelet Technical Committee, Pacific Seabird Group, Corvallis, Oregon.

Ralph, C. J., S. K. Nelson, M. M. Shaughnessy, and S. L. Miller, comps. 1993. *Methods for Surveying Marbled Murrelets in Forests*. Marbled Murrelet Technical Committee, Pacific Seabird Group, Corvallis, Oregon.

Ralph, C. J., S. K. Nelson, M. M. Shaughnessy, S. L. Miller, and T. E. Hamer. 1994. *Methods for Surveying Marbled Murrelets in Forests: A Protocol for Land Management and Research*. Marbled Murrelet Technical Committee, Pacific Seabird Group, Corvallis, Oregon.

Ralph, C. J., G. L. Hunt Jr., M. G. Raphael, and J. F. Piatt, eds. 1995. *Ecology and Conservation of the Marbled Murrelet*. General Technical Report PSW-GTR-152. USDA Forest Service, Albany, California.

Raphael, M. G. 2006. Conservation of the Marbled Murrelet under the Northwest Forest Plan. *Conservation Biology* 20(2):297–305.

Raphael, M. G., D. Evans-Mack, J. M. Marzluff, and J. M. Luginbuhl. 2002. Effects of forest fragmentation on populations of the Marbled Murrelet. *Studies in Avian Biology* 25:221–35.

Raphael, M. G., J. Baldwin, G. A. Falxa, M. H. Huff, M. Lance, S. L. Miller, S. F. Pearson, C. J. Ralph, C. Strong, and C. Thompson. 2007. *Regional Population Monitoring of the Marbled Murrelet: Field and Analytical Methods*. General Technical Report PNW-GTR-716. USDA Forest Service, Portland, Oregon.

Raphael, M. G., G. A. Falxa, and A. E. Burger. 2018. Marbled Murrelet. Pages 301–50 in T. A. Spies, P. A. Stine, et al., eds. *Synthesis of Science to Inform Land Management within the Northwest Forest Plan Area*, vol. 1. General Technical Report PNW-GTR-966. US Forest Service, Pacific Northwest Research Station, Portland, Oregon.

Ratti, J. T. 1985. A test of water depth niche partitioning by Western Grebe color morphs. *Auk* 102:635–37.

Rattner, B. A., J. C. Franson, S. R. Sheffield, C. I. Goddard, N. J. Leonard, D. Stang, and P. J. Wingate. 2008. *Sources and Implications of Lead Ammunition and Fishing Tackle on Natural Resources*. Wildlife Society Technical Review 08-01. The Wildlife Society, Bethesda, Maryland.

Redig, P. T., E. M. Lawler, S. Schwartz, J. L. Dunnette, B. Stephenson, and G. E. Duke. 1991. Effects of chronic exposure to sublethal concentrations of lead acetate on heme synthesis and immune function in red-tailed hawks. *Archives of Environmental Contamination and Toxicology* 21:72–77.

Redmond, L. J., M. T. Murphy, and A. C. Dolan. 2007. Nest reuse by Eastern Kingbirds: adaptive behavior or ecological constraint? *Condor* 109:463–68.

Redmond, L. J., M. T. Murphy, A. C. Dolan, and K. Sexton. 2009. Public information facilitates habitat selection of a territorial species: the eastern kingbird. *Animal Behaviour* 77:457–63.

Redmond, L. J., M. T. Murphy, N. W. Cooper, and K. M. O'Reilly. 2016. Testosterone secretion in a socially monogamous but sexually promiscuous migratory passerine. *General and Comparative Endocrinology* 228:24–32.

Remington, T. A., P. A. Deibert, S. E. Hanser, D. M. Davis, L. A. Robb, and J. L. Welty. 2021. *Sagebrush Conservation Strategy—Challenges to Sagebrush Conservation*. Open-File Report 2020-1125. Western Association of Fish and Wildlife Agencies, Bureau of Land Management, and US Fish and Wildlife Service, Washington, DC.

Remsen, J. V. 1995. The importance of continued collecting of bird specimens to ornithology and bird conservation. *Bird Conservation International* 5:145–80.

Reynolds, R. T., J. M. Scott, and R. A. Nussbaum. 1980. A variable circular-plot method for estimating bird numbers. *Condor* 82:309–13.

Rich, T. D., C. J. Beardmore, H. Berlanga, P. J. Blancher, M. S. W. Bradstreet, et al. 2004. *Partners in Flight North American Landbird Conservation Plan*. Cornell Lab of Ornithology, Ithaca, New York.

Richmond, M. L., C. J. Henny, R. L. Floyd, R. W. Mannan, D. M. Finch, and L. R. DeWeese. 1979. *Effects of Sevin-4-Oil, Dimilin, and Orthene on Forest Birds in Northeastern Oregon*. Research Paper PSW-148. US Forest Service, Washington, DC.

Ricketts, T. H. 1999. *Terrestrial Ecoregions of North America: A Conservation Assessment*. Island Press, Washington, DC.

Rivers, J., M. Johnson, S. M. Haig, C. J. Schwarz, J. W. Glendening, L. J. Burnett, J. Brandt, D. George, and J. Grantham. 2014. A quantitative analysis of monthly home range size across the annual cycle in the critically endangered California Condor *Gymnogyps californianus. Bird Conservation International* 24:492–504.

Rivers, J., J. Verschuyl, C. J. Schwartz, A. J. Kroll, and M. G. Betts. 2019. No evidence of a demographic response to experimental herbicide treatments by the White-crowned Sparrow, an early successional forest songbird. *Condor* 121(2):1–13.

Robert, M., L. Cloutier, and P. Laporte. 1997. The summer diet of the Yellow Rail in southern Quebec. *Wilson Bulletin* 109:702–10.

Roby, D. D., K. Collis, D. E. Lyons, D. P. Craig, J. Y. Adkins, A. M. Myers, and R. M. Suryan. 2002. Effects of colony relocation on diet and productivity of Caspian terns. *Journal of Wildlife Management* 66:662–73.

Roby, D. D., D. E. Lyons, D. P. Craig, K. Collis, and G. H. Visser. 2003a. Quantifying the effect of predators on endangered species using a bioenergetics approach: Caspian terns and juvenile salmonids in the Columbia River estuary. *Canadian Journal of Zoology* 81:250–65.

Roby, D. D., K. Collis, D. E. Lyons, D. P. Craig, and M. Antolos. 2003b. Caspian Tern *Sterna caspia*. Pages 277–79 in D. B. Marshall, M. G. Hunter, and A. L. Contreras, eds. *Birds of Oregon: A*

*General Reference.* Oregon State University Press, Corvallis.

Roby, D. D., T. J. Lawes, D. E. Lyons, K. Collis, A. F. Evans, et al. 2021a. Caspian tern management in the Columbia River estuary. Pages 21–113 in D. D. Roby, A. F. Evans, and K. Collis, eds. *Avian Predation on Salmonids in the Columbia River Basin: A Synopsis of Ecology and Management.* Synthesis report submitted to the US Army Corps of Engineers, Walla Walla, Washington; Bonneville Power Administration, Portland, Oregon; Grant County Public Utility District / Priest Rapids Coordinating Committee, Ephrata, Washington; and Oregon Department of Fish and Wildlife, Salem.

Roby, D. D., A. F. Evans, and K. Collis, eds. 2021b. *Avian Predation on Salmonids in the Columbia River Basin: A Synopsis of Ecology and Management.* Synthesis report submitted to the US Army Corps of Engineers, Walla Walla, Washington; Bonneville Power Administration, Portland, Oregon; Grant County Public Utility District / Priest Rapids Coordinating Committee, Ephrata, Washington; and Oregon Department of Fish and Wildlife, Salem.

Rockwell, S. M., J. D. Alexander, J. L. Stephens, R. I. Frey, and C. J. Ralph. 2017. Spatial variation in songbird demographic trends from a regional network of banding stations in the Pacific Northwest. *Condor* 119:732–44.

Rockwell, S. M., J. L. Stephens, and B. Altman. 2021. *Population and Habitat Objectives for Landbirds in Grassland, Oak, and Riparian Habitats in the Puget Lowlands, Willamette Valley, and Klamath Mountains Ecoregions.* Version 2.0. Klamath Bird Observatory and American Bird Conservancy. Prepared for Oregon-Washington Partners in Flight, Pacific Birds Habitat Joint Venture, Bureau of Land Management, and US Forest Service. Corvallis, Oregon.

Rocque, D. A., and K. Winker. 2005. Use of bird collections in contaminant and stable-isotope studies. *Auk* 122:990–94.

Rodway, M. S. 1990. *Status Report on the Marbled Murrelet* Brachyramphus marmoratus. Committee on the Status of Endangered Wildlife in Canada, Ottawa.

Rohwer, S., and V. G. Rohwer. 2018. Breeding and multiple waves of primary molt in common ground doves of coastal Sinaloa. *PeerJ* 6:e4243.

Rose, C. L., B. G. Marcot, T. K. Mellen, J. L. Ohmann, K. L. Waddell, D. L. Lindley, and B. Schreiber. 2001. Decaying wood in Pacific Northwest forests: concepts and tools for habitat management. Pages 580–623 in *Wildlife-Habitat Relationships in Oregon and Washington.* Oregon State University Press, Corvallis.

Rosenberg, K. V., J. A. Kennedy, R. Dettmers, R. P. Ford, D. Reynolds, et al. 2016. *Partners in Flight Landbird Conservation Plan: 2016 Revision for Canada and Continental United States.* Partners in Flight Science Committee.

Rosenberg, K. V., A. M. Dokter, P. J. Blancher, J. R. Sauer, A. C. Smith, et al. 2019. Decline of the North American avifauna. *Science* 366:120–24.

Rowe, D. L., M. T. Murphy, R. C. Fleischer, and P. G. Wolf. 2001. High frequency of extra-pair paternity in Eastern Kingbirds. *Condor* 103:845–51.

Ruggerone, G. T. 1986. Consumption of migrating juvenile salmonids by gulls foraging below a Columbia River dam. *Transactions of the American Fisheries Society* 115:736–42.

Ruggiero, L. F., K. B. Aubry, A. B. Carey, and M. H. Huff. 1991. *Wildlife and Vegetation of Unmanaged Douglas-Fir Forests.* General Technical Report PNW-GTR-285. USDA Forest Service, Pacific Northwest Forest and Range Experiment Station. Corvallis, Oregon.

Ruth, M. M. 2005. *Rare Bird: Pursuing the Mystery of the Marbled Murrelet.* Mountaineers Books, Seattle, Washington.

Ryder, G. R., R. W. Campbell, H. R. Carter, and S. G. Sealy. 2012. Earliest well-described tree nest of the Marbled Murrelet: Elk Creek, British Columbia, 1955. *Wildlife Afield* 9:49–58.

Sallabanks, R., E. Arnett, and J. Marzluff. 2000. An evaluation of research on the effects of timber harvest on bird populations. *Wildlife Society Bulletin* 28:1144–55.

Sallabanks, R., R. A. Riggs, and L. E. Cobb. 2002. Bird use of forest structural classes in grand fir forests of the Blue Mountains, Oregon. *Forest Science* 48:311–21.

Sanzenbacher, P. M., and S. M. Haig. 2001. Killdeer population trends in North America. *Journal of Field Ornithology* 72:160–69.

Sanzenbacher, P. M., and S. M. Haig. 2002a. Residency and movement patterns of wintering Dunlin (*Calidris alpina*) in the Willamette Valley, Oregon. *Condor* 104:271–80.

Sanzenbacher, P. M., and S. M. Haig. 2002b. Regional fidelity and movement patterns of wintering Killdeer in an agricultural landscape. *Waterbirds* 25:16–25.

Sapozhnikova, Y., O. Bawardi, and D. Schlenk. 2004. Pesticides and PCBs in sediments and fish from the Salton Sea, California, USA. *Chemosphere* 55(6):797–809.

Sauer, J. R., and W. A. Link. 2011. Analysis of the North American Breeding Bird Survey using hierarchical models. *Auk* 128:87–98.

Sauer, J. R., J. E. Hines, and J. Fallon. 2008. *The North American Breeding Bird Survey, Results and Analysis 1966–2004*. Version 2005.2. US Geological Survey Patuxent Wildlife Research Center, Laurel, Maryland.

Sauer, J. R., K. L. Pardieck, D. J. Ziolkowski Jr., A. C. Smith, M.-A. R. Hudson, R. Rodriguez, H. Berlanga, D. K. Niven, and W. A. Link. 2017. The first 50 years of the North American Breeding Bird Survey. *Condor: Ornithological Applications* 119:576–93.

Schmalz, H. J., R. V. Taylor, T. N. Johnson, P. L. Kennedy, S. J. DeBano, B. Newingham, and P. A. McDaniel. 2013. Soil morphologic properties and cattle stocking rate affect dynamic soil properties. *Range Ecology and Management* 66:445–53.

Schreck, C. B., T. P. Stahl, L. E. Davis, D. D. Roby, and B. J. Clemens. 2006. Mortality estimates of juvenile spring-summer Chinook salmon in the lower Columbia River and estuary, 1992–1998: evidence for delayed mortality? *Transactions of the American Fisheries Society* 135:457–75.

Schreiber, B., and D. S. deCalesta. 1992. The relationship between cavity-nesting birds and snags on clearcuts in western Oregon. *Forest Ecology and Management* 50(3–4):299–316.

Schwarzbach, S. E., J. D. Albertson, and C. M. Thomas. 2006. Effects of predation, flooding, and contamination on reproductive success of California Clapper Rails (*Rallus longirostris obsoletus*) in San Francisco Bay. *Auk* 123:45–60.

Scott, J. M. 1971. Interbreeding of the Glaucous-winged Gull and Western Gull in the Pacific Northwest. *California Birds* 2:129–33.

Scott, J. M., J. A. Wiens, and R. R. Claeys. 1975. Organochlorine levels associated with a common murre die-off in Oregon. *Journal of Wildlife Management* 39:310–20.

Seavy, N. E. 2006. Effects of disturbance on animal communities: fire effects on birds in mixed-conifer forest. Ph.D. dissertation. University of Florida, Gainesville.

Seavy, N. E., and J. D. Alexander. 2011. Interactive effects of vegetation structure and composition describe bird habitat associations in mixed broadleaf-conifer forest. *Journal of Wildlife Management* 75:344–52.

Seavy, N. E., and J. D. Alexander. 2014. Songbird response to wildfire in mixed-conifer forest in southwestern Oregon. *International Journal of Wildland Fire* 23:246–58.

Seavy, N. E., J. D. Alexander, and P. E. Hosten. 2008. Bird community composition after mechanical mastication fuel treatments in southwest Oregon oak woodland and chaparral. *Forest Ecology and Management* 256:774–78.

Sebring, S. H., M. C. Carper, R. D. Ledgerwood, B. P. Sandford, G. M. Matthews, and A. F. Evans. 2013. Relative vulnerability of PIT-tagged sub-yearling fall Chinook salmon to predation by Caspian terns and double-crested cormorants in the Columbia River Estuary. *Transactions of the American Fisheries Society* 142:1321–34.

Senner, N. R., J. N. Moore, S. T. Seager, K. Kreuz, S. Dougill, and S. E. Senner. 2018. A salt lake under stress: the relationship between birds, water levels, and invertebrates at a Great Basin saline lake. *Biological Conservation* 220:320–29.

Severson J. P., C. A. Hagen, J. D. Tack, J. D. Maestas, D. E. Naugle, J. T. Forbes, and K. P. Reese. 2017a. Restoring sage-grouse nesting habitat through removal of early successional conifer. *Restoration Ecology* 25:126–34.

Severson, J. P., C. A. Hagen, J. D. Tack, J. D. Maestas, D. E. Naugle, J. T. Forbes, and K. P. Reese. 2017b.

Better living through conifer removal: a demographic analysis of sage-grouse vital rates. *PLoS ONE* 12(3):e0174347. https://doi.org/10.1371/journal.pone.0174347.

Seymour, R., and M. Hunter. 1999. Principles of ecological forestry. Pages 22–62 in M. L. Hunter, ed. *Maintaining Biodiversity in Forest Ecosystems*. Cambridge University Press, Cambridge.

Sheehan, T., T. Bachelet, and K. Ferschweiler. 2015. Projected major fire and vegetation changes in the Pacific Northwest of the conterminous United States under selected CMIP5 climate futures. *Ecological Modelling* 317:16–29.

Sherker, Z. T., K. Pellet, J. Atkinson, J. Damborg, and A. W. Trites. 2021. Pacific great blue herons (*Ardea herodias fannini*) consume thousands of juvenile salmon (*Oncorhynchus* spp.). *Canadian Journal of Zoology* 99:349–61.

Sherman, L., and J. Hagar. 2020. The snag's the limit: habitat selection modeling for the western purple martin in a managed forest landscape. *Forest Ecology and Management* 480:118689.

Shipley, A. A., M. T. Murphy, and A. H. Elzinga. 2013. Residential edges as ecological traps: postfledging survival of Spotted Towhees in an urban park. *Auk* 130:501–11.

Shirley, S. M., Z. Yang, R. A. Hutchinson, J. D. Alexander, K. McGarigal, and M. G. Betts. 2013. Species distribution modelling for the people: unclassified landsat TM imagery predicts bird occurrence at fine resolutions. *Diversity and Distributions* 19:855–66.

Shuford, W. D., R. E. Gill Jr., and C. M. Handel. 2018. *Trends and Traditions: Avifaunal Change in Western North America*. Studies of Western Birds 3. Western Field Ornithologists, Camarillo, California.

Shuford, W. D., L. D. L. Thomson, D. M. Mauser, and J. Beckstrand. 2004. *Abundance, Distribution and Phenology of Nongame Waterbirds in the Klamath Basin of Oregon and California in 2003*. Point Blue Conservation Science. Final Report to US Fish and Wildlife Service, Klamath Basin National Wildlife Refuge, Point Reyes, California.

Shunk, S. A. 2011a. A woodpecker to make Meriwether Lewis proud. *Living Bird* (Winter 2011).

Shunk, S. A. 2011b. The ABCs of Lewis's Woodpecker conservation. *Living Bird* (Winter 2011).

Sickinger, T. 2019. Failing forestry: with $1 billion timber lawsuit, not all 14 counties are big winners. *Oregon Live*, 26 November 2019, https://www.oregonlive.com/politics/2019/11/failing-forestry-with-1-billion-timber-lawsuit-not-all-14-counties-are-big-winners.html.

Simons, D. D. 1983. Interactions between California Condors and humans in prehistoric far western North America. Pages 470–94 in S. R. Wilbur and J. A. Jackson, eds. *Vulture Biology and Management*. University of California Press, Berkeley.

Sims, M. D. 1983. Breeding success and nest site characteristics of the Western Bluebird on Parrett Mountain. M.S. thesis. Portland State University, Portland, Oregon.

Smith, J. T., B. W. Allred, C. Boyd, K. Davies, M. Jones, J. D. Maestas, J. Moford, and D. E. Naugle. 2022. The elevational ascent and spread of exotic annual grasses in the Great Basin, USA. *Diversity and Distributions* 28:83–96.

Snyder, N. F. R., and J. A. Hamber. 1985. Replacement-clutching and annual nesting of California Condors. *Condor* 87:374–78.

Snyder, N. F. R., and H. A. Snyder. 2000. *The California Condor: A Saga of Natural History and Conservation*. Academic Press, San Diego, California.

Sovern, S. G., E. D. Forsman, K. M. Dugger, and M. Taylor. 2015. Roosting habitat use and selection by northern spotted owls during natal dispersal. *Journal of Wildlife Management* 79:254–62.

Sovern, S. G., D. B. Lesmeister, K. M. Dugger, M. S. Pruett, R. J. Davis, and J. M. Jenkins. 2019. Activity center selection by northern spotted owls. *Journal of Wildlife Management* 83:714–27. doi:10.1002/jwmg.21632.

Spiegel, C. S., S. M. Haig, M. I. Goldstein, and M. Huso. 2012. Factors affecting incubation patterns and sex roles of Black Oystercatchers in Alaska. *Condor* 114:123–34.

Spies, T. A., K. N. Johnson, K. M. Burnett, J. L. Ohmann, B. C. McComb, G. H. Reeves, P. Bettinger, J. D. Kline, and B. Garber-Yonts. 2007. Cumulative ecological and socioeconomic effects of forest policies in coastal Oregon. *Ecological Applications* 17:5–17.

Sprunt, A., IV, W. B. Robertson Jr., S. Postupalsky, R. J. Hensel, C. E. Knoder, and F. J. Ligas. 1973. Comparative productivity of six bald eagle populations. *Transactions of the Thirty-Eighth North American Wildlife and Natural Resources Conference* 38:96–106.

Stalmaster, M. V. 1987. *The Bald Eagle*. Universe Books, New York.

Stauber, E., N. Finch, P. A. Talcott, and J. M. Gay. 2010. Lead poisoning of Bald (*Haliaeetus leuco-cephalus*) and Golden (*Aquila chrysaetos*) Eagles in the US Inland Pacific Northwest region—an 18-year retrospective study: 1991–2008. *Journal of Avian Medical Surgery* 24:279–87.

Steenhof, K., L. Bond, K. K. Bates, and L. L. Leppert. 2002. Trends in midwinter counts of Bald Eagles in the contiguous United States, 1986–2000. *Bird Populations* 6:21–32.

Stephens, J. L. 2005. A comparison of bird abundance and nesting in harvest units, habitat islands, and mature coniferous forests in southwestern Oregon. M.S. thesis, Southern Oregon University, Ashland.

Stephens, J. L. 2016. *Grasshopper Sparrow Abundance on the Imperatrice Property: Results from 2016 Surveys*. Report KBO-2016-0009. Klamath Bird Observatory, Ashland, Oregon.

Stephens, J. L., and C. R. Gillespie. 2016. *Chaparral Patch Size and Nearby Chaparral Amount Influence Songbird Occupancy*. Report KBO-2016-0017. Klamath Bird Observatory, Ashland, Oregon.

Stephens, J. L., and S. M. Rockwell. 2020. *Limiting Factors for the Oregon Vesper Sparrow Population in the Rogue Basin, Oregon: 2018–2020 Summary Report*. Klamath Bird Observatory, Ashland, Oregon.

Stephens, J. L., I. J. Ausprey, N. E. Seavy, and J. D. Alexander. 2015a. Fire severity affects mixed broadleaf-conifer forest bird communities: results for 9 years following fire. *Condor: Ornithological Applications* 117:430–46.

Stephens, J. L., S. M. Rockwell, C. J. Ralph, and J. D. Alexander 2015b. Decline of the Black Tern (*Chlidonias niger*) population in the Klamath Basin, Oregon, 2001–2010. *Northwestern Naturalist* 96(3):196–204.

Stephens, J. L., E. C. Dinger, J. D. Alexander, S. R. Mohren, C. J. Ralph, and D. A. Sarr. 2016. Bird communities and environmental correlates in southern Oregon and northern California, USA. *PLoS ONE* 11:e0163906.

Stephens, J. L., E. C. Dinger, and J. D. Alexander. 2019. Established and empirically derived landbird focal species lists correlate with vegetation and avian metrics. *Ecological Applications* 29(3):e01865.

Stephens, J. L., C. R. Gillespie, and J. D. Alexander. 2021. Restoration treatments reduce threats to oak ecosystems and provide immediate subtle benefits for oak-associated birds. *Restoration Ecology* 29:e13298.

Stephensen, S. 2020. Oregon Coast NWRC: Surface Nesting Seabirds-Aerial Photographic Colony Surveys-Annual Count Summary Reports. Unpublished report, US Fish and Wildlife Service, Newport, Oregon.

Stern, M. A., and R. L. Jarvis. 1991. Sexual dimorphism and assortative mating in Black Terns. *Wilson Bulletin* 103:266–71.

Stern, M. A., G. J. Pampush, K. Kristensen, and R. S. Del Carlo. 1986. Survivorship, causes of mortality and movements of juvenile Sandhill Cranes at Sycan Marsh, Oregon, 1984–1985. Unpublished report. Oregon Department of Fish and Wildlife Nongame Program, Portland.

Stern, M. A., G. J. Pampish, and R. S. Del Carlo. 1987a. Nesting ecology and productivity of Greater Sandhill Cranes at Sycan Marsh, Oregon. Pages 249–56 in J. C. Lewis, ed. *Proceedings of the 1985 Crane Workshop*. Platte River Whooping Crane Maintenance Trust, Grand Island, Nebraska.

Stern, M. A., R. Del Carlo, M. Smith, and K. Kristensen. 1987b. Birds of Sycan Marsh, Lake Co., Oregon. *Oregon Birds* 13:184–92.

Stern, M. A., J. F. Morawski, and G. A. Rosenberg. 1993. Rediscovery and status of a disjunct population of breeding Yellow Rails in southern Oregon. *Condor* 95:1024–27.

Stinson, D. W. 2005. *Status Report for the Mazama Pocket Gopher, Streaked Horned Lark, and Taylor's Checkerspot*. Washington Department of Fish and Wildlife, Olympia.

Stockenberg, E., B. Altman, M. Green, and J. D. Alexander. 2008. Using GAP in landbird biological objective setting: process and examples from oak habitats in the Pacific Northwest. *Gap Analysis Bulletin* 15:34.

Stocking, J., E. Elliott-Smith, N. Holcomb, and S. M. Haig. 2010, *Long-billed Curlew Breeding Success on Mid-Columbia River National Wildlife Refuges, South-Central Washington and North-Central Oregon, 2007–08*. Open-File Report 2010-1089. US Geological Survey, Reston, Virginia.

Stralberg, D., D. Jongsomjit, C. A. Howell, M. A. Snyder, J. D. Alexander, J. D. Wiens, and T. L. Root. 2009. Re-shuffling of species with climate disruption: a no-analog future for California birds? *PLoS ONE* 4(9):e6825.

Strong, C. S. 2020. *Marbled Murrelet Population Monitoring in Conservation Zone 3, Oregon, during 2020*. US Fish and Wildlife Service, Portland, Oregon.

Suckley, G., and J. G. Cooper. 1860. *The Natural History of Washington Territory and Oregon: With Much Relating to Minnesota, Nebraska, Kansas, Utah, and California, between the Thirty-Sixth and Forty-Ninth Parallels of Latitude, Being Those Parts of the Final Reports on the Survey of the Northern Pacific Railroad Route, Relating to the Natural History of the Regions Explored, with Full Catalogues and Descriptions of the Plants and Animals Collected from 1853 to 1860*. Ballière Brothers, New York.

Summers, S. D. 1993. *A Birder's Guide to the Klamath Basin*. Klamath Basin Audubon Society, Klamath Falls, Oregon.

Suryan, R. M., D. P. Craig, D. D. Roby, N. D. Chelgren, K. Collis, W. D. Shuford, and D. E. Lyons. 2004. Redistribution and growth of the Caspian Tern population in the Pacific Coast region of North America, 1981–2000. *Condor* 106:777–90.

Suryan, R. M., K. J. So, E. M. Phillips, J. E. Zamon, R. W. Lowe, and S. W. Stephensen. 2012. *Seabird Colony and At-Sea Distribution along the Oregon Coast: Implications for Offshore Energy Facility Placement and Information Gap Analysis*. Report to the Northwest National Marine Renewable Energy Center, Oregon State University, Corvallis.

Suryan, R., C. Horton, S. Wheeler, C. Alexander, and E. Nelson. 2014. *Yaquina Head Seabird Colony Monitoring 2014 Season Summary*. Oregon State University, Hatfield Marine Science Center, Newport.

Suryan, R. M., R. Albertani, and B. Polagye. 2016. *A Synchronized Sensor Array for Remote Monitoring of Avian and Bat Interactions with Offshore Renewable Energy Facilities*. Final Report to the Department of Energy for Project DOE_EE0005363. Department of Energy, Washington, DC.

Suzuki, Y., D. D. Roby, D. E. Lyons, K. N. Courtot, and K. Collis. 2015. Developing nondestructive techniques for managing conflicts between fisheries and double-crested cormorant colonies. *Wildlife Society Bulletin* 39:764–71. doi:10.1002/wsb.595.

Suzuki, Y., J. Heinrichs, D. E. Lyons, D. D. Roby, and N. Schumaker. 2018. *Modeling the Pacific Flyway Population of Caspian Terns to Investigate Current Population Dynamics and Evaluate Future Management Options*. Final report submitted to Bonneville Power Administration and Northwest Power and Conservation Council, Portland, Oregon.

Syverson, V. J., and D. R. Prothero. 2010. Evolutionary patterns in Late Quaternary California Condors. *PalArch's Journal of Vertebrate Paleontology* 7:1–18.

Taft, O. W., and S. M. Haig. 2003. Historical wetlands in Oregon's Willamette Valley: implications for restoration of winter waterbird habitat. *Wetlands* 23:51–64.

Taft, O. W., and S. M. Haig. 2005. The value of agricultural wetlands as invertebrate resources for wintering shorebirds. *Agriculture, Ecosystems, and Environment* 110:249–56.

Taft, O. W., and S. M. Haig. 2006a. Landscape context mediates influence of local food abundance on wetland use by wintering shorebirds in an agricultural valley. *Biological Conservation* 128:298–307.

Taft, O. W., and S. M. Haig. 2006b. Importance of wetland landscape structure to shorebirds wintering in an agricultural valley. *Landscape Ecology* 21:169–84.

Taft, O. W., S. M. Haig, and C. Kiilsgaard. 2004. Use of radar remote sensing (RADARSAT) to map winter wetland habitat for shorebirds in an agricultural landscape. *Environmental Management* 33:749–62.

Taft, O. W., P. M. Sanzenbacher, and S. M. Haig. 2008. Movements of wintering Dunlin and changing habitat availability in an agricultural wetland landscape. *Ibis* 150:541–49.

Tallis, H., P. S. Levin, M. Ruckelshaus, S. E. Lester, K. L. McLeod, D. L. Fluharty, and B. S. Halpern. 2009. The many faces of ecosystem-based management: making the process work today in real places. *Marine Policy* 34:340–48.

Tarof, S., and C. R. Brown. 2013. Purple Martin (*Progne subis*). In A. Poole, ed. *The Birds of North America*. Cornell Lab of Ornithology, Ithaca, New York.

Taylor, A. L., and E. D. Forsman. 1976. Recent range extensions of the Barred Owl in western North America, including the first records for Oregon. *Condor* 78:560–61.

Taylor, W. P. 1921. The Marbled Murrelet mystery. *Murrelet* 2:8.

Tessler, D. F., J. A. Johnson, B. A. Andres, S. Thomas, and R. B. Lanctot. 2014. A global assessment of the conservation status of the Black Oystercatcher *Haematopus bachmani*. *International Wader Studies* 20:83–96.

Thaxter, C. B., G. M. Buchanan, J. Carr, S. H. M. Butchart, T. Newbold, R. E. Green, J. A. Tobias, W. B. Foden, S. O'Brien, and J. W. Pearce-Higgins. 2017. Bird and bat species' global vulnerability to collision mortality at wind farms revealed through a trait-based assessment. *Proceedings of the Royal Society B* 284(1862).

The Nature Conservancy and J. Krueger. 2013. *Umpqua Prairie and Oak Partnership Conservation Strategy*. The Nature Conservancy and Jeff Krueger Environments, LLC, Roseburg, Oregon.

Thomas, J. W. 1979. *Wildlife Habitats in Managed Forests the Blue Mountains of Oregon and Washington*. Agriculture Handbook 553. USDA Forest Service, Washington, DC.

Thomas, J. W., E. D. Forsman, J. B. Lint, E. C. Meslow, B. R. Noon, and J. Verner. 1990. *A Conservation Strategy for the Northern Spotted Owl: Report of the Interagency Scientific Committee to Address the Conservation of the Northern Spotted Owl*. USDA Forest Service, US Department of the Interior Bureau of Land Management, Fish and Wildlife Service, and National Park Service, Portland, Oregon.

Thomas, J. W., J. F. Franklin, J. Gordon, and K. N. Johnson. 2006. The Northwest Forest Plan: origins, components, implementation experience, and suggestions for change. *Conservation Biology* 20:277–87. https://doi.org/10.1111 /j.1523-1739.2006.00385.x.

Thompson, A. R., I. D. Schroeder, S. J. Bograd, E. L. Hazen, M. G. Jacox, et al. 2019. State of the California Current 2018–19: a novel anchovy regime and a new marine heat wave? *CalCOFI Reports* 60:1–66.

Tilman, D., R. M. May, C. L. Lehman, and M. A. Nowak. 1994. Habitat destruction and the extinction debt. *Nature* 371:65–66.

Torgersen, T. R., J. W. Thomas, R. R. Mason, and D. Van Horn. 1984. Avian predators of Douglas-fir tussock moth, *Orgyia pseudotsugata* (McDunnough) (Lepidoptera: Lymantriidae) in southwestern Oregon. *Environmental Entomology* 13:1018–22.

Townsend, J. K. 1839. *Narrative of a Journey across the Rocky Mountains to the Columbia River*. Edited by H. Perkins. Philadelphia, Pennsylvania. [Republished in 1978 as *Across the Rockies to the Columbia*. University of Nebraska Press, Lincoln.]

Trail, P. W. 2017. Identifying bald versus golden eagle bones: a primer for wildlife biologists and law enforcement officers. *Journal of Fish and Wildlife Management* 8(2):596–610.

Trail, P. W., R. Cooper, and D. Vroman. 1997. The breeding birds of the Klamath/Siskiyou region. In *Proceedings of the First Conference on Siskiyou Ecology*. Siskiyou Project and Nature Conservancy, Cave Junction, Oregon.

USACE. US Army Corps of Engineers. 2006. *Caspian Tern Management to Reduce Predation of Juvenile Salmonids in the Columbia River Estuary: Record of Decision, November 2006*. Portland District, Northwestern Division, Portland, Oregon.

USACE. US Army Corps of Engineers. 2014. *Inland Avian Predation Management Plan Environmental Assessment*. Walla Walla District, Northwestern Division. Walla, Walla, Washington.

USACE. US Army Corps of Engineers. 2015. *Double-crested Cormorant Management Plan to Reduce Predation on Juvenile Salmonids in the Columbia River Estuary: Final Environmental Impact Statement, February 2015*. Portland District, Portland, Oregon.

US Department of Agriculture and US Department of the Interior. 1994. *Final Supplemental Environmental Impact Statement on Management of Habitat for Late-Successional and Old-Growth Forest Related Species within the Range of the Northern Spotted Owl*. Portland, Oregon.

US Forest Service. 1977. *Final Environmental Statement for Oregon Dunes National Recreation Area Management Plan*. Siuslaw National Forest, Pacific Northwest Region, Corvallis, Oregon.

US Forest Service and Bureau of Land Management. 1994a. *Record of Decision for Amendments to Forest Service and Bureau of Land Management Planning Documents within the Range of the Northern Spotted Owl*. Washington, DC.

US Forest Service and Bureau of Land Management. 1994b. *Final Supplemental Environmental Impact Statement on Management of Habitat for Late-Successional and Old-Growth Forest Related Species within the Range of the Northern Spotted Owl*. Portland, Oregon.

USFWS. US Fish and Wildlife Service. 1967. Endangered species. *Federal Register* 32:4001.

USFWS. US Fish and Wildlife Service. 1975. *California Condor Recovery Plan*. Portland, Oregon.

USFWS. US Fish and Wildlife Service. 1980. *California Condor Recovery Plan*, 1st rev. Portland, Oregon.

USFWS. US Fish and Wildlife Service. 1984. *California Condor Recovery Plan*, 2nd rev. Portland, Oregon.

USFWS. US Fish and Wildlife Service. 1986. *Recovery Plan for the Pacific Bald Eagle*. Portland, Oregon.

USFWS. US Fish and Wildlife Service. 1990. Endangered and threatened wildlife and plants: determination of threatened status for the Northern Spotted Owl. *Federal Register* 55:26, 114–94.

USFWS. US Fish and Wildlife Service. 1992. Endangered and threatened wildlife and plants: determination of threatened status for the Washington, Oregon, and California population of the Marbled Murrelet. *Federal Register* 57:45,328–37.

USFWS. US Fish and Wildlife Service. 1993. Endangered and threatened wildlife and plants: determination of threatened status for the Pacific Coast population of the Western Snowy Plover. *Federal Register* 58:12,864–75.

USFWS. US Fish and Wildlife Service. 1996. *California Condor Recovery Plan*, 3rd rev. Portland, Oregon.

USFWS. US Fish and Wildlife Service. 1997. *Recovery Plan for the Threatened Marbled Murrelet* (Brachyramphus marmoratus) *in Washington, Oregon, and California*. Portland, Oregon.

USFWS. US Fish and Wildlife Service. 2003. *Double-crested Cormorant Management in the United States: Final Environmental Impact Statement*. Division of Migratory Bird Management, Arlington, Virginia.

USFWS. US Fish and Wildlife Service. 2004. *Northern Spotted Owl Five-Year Review: Summary and Evaluation*. Portland, Oregon.

USFWS. US Fish and Wildlife Service. 2005a. *Caspian Tern Management to Reduce Predation of Juvenile Salmonids in the Columbia River Estuary: Final Environmental Impact Statement, January 2005*. Migratory Birds and Habitat Program, Portland, Oregon.

USFWS. US Fish and Wildlife Service. 2005b. Endangered and threatened wildlife and plants: revised designation of critical habitat for the Pacific Coast population of the Western Snowy Plover. *Federal Register* 77(118):36,728–69.

USFWS. US Fish and Wildlife Service. 2006. *Caspian Tern Management to Reduce Predation of Juvenile Salmonids in the Columbia River Estuary: Record of Decision, November 2006*. Migratory Birds and Habitat Programs, Portland, Oregon.

USFWS. US Fish and Wildlife Service. 2007. Endangered and threatened wildlife and plants: removing the Bald Eagle in the lower 48 states from the list of endangered and threatened wildlife; final rule. *Federal Register* 72:37,346–72.

USFWS. US Fish and Wildlife Service. 2009. *Marbled Murrelet (*Brachyramphus marmoratus*) 5-Year Status Review*. Washington Fish and Wildlife Office, Lacey.

USFWS. US Fish and Wildlife Service. 2011. *Revised Recovery Plan for the Northern Spotted Owl (*Strix occidentalis caurina*)*. US Department of the Interior, Portland, Oregon.

USFWS. US Fish and Wildlife Service. 2012a. *Report on Marbled Murrelet Recovery Implementation Team Meeting and Stakeholder Workshop*. Washington Field Office, Lacey.

USFWS. US Fish and Wildlife Service. 2012b. Endangered and threatened wildlife and plants: designation of revised critical habitat for the Northern Spotted Owl. *Federal Register* 77:32,483–93.

USFWS. US Fish and Wildlife Service. 2013a. *Experimental Removal of Barred Owl to Benefit Threatened Northern Spotted Owls—Final Environmental Impact Statement*. Portland, Oregon.

USFWS. US Fish and Wildlife Service. 2013b. Endangered and Threatened wildlife and plants: determination of endangered status for the Taylor's Checkerspot Butterfly and threatened status for the Streaked Horned Lark, final rule. *Federal Register* 78:61,452–503.

USFWS. US Fish and Wildlife Service. 2016a. *Double-crested Cormorant Western Population Status Evaluation: Final Annual 2015 Report*. Report to the US Army Corps of Engineers, USFWS, Migratory Bird and Habitat Programs, Portland, Oregon.

USFWS. US Fish and Wildlife Service. 2016b. Lower Klamath, Clear Lake, Tule Lake, Upper Klamath, and Bear Valley National Wildlife Refuges, Klamath County, OR; Siskiyou and Modoc Counties, CA: Final comprehensive conservation plan / environmental impact statement. *Federal Register* 81(237).

USFWS. US Fish and Wildlife Service. 2018. Endangered and threatened wildlife and plants: 90-day findings for three species. *Federal Register* 83:30,091–94.

USFWS. US Fish and Wildlife Service. 2019a. *Marbled Murrelet (*Brachyramphus marmoratus*) 5-Year Status Review*. Washington Fish and Wildlife Office, Lacey.

USFWS. US Fish and Wildlife Service. 2019b. *Species Biological Report for the Streaked Horned Lark (*Eremophila alpestris strigata*)*. Version 1.0. Portland, Oregon.

USFWS. US Fish and Wildlife Service. 2019c. *Biological Opinion on the Effects of Proposed Klamath Project Operations from April 1, 2019, through March 31, 2024, on the Lost River Sucker and the Shortnose Sucker*. Klamath Falls Fish and Wildlife Office, Klamath Falls, Oregon.

USFWS. US Fish and Wildlife Service. 2019d. Establishment of a nonessential experimental population of the California Condor in the Pacific Northwest; proposed rule. *Federal Register* 84:13,587–603.

USFWS. US Fish and Wildlife Service. 2020a. *Final Report: Bald Eagle Population Size*. 2020 Update. Division of Migratory Bird Management, Washington, DC.

USFWS. US Fish and Wildlife Service. 2020b. *Double-crested Cormorant Western Population Status Evaluation: Final Annual 2019 Report*. Migratory Birds and Habitat Programs, Portland, Oregon.

USFWS. US Fish and Wildlife Service. 2020c. *Final Environmental Impact Statement for the Management of Conflicts Associated with Double-crested Cormorants*. USFWS, Falls Church, Virginia.

USFWS. US Fish and Wildlife Service. 2020d. Endangered and threatened wildlife and plants: 12-month finding for the Northern Spotted Owl. *Federal Register* 85:81,144–52.

USFWS. US Fish and Wildlife Service. 2021. Establishment of a nonessential experimental population of the California Condor in the Pacific Northwest: final rule. *Federal Register* 86:15,602–23.

Valente, J. J., S. K Nelson, J. W. Rivers, D. D. Roby, and M. G. Betts. 2021. Experimental evidence that social information affects habitat selection in Marbled Murrelets. *Ornithology* 138:1–13.

Varva, M., and S. D. Wood. 2000. *Oregon State of the Environment Report 2000*. State of Oregon, Salem.

Veloz, S., L. Salas, B. Altman, J. D. Alexander, D. Jongsomjit, N. Elliott, and G. Ballard. 2015. Improving effectiveness of systematic conservation planning with density data. *Conservation Biology* 29:1217–27.

Viste-Sparkman, K. 2005. White-breasted Nuthatch density and nesting ecology in oak woodlands of the Willamette Valley, Oregon. M.S. thesis, Oregon State University, Corvallis.

Walters, J. R., S. R. Derrickson, D. M. Fry, S. M. Haig, J. M. Marzluff, J. M. Wunderle Jr., B. B. Bernstein, and K. L. Velas. 2010. Status of the California Condor (*Gymnogyps californianus*) and efforts to achieve its recovery. *Auk* 127:969–1001.

Wan, H. Y., S. A. Cushman, and L. L. Ganey. 2019. Recent and projected future wildfire trends across the ranges of three Spotted Owl subspecies under climate change. *Frontiers in Ecology and Evolution* 7:37.

Wang, B., X. Luo, Y.-M. Yang, W. Sun, M. A. Cane, W. Cai, S.-W. Yeh, and J. Liu. 2019. Historical change of El Niño properties sheds light on future changes of extreme El Niño. *Proceedings of the National Academy of Sciences* 116:22,512–17. https://doi.org/10.1073/pnas.1911130116.

Warnock, N. D., S. M. Haig, and L. W. Oring. 1998. Monitoring species richness and abundance of shorebirds in arid interior regions: an example from the western Great Basin. *Condor* 100:589–600.

Warrick, D. R., B. W. Tobalske, and D. R. Powers. 2009. Lift production in the hovering hummingbird. *Proceedings of the Royal Society B* 27:63,747–52.

Watkins, W. W. 1988. *Population Density and Habitat Use by Red-necked Grebes on Upper Klamath Lake, Klamath County, Oregon*. Technical Report 88-3-04. Nongame Wildlife Program, Oregon Department of Fish and Wildlife, Portland.

Watson, R. T., P. S. Kolar, M. Ferrer, T. Nygard, N. Johnston, W. G. Hunt, H. A. Smit-Robinson, C. Farmer, M. M. Huso, and T. E. Katzner. 2018. Raptor interactions with wind energy—case studies from around the world. *Journal of Raptor Research* 52:1–18. https://doi.org/10.3356/JRR-16-100.1.

Webster, M. S. 2017. The extended specimen 1. Pages 1–10 in *The Extended Specimen*. CRC Press, Boca Raton, Florida.

Weikel, J. M., and J. P. Hayes. 1999. The foraging ecology of cavity-nesting birds in young forests of the northern Coast Range of Oregon. *Condor* 101:58–66.

Weiser, A., and D. Lepofsky. 2009. Ancient land use and management of Ebey's Prairie, Whidbey Island, Washington. *Journal of Ethnobiology* 29:184–212.

Wells, B. K., J. A. Santora, M. J. Henderson, P. Warzybok, J. Jahncke, et al. 2017. Environmental conditions and prey switching by a seabird predator impact juvenile salmon survival. *Journal of Marine Systems* 174:54–63.

West, C. J., J. D. Wolfe, A. Wiegardt, and T. Williams-Claussen. 2017. Feasibility of California Condor recovery in northern California, USA: contaminants in surrogate Turkey Vultures and Common Ravens. *Condor* 119(4):720–31.

Western Purple Martin Working Group. 2018. Interim population objective for the Pacific population of the Western Purple Martin (*Progne subis arboricola*). Point Blue Conservation Science, Point Reyes, California.

Whitlock, C. 1992. Vegetational and climatic history of the Pacific Northwest during the last 20,000 years: implications for understanding present-day biodiversity. *Northwest Environmental Journal* 8:5–28.

Whitlock, C., and P. J. Bartlein. 1997. Vegetation and climate change in northwest America during the past 125 kyr. *Nature* 388:57–61.

Whittaker, R. H. 1960. Vegetation of the Siskiyou Mountains, Oregon and California. *Ecological Monographs* 30:279–338.

Wiedemann, A. M. 1987. *The Ecology of European Beachgrass* (Ammophila arenaria *[L.] Link): A Review of the Literature*. Technical Report 87-1-01. Nongame Wildlife Program, Oregon Department of Fish and Wildlife, Corvallis.

Wiedemann, A. M., and A. Pickart. 1996. The *Ammophila* problem on the northwest coast of North America. *Landscape and Urban Planning* 34(3–4):287–99.

Wiegardt, A. K., D. C. Barton, and J. D. Wolfe. 2017a. Post-breeding dynamics indicate upslope molt-migration by Wilson's Warblers. *Journal of Field Ornithology* 88:47–52.

Wiegardt, A. K., J. Wolfe, C. J. Ralph, J. L. Stephens, and J. Alexander. 2017b. Postbreeding elevational movements of western songbirds in northern California and southern Oregon. *Ecology and Evolution* 7(19):7750–64.

Wiens, J. A., and R. A. Nussbaum. 1975. Model estimation of energy flow in northwestern coniferous forest bird communities. *Ecology* 56:547–61.

Wiens, J. A., and J. T. Rotenberry. 1981. Habitat associations and community structure of birds in shrubsteppe environments. *Ecological Monographs* 51:21–41.

Wiens, J. A., and J. M. Scott. 1975. Model estimation of energy flow in Oregon coastal seabird populations. *Condor* 77:439–52.

Wiens, J. D., R. G. Anthony, and E. D. Forsman. 2011. Barred owl occupancy surveys within the range of the northern spotted owl. *Journal of Wildlife Management* 75:531–38.

Wiens, J. D., R. G. Anthony, and E. D. Forsman. 2014. Competitive interactions and resource partitioning between northern spotted owls and barred owls in western Oregon. *Wildlife Monographs* 185:1–50.

Wiens, J. D., K. M. Dugger, K. E. Dilione, and D. C. Simon. 2019a. *Effects of Barred Owl* (Strix varia) *Removal on Population Demography of Northern Spotted Owls* (Strix occidentalis caurina) *in Washington and Oregon, 2015–18*. Open-File Report 2019-1074. US Geological Survey, Reston, Virginia.

Wiens, J. D., K. E. Dilione, C. A. Eagles-Smith, G. Herring, D. B. Lesmeister, M. W. Gabriel, G. M. Wengert, and D. C. Simon. 2019b. Anticoagulant rodenticides in *Strix* owls indicate widespread exposure in West Coast forests. *Biological Conservation* 238:108238.

Wiens, J. D., K. M. Dugger, J. M. Higley, D. B. Lesmeister, A. B. Franklin, et al. 2021. Invader removal triggers competitive release in a threatened avian predator. *Proceedings of the National Academy of Science* 118(31): e2102859118.

Willamette Valley Oak and Prairie Cooperative. 2020. Willamette Valley oak and prairie cooperative strategic action plan. Willamette Partnership, Portland, Oregon.

Wilson-Jacobs, R., and E. C. Meslow. 1984. Distribution, abundance, and nesting characteristics of Snowy Plovers on the Oregon coast. *Northwest Science* 58:40–48.

Wilson-Jacobs, R., and G. L. Dorsey. 1985. Snowy Plover use of Coos Bay north spit, Oregon. *Murrelet* 66:75–81.

Wires, L. R. 2014. *The Double-crested Cormorant, Plight of a Feathered Pariah*. Yale University Press, New Haven, Connecticut.

Wires, L. R., and F. J. Cuthbert. 2000. Trends in Caspian Tern numbers and distribution in North America: a review. *Waterbirds* 23:388–404.

Wires, L. R., and F. J. Cuthbert. 2006. Historic populations of the Double-crested Cormorant (*Phalacrocorax auritus*): implications for conservation and management in the 21st century. *Waterbird*s 29:9–37.

Wolf, A. 2012. South Puget Sound Streaked Horned Lark (*Eremophila alpestris strigata*) genetic rescue study report for year 2. Unpublished report, Center for Natural Lands Management, Olympia, Washington.

Wolfe, J. D., J. D. Alexander, J. L. Stephens, and C. J. Ralph. 2019. A novel approach to understanding bird communities using informed diversity estimates at local and regional scales in northern California and southern Oregon. *Ecology and Evolution* 9:4431–42.

Wood, C. C. 1987. Predation of juvenile Pacific salmon by the common merganser (*Mergus merganser*) on eastern Vancouver Island I: predation during the seaward migration. *Canadian Journal of Fisheries and Aquatic Sciences* 44:941–49.

Wood, H. B. 1945. The history of bird banding. *Auk* 62:256–65.

Wood, W. E., and S. M. Yezerinac. 2006. Song Sparrow (*Melospiza melodia*) song varies with urban noise. *Auk* 123:650–59.

Woodcock, A. R. 1902. *An Annotated List of the Birds of Oregon*. Bulletin 68. Oregon Agricultural Experiment Station, Corvallis.

Wright, S. K., D. D. Roby, and R. G. Anthony. 2007. Responses of California Brown Pelicans to disturbances at a large Oregon roost. *Waterbirds* 30:479–87.

Wyss, L. A., B. D. Dugger, A. T. Herlihy, W. J. Gerth, and J. L. Li. 2013. Effects of grass seed agriculture on aquatic invertebrate communities inhabiting seasonal wetlands of the southern Willamette Valley, Oregon. *Wetlands* 33:921–37.

Yackulic, C. B., J. Reid, J. D. Nichols, J. E. Hines, R. J. Davis, and E. Forsman. 2014. The roles of competition and habitat in the dynamics of populations and species distributions. *Ecology* 95:265–79.

Yackulic, C. B., L. L. Bailey, K. M. Dugger, R. J. Davis, A. Franklin, et al. 2019. The past and future roles of competition and habitat in the range-wide occupancy dynamics of Northern Spotted Owls. *Ecological Applications* 29(3): e01861.

Yegorova, S., M. G. Betts, J. Hagar, and K. J. Puettmann. 2013. Bird-vegetation associations in thinned and unthinned young Douglas-fir forests 10 years after thinning. *Forest Ecology and Management* 310:1057–70.

Zenk, H. B. 1976. Contributions to Tualatin ethnography: subsistence and ethnobiology. MA thesis. Portland State University, Portland, Oregon.

Zipkin, E. F., S. Rossman, C. B. Yackulic, J. D. Wiens, J. T. Thorson, R. J. Davis, and E. H. C. Grant. 2017. Integrating count and detection–non-detection data to model population dynamics. *Ecology* 98:1640–50.

Zusi, R. L. 2013. Introduction to the skeleton of hummingbirds (Aves: Apodiformes, Trochilidae) in functional and phylogenetic contexts. *Ornithological Monographs* 77(1):1–94.

Zwickel, F., and M. A. Schroeder. 2003. Grouse of the Lewis and Clark expedition. *Northwestern Naturalist* 84:1–19.

# CONTRIBUTORS

JOHN ALEXANDER is the Executive Director of the Klamath Bird Observatory in Ashland, Oregon.

BOB ALTMAN retired from the American Bird Conservancy as their Pacific Northwest Conservation Officer in Corvallis, Oregon.

MATTHEW BETTS is a Professor of Forest Ecosystems and Society at Oregon State University in Corvallis, Oregon.

M. RALPH BROWNING retired from the Biological Survey of the US Department of the Interior at the Division of Birds, National Museum of Natural History (Smithsonian) and now resides in Medford, Oregon.

CHARLIE BRUCE retired as the Threatened and Endangered Species Coordinator for the Oregon Department of Fish and Wildlife in Corvallis, Oregon.

ALAN CONTRERAS is retired from a career in higher education oversight and has written and edited several books on Pacific Northwest natural history. He lives in Eugene, Oregon.

JESSE D'ELIA is the Acting Program Manager for Conservation Planning and Decision Support in the Ecological Services Office of the US Fish and Wildlife Service, Portland, Oregon.

KATIE DUGGER is the Assistant Leader at the USGS Oregon Cooperative Fish and Wildlife Research Unit and a Professor of Wildlife Ecology in the Department of Fisheries, Wildlife, and Conservation Sciences at Oregon State University in Corvallis, Oregon.

ELISE ELLIOTT-SMITH is a Wildlife Ecologist at the USGS Forest and Rangeland Ecosystem Science Center, Corvallis, Oregon.

GREG GREEN is a Wildlife Ecologist at Owl Ridge Natural Resource Consultants, Bothell, Washington.

JOAN HAGAR is a Forest Ecologist at the USGS Forest and Rangeland Ecosystem Science Center and an Associate Professor of Forest Ecosystems and Society at Oregon State University in Corvallis, Oregon.

CHRISTIAN HAGEN is an Associate Professor of Wildlife Ecology in the Department of Fisheries, Wildlife, and Conservation Sciences at Oregon State University in Corvallis, Oregon.

SUSAN HAIG is a Professor of Wildlife Ecology in the Department of Fisheries, Wildlife, and Conservation Sciences at Oregon State University, and retired Senior Scientist at the USGS Forest and Rangeland Ecosystem Science Center in Corvallis, Oregon.

TASHI HAIG is the Interim Coordinator of Museum Programs and Partnerships, Rubin Museum, New York, NY.

CHARLES HENNY is a Research Ecologist Emeritus at the USGS Forest and Rangeland Ecosystem Science Center in Corvallis, Oregon.

FRANK ISAACS was a Senior Faculty Research Assistant in the Department of Fisheries, Wildlife, and Conservation Sciences at Oregon State University in Corvallis and is a cofounder and Project Leader for Oregon Eagle Foundation, Inc., in Klamath Falls, Oregon.

GARY IVEY is a Research Associate of the International Crane Foundation in Bend, Oregon.

PATRICIA KENNEDY is a Professor of Wildlife Ecology Emerita in the Department of Fisheries, Wildlife, and Conservation Sciences, at Oregon State University and resides in Union, Oregon.

DONALD KROODSMA retired as an Ornithologist at the University of Massachusetts Amherst.

DAVID LAUTEN is a Faculty Research Assistant at Portland State University's Institute of Natural Resources based in Bandon, Oregon.

JOE LIEBEZEIT is the Staff Scientist and Avian Conservation Program Manager for Portland Audubon, Portland, Oregon.

ROY LOWE retired as Manager of the Oregon Coast National Wildlife Refuge Complex for the US Fish and Wildlife Service in Newport, Oregon.

DONALD LYONS is Director of Conservation Science with the National Audubon Society and an Assistant Professor Senior Research in the Department of Fisheries, Wildlife, and Conservation Sciences at Oregon State University in Corvallis, Oregon.

JEFFREY MARKS is the Executive Director of Montana Bird Advocacy and resides in Portland, Oregon.

CHRISTOPHER MATHEWS is a Distinguished Professor Emeritus and retired Chair of Biochemistry at Oregon State University and former President of the Audubon Society of Corvallis in Oregon.

BRENDA MCCOMB retired as Dean of the Graduate School at Oregon State University in Corvallis, Oregon.

RANDY MOORE is a Senior Instructor in the Department of Fisheries, Wildlife, and Conservation Sciences at Oregon State University in Corvallis, Oregon.

MICHAEL T. MURPHY is a Professor of Biology at Portland State University, Portland, Oregon.

S. KIM NELSON retired as a Senior Faculty Research Assistant in the Department of Fisheries, Wildlife, and Conservation Sciences at Oregon State University in Corvallis, Oregon.

M. CATHY NOWAK is a Wildlife Biologist for the Oregon Department of Fish and Wildlife in Union, Oregon.

LEWIS ORING is a Professor Emeritus of Ecology, Evolution, and Conservation Biology at the University of Nevada, Reno, living in Susanville, California.

RAM PAPISH is a natural historian and artist from Toldeo, Oregon.

DANIEL ROBY is a Professor of Wildlife Ecology (retired) in the Department of Fisheries, Wildlife, and Conservation Sciences at Oregon State University and the former Leader-Wildlife of the USGS Oregon Cooperative Fish and Wildlife Research Unit in Corvallis, Oregon.

PETER SANZENBACHER is a Wildlife Biologist for the US Fish and Wildlife Service in Palm Springs, California.

MARK STERN retired as Director of Forest Conservation for The Nature Conservancy in Portland, Oregon.

ORIANE TAFT is the Proposal Coordinator at NV5 Geospatial in Corvallis, Oregon.

PEPPER TRAIL retired as the Senior Ornithologist at the US Fish and Wildlife Service National Wildlife Forensic Laboratory in Ashland, Oregon.

DAVID WIENS is a Raptor Ecologist with the USGS Forest and Rangeland Ecosystem Science Center and an Assistant Professor of Wildlife Ecology in the Department of Fisheries, Wildlife, and Conservation Sciences at Oregon State University in Corvallis, Oregon.

JOHN WIENS is an Avian Ecologist who once chaired Oregon State University's Zoology Department, is a Distinguished Professor Emeritus at Colorado State University, and after many ornithological adventures has retired to Corvallis, Oregon.

CONTRIBUTING BIRD PHOTOGRAPHERS:

Scott Carpenter: pp. 132, 146, 192, 250, 252
Mel Clements: pp. 163, 196, 210
Dan Cushing: pp. 86, 261
Alan Dyck: pp. xx, 139
Rod Gilbert: p. 156
Susan Haig: pp. 206, 213, 259
Jared Hobbs: pp. 5, 46, 174, 178, 186, 209, 215, 226, 228
Tim Huntington: pp. ii, xxii, 44, 241
Timothy Lawes: pp. 49, 69, 79, 113, 183
David Leonard: pp. 43, 136, 160, 166, 176, 233
Brett Lovelace: p. 86
Roy Lowe: pp. 27, 53, 64, 74, 75, 76, 82, 119, 124
Kim Nelson: p. 86
Ken Popper: p. 202
Dan Roby: pp. 34, 38, 40, 50, 57, 80, 83, 97, 99, 103, 108, 111, 115, 123, 128, 135, 150, 152, 169, 170, 189, 219, 239, 247
Mick Thompson: pp. xvii, 199, 225, 253, 260
Jeffrey M. Wells: p. 142

# INDEX